Carbon-13 NMR Shift Assignments of Amines and Alkaloids

Carbon-13 NMR Shift Assignments of Amines and Alkaloids

Maurice Shamma and
David M. Hindenlang

The Pennsylvania State University
University Park, Pennsylvania

PLENUM PRESS • NEW YORK AND LONDON

Library of Congress Cataloging in Publication Data

Shamma, Maurice, 1926-
Carbon-13 nmr shift assignments of amines and alkaloids.

Bibliography: p.
Includes index.
1. Amines – Spectra. 2. Alkaloids – Spectra. 3. Carbon – Isotopes – Spectra. 4. Nuclear magnetic resonance spectroscopy. I. Hindenlang, David M., joint author. II. Title.
QC463.A8S5 547'.042 78-10117
ISBN 0-306-40107-X

A Division of Plenum Publishing Corporation
227 West 17th Street, New York, N.Y. 10011

Printed in the United States of America

PREFACE

The aim of this book is to gather under one cover most of the data presently available on the carbon-13 nuclear magnetic resonance (cmr) spectra of alkaloids. The term "alkaloids" is used here in a very broad sense to include synthetic analogues of the natural products. Simple model amines are also incorporated since these often supply the basic information required in the assignment of chemical shifts for the more complex compounds. The literature on alkaloid cmr spectroscopy has been covered through 1977, but the collection of compounds presented here is illustrative rather than exhaustive. The papers included in the reference list afford further information not only on the cmr assignments of the particular compounds provided here, but also incorporate data on additional related structures. Only a few dimeric indole alkaloids are included since to a large extent their cmr spectra can be correlated directly with those of their monomeric analogues. The present volume is thus a representative empirical compendium of cmr assignments focusing upon alkaloids and model amines, and is intended to aid cmr research in heterocyclic and alkaloid chemistry.

The compounds and data presented in this book are classified and organized according to structural similarity. The purpose of such a presentation is to demonstrate the common cmr characteristics of a given structural type, while also facilitating an empirical evaluation of the cmr spectral changes specifically resulting from relatively minor variations in oxidation level, substitution, or stereochemistry. The basic ordering within chapters and subsections is intended to progress from simpler to more complex structural skeletons and substitution patterns.*

The early chapters introduce generally simple single-ring heterocycles and other compounds which serve as models for the more complex structures. The isoquinoline alkaloids and indole alkaloids are particularly noteworthy in terms of both the wealth of structural variations and the extent of cmr data available. The fact that these two groups of natural products, along with some

* Compounds and data that appeared in the literature after the manuscript had been completed and sent off to the publisher, and therefore too late for incorporation in the main body of the text, are gathered in Chapter 23.

closely related structures, comprise nearly half the total number of compounds presented in this book reflects the magnitude of the research effort in these areas.

In most cases the structures included here have been named, but several synthetic analogues and model compounds remain unnamed. The compounds are listed in the Index according to name, or by key structural characteristics when unnamed, or by both in the case of some natural products whose common names are more descriptive of origin than of chemical structure.

It is not the purpose of this book to elaborate upon the theory or the more advanced applications of cmr. Several excellent reference books are available to the organic chemist on the principles, techniques, and interpretations of cmr, such as: G. C. Levy and G. L. Nelson, *Carbon-13 Nuclear Magnetic Resonance for Organic Chemists*, Wiley-Interscience, New York (1972); J. B. Stother, *Carbon-13 N.M.R. Spectroscopy*, Academic Press, New York (1972); and F. W. Wehrli and T. Wirthlin, *Interpretation of Carbon-13 NMR Spectra*, Heyden, London (1976). Moreover, there exist review books devoted to nuclear magnetic resonance in general and cmr in particular, which explicate current advances in techniques and new applications. Notable in this respect is G. C. Levy (ed.), *Topics in Carbon-13 NMR Spectroscopy*, Vols. 1 and 2, Wiley-Interscience, New York (1974 and 1976). In addition, L. F. Johnson and W. C. Jankowski, *Carbon-13 NMR Spectra*, Wiley-Interscience, New York (1972), provides a useful indexed compilation of 500 assigned cmr spectra selected to cover the broad range of types of organic molecules.

The cmr values presented here are the noise-decoupled chemical shifts. However, further information can be obtained from various cmr techniques other than the mere observation of one resonance for each type of carbon nucleus in the sample molecule. Likewise, more is involved in making the reported assignments than simple chemical shift theory and comparison of model systems. Thus, there are varying degrees of certainty regarding assignments of cmr resonances, depending upon the extent of the investigation. The vast majority of references cited here have, at a minimum, employed off-resonance proton decoupling to determine the first-order multiplicities and hence the number of protons bound to each carbon nucleus in order to substantiate assignments. The first-order splitting rules state that a singlet indicates an unprotonated center; a doublet, one proton; a triplet, two protons; and a quartet, three protons, all directly bonded to the carbon nucleus observed. Gated decoupling experiments with the decoupler off only during data acquisition afford fully coupled spectra which also allow the observation of long-range couplings on quaternary carbons, resulting in improved accuracy of the shift assignments.

In a minority of the studies cited here, greater certainty of assignments was achieved by the application of more elaborate techniques. For example,

T_1 relaxation studies can determine the number of protons within two bond distances of a quaternary carbon nucleus. Other spectroscopic techniques applied to establish more rigorously the cmr assignments include selective single-frequency proton decoupling and nuclear Overhauser enhancement experiments. Chemical methods such as selective deuteration, selective C-13 incorporation, and shift reagent experiments have also been applied in some of the studies reported here.

The reader should be cautioned that since cmr spectroscopy is a relatively new tool for organic chemists, a few of the chemical shift assignments may in the future have to be modified or corrected. In cases where accurate assignments are particularly difficult, interchangeable chemical shifts are denoted by superior letters (*a*, *b*, *c*, . . .). For the most part, these interchangeable assignments are marked only as presented in the original reports. However, in some cases the present authors have added such caveats when deemed appropriate in light of other data, usually unavailable at the time of writing of the original paper.

It should also be pointed out that in certain instances some carbon atoms shown have not been assigned chemical shifts. This is simply a reflection of the limited data provided in the original literature. Nevertheless, these partially assigned structures are sufficiently instructive to merit inclusion.

As with the convention for proton nuclear magnetic resonance, the cmr shifts are reported in parts per million (ppm) downfield from standard tetramethylsilane (TMS). If the chemical shifts were originally reported relative to some other internal standard, these values were converted to a TMS reference for uniformity and ease of comparison. The equation used was: $\delta_{TMS} = \delta_{CDCl_3} + 77.0 = 192.4 - \delta_{CS_2} = \delta_{C_6H_{12}} + 26.9$. Deuteriochloroform ($CDCl_3$) is now the preferred and most common solvent for the measurement of cmr spectra. When the data were not recorded in $CDCl_3$, the appropriate solvent is indicated in parentheses below the structure.

In a historical context, each decade since the 1930s has witnessed the development of an important branch of spectroscopy used by the organic chemist. In the 1930s ultraviolet spectroscopy was first used, to be followed in the 1940s by infrared spectroscopy. The 1950s saw the development of proton nuclear magnetic resonance spectroscopy, and the 1960s the development of mass spectroscopy. Carbon-13 nuclear magnetic resonance spectroscopy has developed in the 1970s as a powerful technique for organic analysis since the advent of the Fourier transform method. The recent flowering of cmr in the realm of natural products chemistry in general, and alkaloids in particular, is due in large part to the prolific contributions of Professor Ernest Wenkert and his colleagues.

The authors are grateful to Dr. Sidney Teitel of Hoffman-La Roche, Inc., for a generous gift of compounds which made possible spectral assignments

318, 319, 324, 353, 413, 443, 444, and 445. The authors also wish to thank Dr. Jerome L. Moniot and Mr. Alan J. Freyer for useful discussions concerning various aspects of this collection of cmr data, as well as Senior Editor, Mr. Ellis H. Rosenberg, and Production Editor, Mr. Robert Golden, for continuous and unfailing interest and assistance.

CONTENTS

Guide to the Presentation of Data in the Diagrams

The carbon-13 nuclear magnetic resonance (cmr) shift spectra are given in the form of numbers adjacent to the carbons in the chemical structures.

The superscript italic letters *a*, *b*, *c*, . . . on the cmr numbers denote interchangeable chemical shifts.

The number in the top left corner of each box is simply a serial number, referred to as the spectral assignment diagram number, used for identification purposes.

The name (when it exists) of the compound is given immediately after the spectral assignment diagram number.

The solvent, when it is other than $CDCl_3$, is given in smaller type in parentheses under the name. Other pertinent comments are given in the same form.

The references are given in the bottom right corner of the box.

With regard to the nomenclature for the *trans*-decahydroquinoline series, the ring system is always taken to be

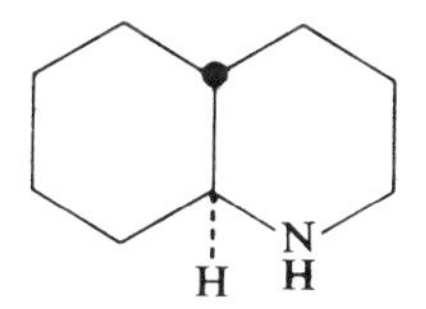

rather than its mirror image.

1. PYRROLES[1]

1 Pyrrole

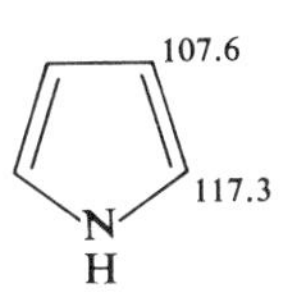

Ref. 2;
see also Ref. 3

2 1-Methylpyrrole

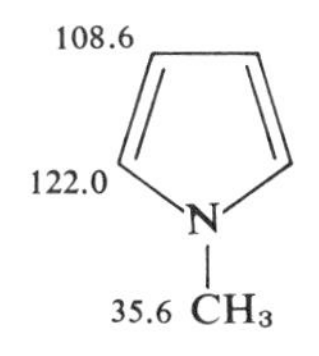

Ref. 3

3 2-Formylpyrrole

110.8 121.6
132.4
126.8 CHO
N 178.8
H

Ref. 2

4 2-Acetylpyrrole

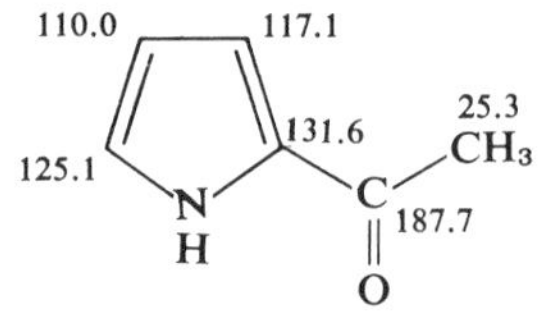

110.0
117.1
25.3
CH_3
131.6
125.1
N
H
C
187.7
O

Ref. 2

5 2-Carbomethoxypyrrole

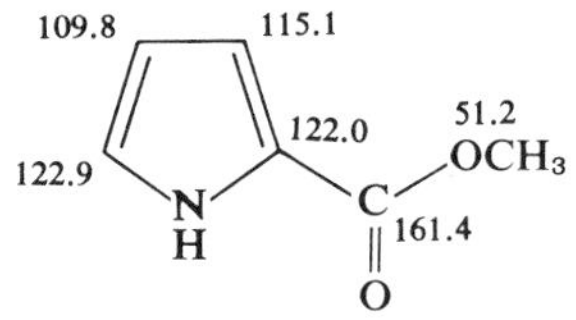

109.8
115.1
51.2
122.0
OCH_3
122.9
N
H
C
161.4
O

Ref. 2

6 2-Methyl-5-carboethoxypyrrole

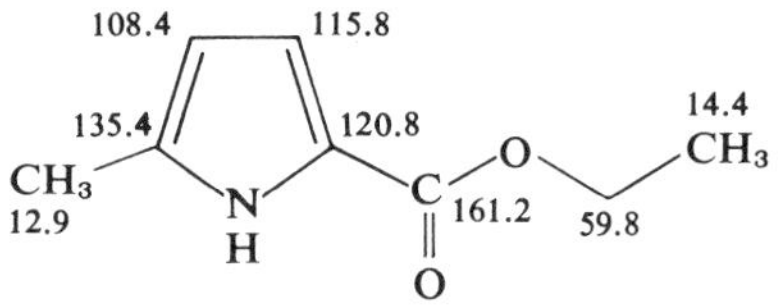

108.4
115.8
14.4
135.4
120.8
CH_3
O
CH_3
12.9
N
H
C
161.2
59.8
O

Ref. 2

7 2-Carboethoxy-3-methylpyrrole

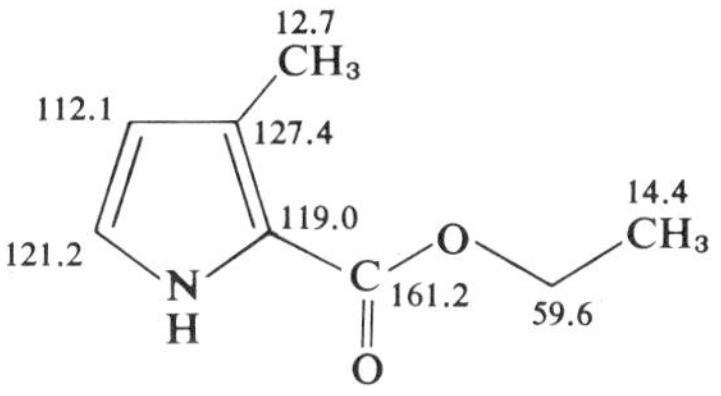

12.7 CH_3
112.1
127.4
119.0
14.4 CH_3
121.2
N H
C 161.2
O
59.6
O

Ref. 2

8 2-Carboethoxy-3,5-dimethylpyrrole

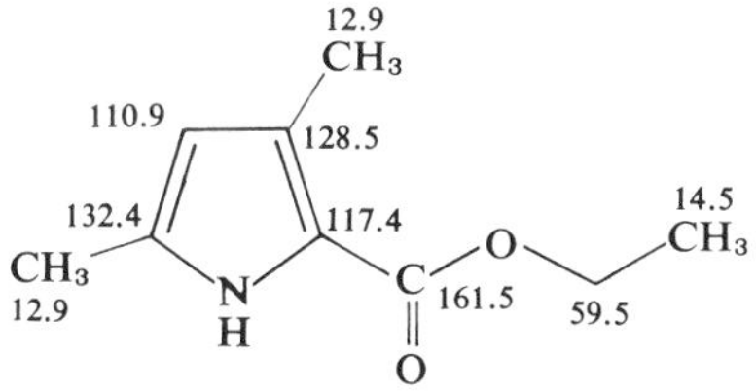

12.9 CH_3
110.9
128.5
132.4
117.4
14.5 CH_3
CH_3 12.9
N H
C 161.5
O
59.5
O

Ref. 2

2. PYRROLIDINES

9 Pyrrolidine

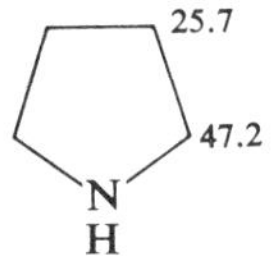

25.7
47.2
N
H

Ref. 4

10 1-Methylpyrrolidine

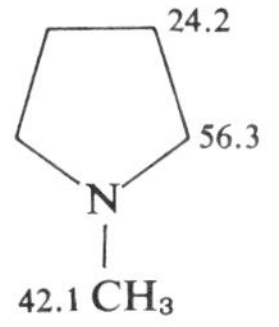

24.2
56.3
N
42.1 CH_3

Ref. 4

11 1,3-Dimethylpyrrolidine

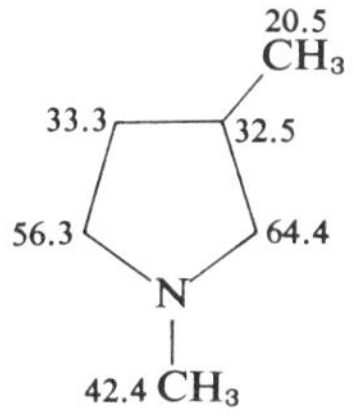

20.5
CH_3
33.3
32.5
56.3
64.4
N
42.4 CH_3

Ref. 4

12 1,3,3-Trimethylpyrrolidine

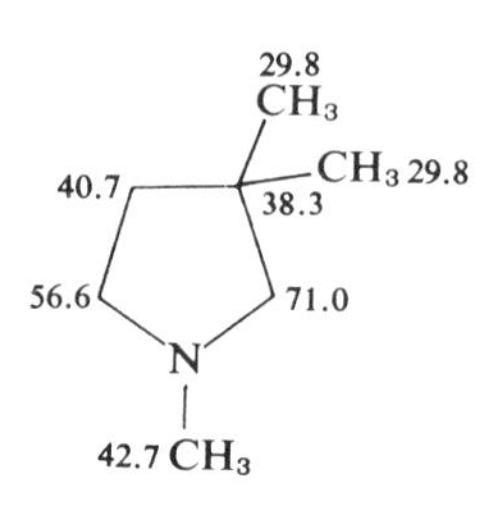

Ref. 4

13 Scabrosin 4-acetate-4′-butyrate

(a lichen metabolite)

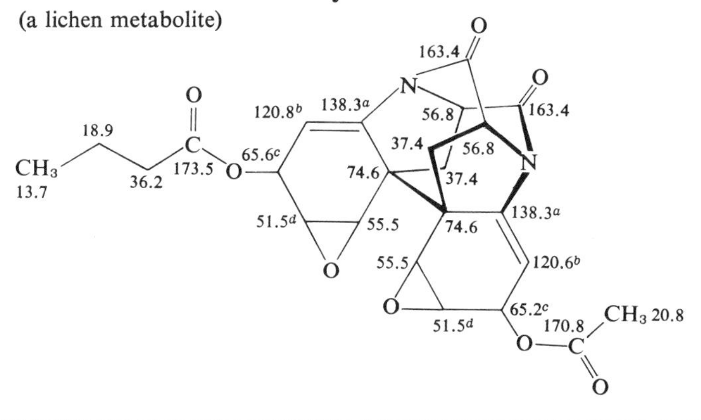

Ref. 5

3. RETRORSINE: A PYRROLIZIDINE ALKALOID

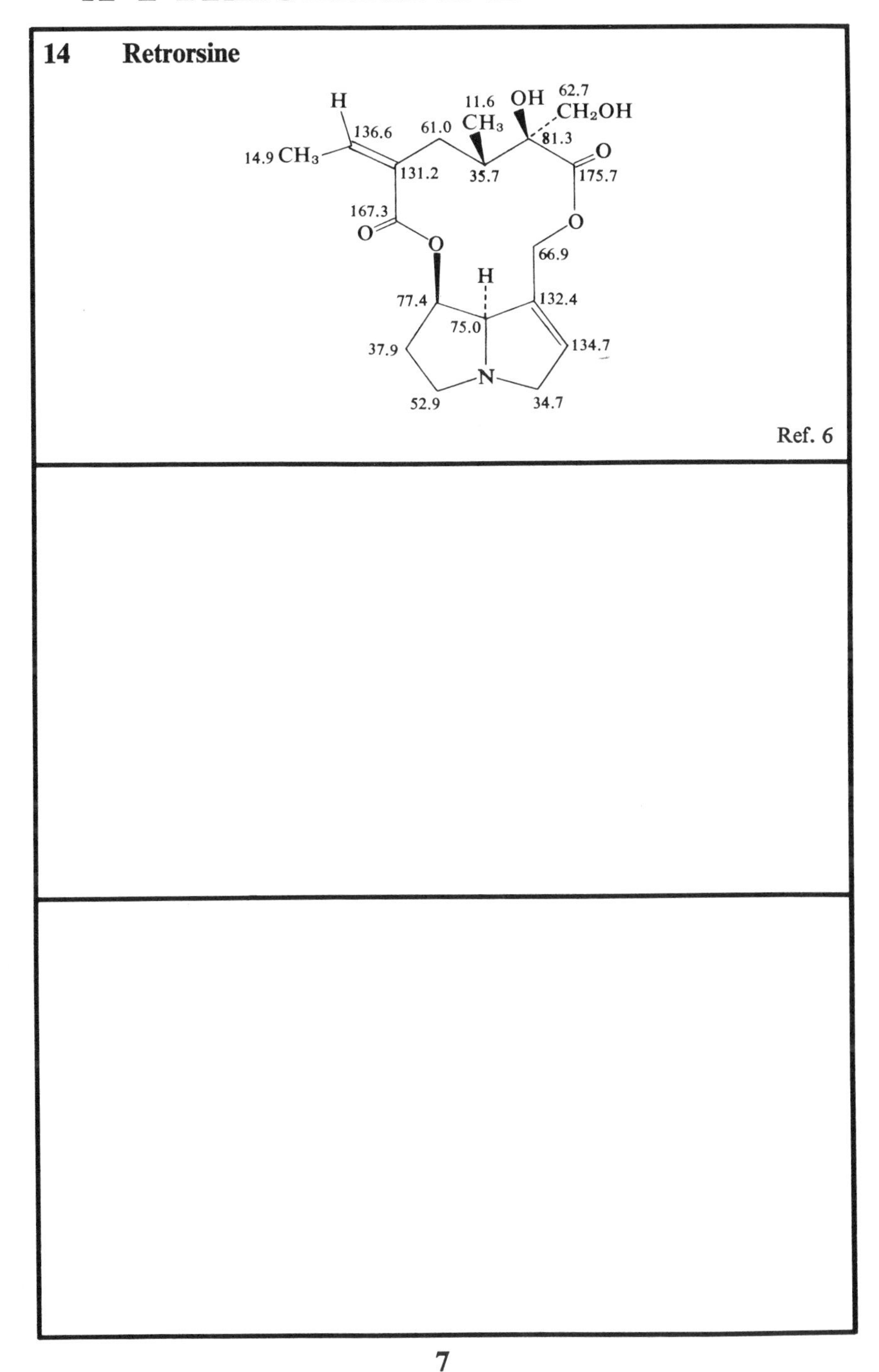

4. PYRIDINES[7]

15 Pyridine

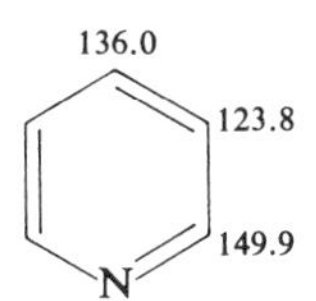

Ref. 8

16 Pyridine *N*-oxide

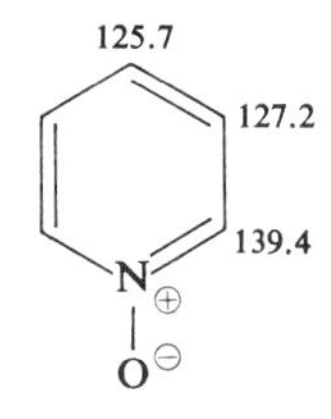

Ref. 8

17 Pyridinium trifluoroacetate
(TFA)

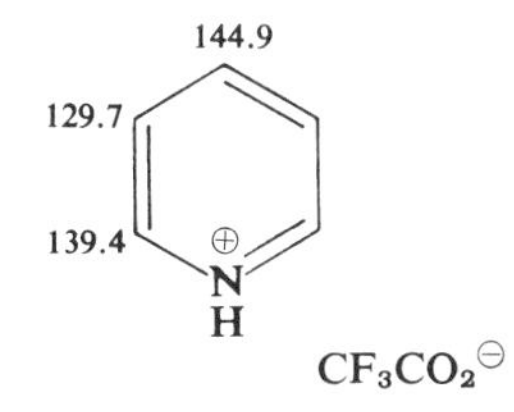

Ref. 9

18 **Pyridine methiodide**
(EtOH)

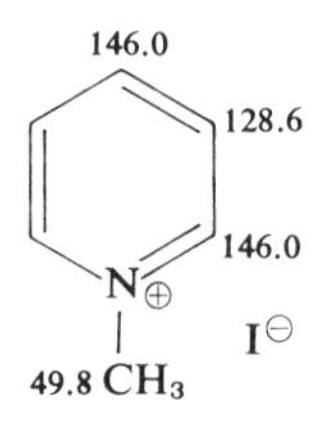

Ref. 8;
see also Ref. 9

19 **2-Methylpyridine**

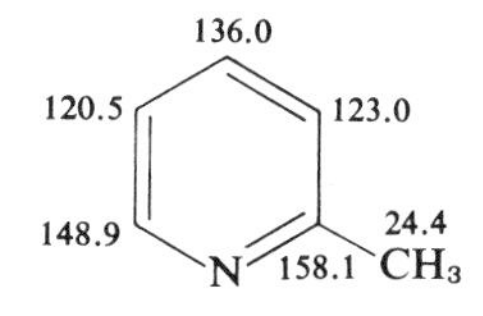

Ref. 10

20 **2-Methylpyridine *N*-oxide**

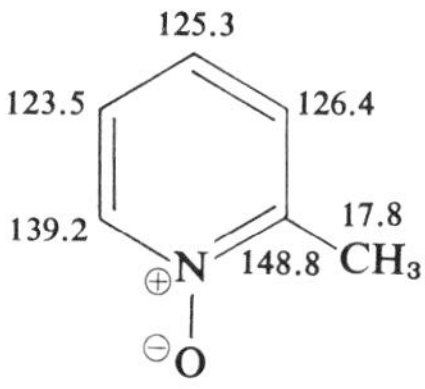

125.3
123.5
126.4
139.2
17.8
⊕N
148.8
CH_3
⊖O

Ref. 10

21 3-Methylpyridine

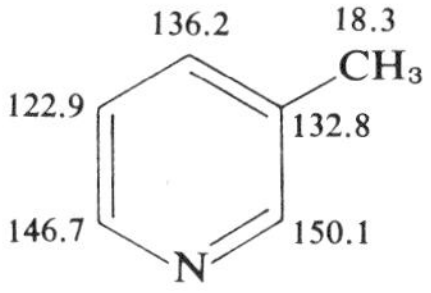

136.2 18.3
CH_3
122.9 132.8
146.7 150.1
N

Ref. 10

22 3-Methylpyridine *N*-oxide

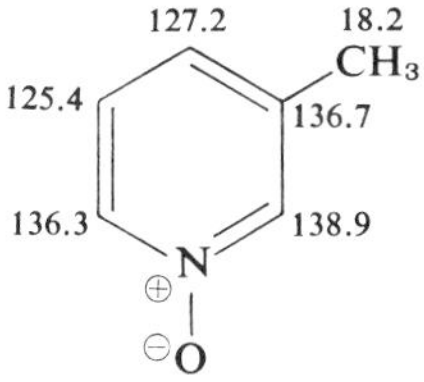

127.2 18.2
CH_3
125.4 136.7
136.3 138.9
$N^{\oplus}$
$O^{\ominus}$

Ref. 10

23 4-Methylpyridine

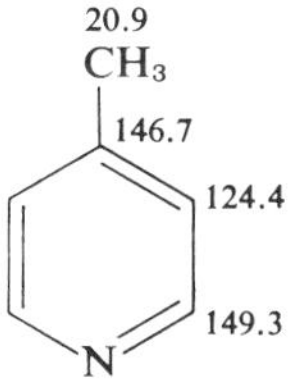

20.9
CH_3
146.7
124.4
149.3
N

Ref. 10

24 4-Methylpyridine *N*-oxide

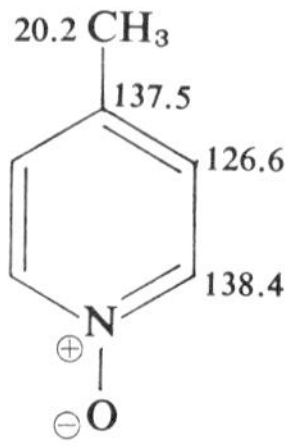

20.2 CH_3
137.5
126.6
138.4
$N^{\oplus}$
$^{\ominus}O$

Ref. 10

25 1-Methylpyridinium-3-oxide
(DMSO-d_6)

132.8
$O^{\ominus}$
127.5
168.4
125.7
134.9
$N^{\oplus}$
47.5 CH_3

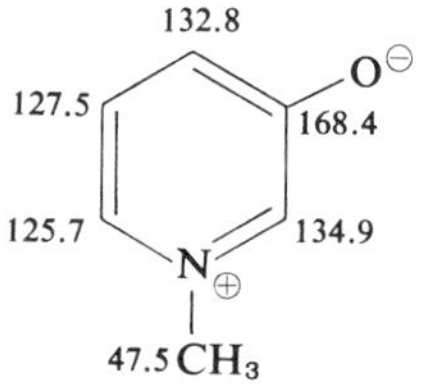

Ref. 11

26 1,6-Dimethylpyridinium-3-oxide
(DMSO-d_6)

131.2
$O^{\ominus}$
128.9
168.0
135.1
134.4
18.5 CH_3
$N^{\oplus}$
45.3 CH_3

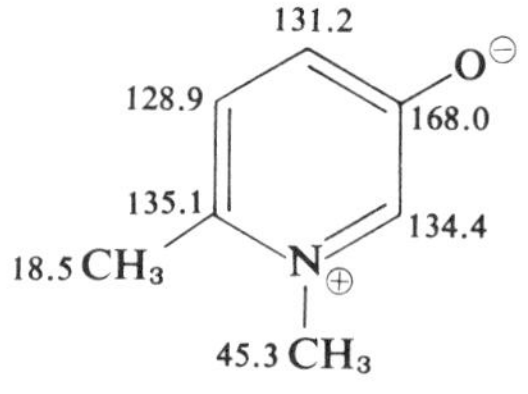

Ref. 11

27 **3-Hydroxypyridine methiodide**
(DMSO-d_6)

131.8, OH, 156.8, 129.1, 136.9, 134.3, $N^{\oplus}$, $I^{\ominus}$, 49.4 CH_3

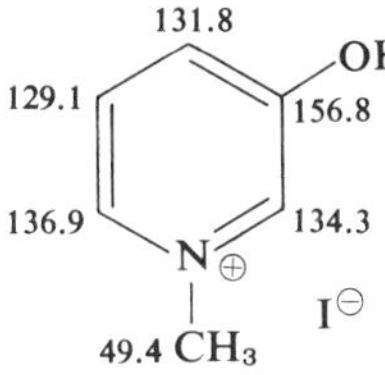

Ref. 11

28 **2-Methyl-5-hydroxypyridine methiodide**
(DMSO-d_6)

132.4, OH, 130.4, 155.3, 146.4, 134.1, 20.7 CH_3, $N^{\oplus}$, $I^{\ominus}$, 47.4 CH_3

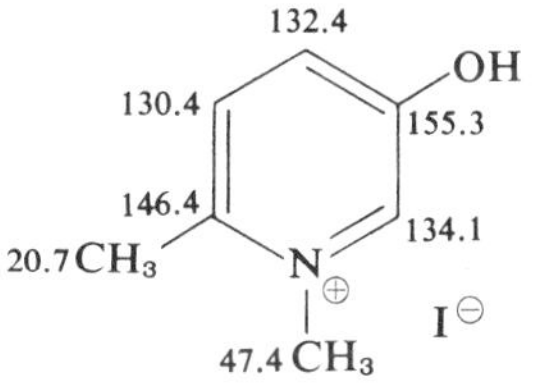

Ref. 11

29 **α-Picoline *N*-metho salt**
(TFA)

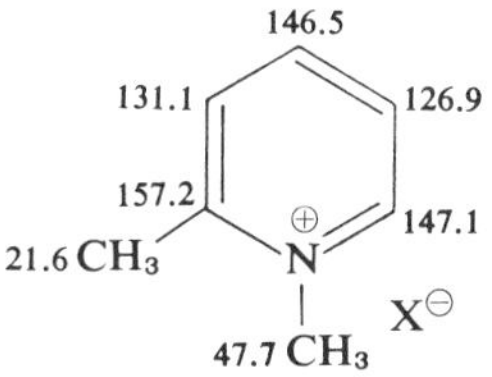

146.5, 131.1, 126.9, 157.2, 147.1, 21.6 CH_3, $N^{\oplus}$, $X^{\ominus}$, 47.7 CH_3

Ref. 9

30 **1,2,4,6-Tetramethylpyridinium salt**
($TFA–CD_2Cl_2$, 4:1)

22.0 CH_3
161.0
129.9
156.7
CH_3 $N^{\oplus}$ CH_3 22.3
$X^{\ominus}$
40.6 CH_3

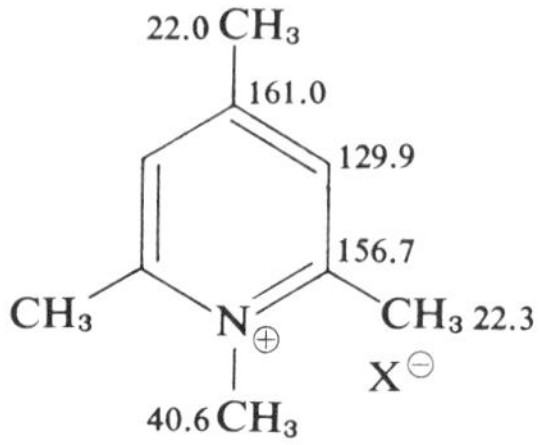

Ref. 12

31 **2,4,6-Trimethylpyridinium salt**
($TFA–CD_2Cl_2$, 4:1)

20.7 CH_3
147.4
121.1
157.4
CH_3 $N^{\oplus}$ H CH_3
24.3
$X^{\ominus}$

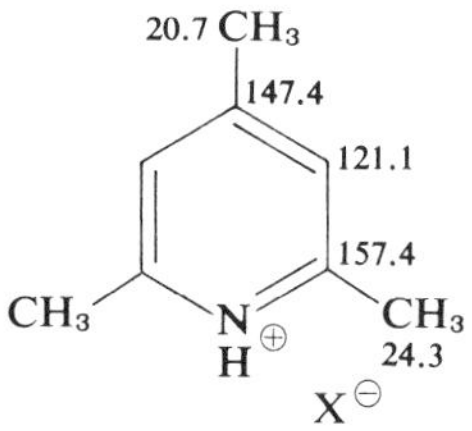

Ref. 12

32 **2-Methoxypyridine**
(DMSO-d_6)

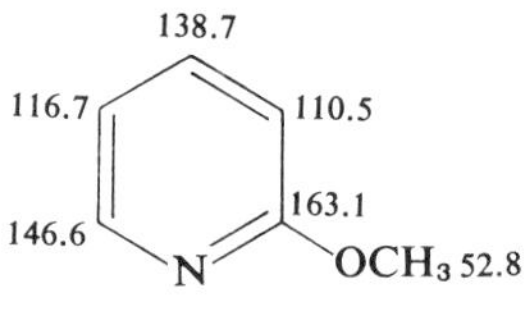

138.7
116.7 110.5
146.6 163.1
N OCH_3 52.8

Ref. 13

33 3-Methoxypyridine
(DMSO-d_6)

120.0
OCH_3 55.3
123.8
155.2
141.4
137.3
N

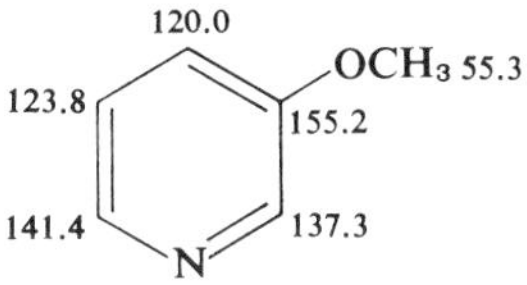

Ref. 13

34 4-Methoxypyridine
(DMSO-d_6)

OCH_3 55.0
164.9
109.8
109.8
150.7
150.7
N

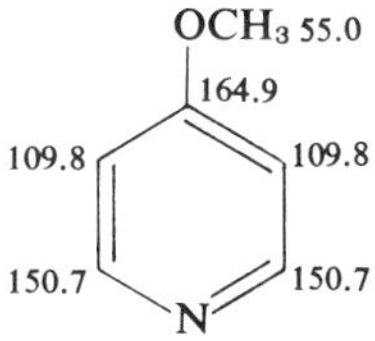

Ref. 13

35 3-Hydroxypyridine
(DMSO-d_6)

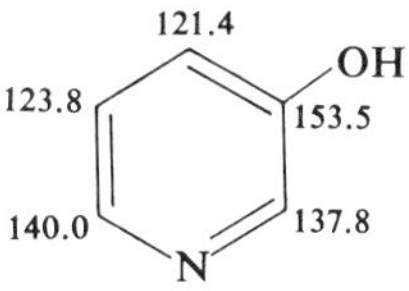

121.4
OH
123.8
153.5
140.0
137.8
N

Ref. 13

36 **3-Acetylpyridine *N*-metho salt**
(TFA)

Ring: 146.1, 130.1, 149.7, 137.0, 147.7; C=O 198.2; CH_3 28.1; N–CH_3 51.2; $X^{\ominus}$

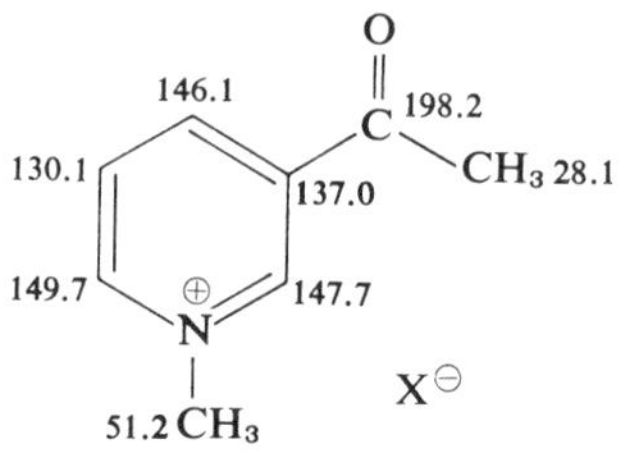

Ref. 9

37 **2-Methyl-5-acetylpyridinium salt**
(TFA)

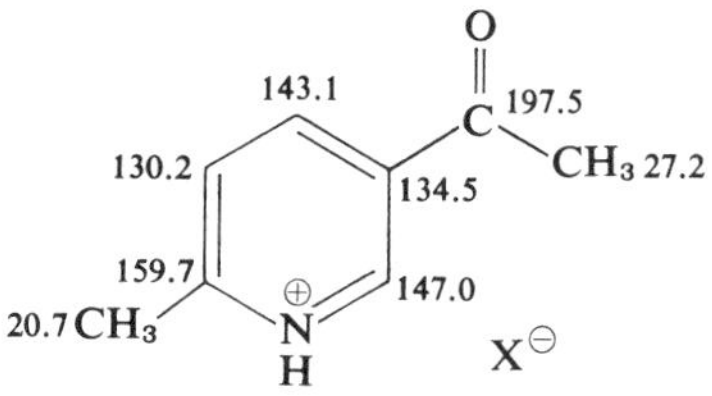

Ring: 143.1, 130.2, 159.7, 134.5, 147.0; C=O 197.5; CH_3 27.2; 2-CH_3 20.7; N–H; $X^{\ominus}$

Ref. 9

38 **2-Methyl-5-acetylpyridine *N*-metho salt**
(TFA)

Ring: 145.3, 131.3, 161.3, 135.6, 148.6; C=O 195.4; CH_3 27.5; 2-CH_3 21.4; N–CH_3 47.7; $CF_3CO_2^{\ominus}$

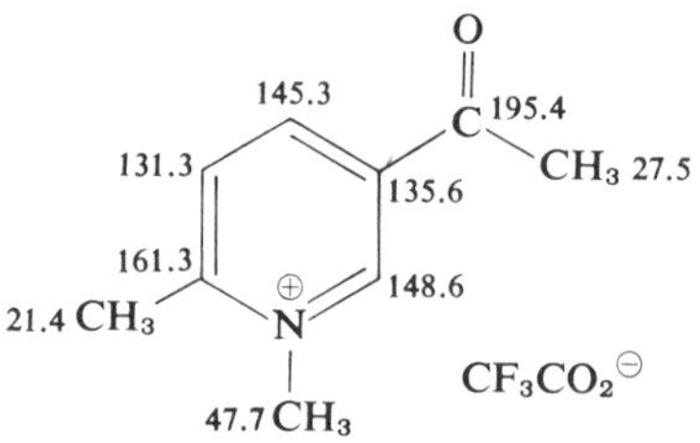

Ref. 9

39 **2-Pyridinecarboxaldehyde**
(neat, values adjusted from CS_2 reference)

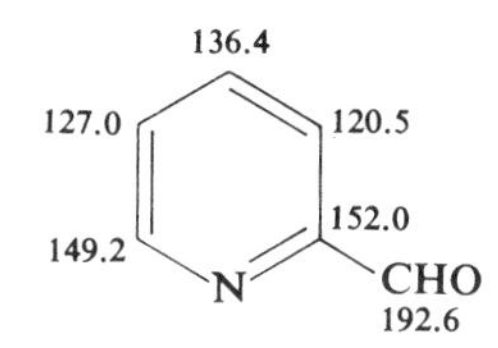

Ref. 14

40 **3-Pyridinecarboxaldehyde**
(neat, values adjusted from CS_2 reference)

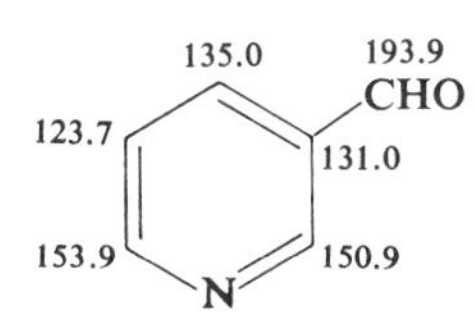

Ref. 14

41 **4-Pyridinecarboxaldehyde**
(neat, values adjusted from CS_2 reference)

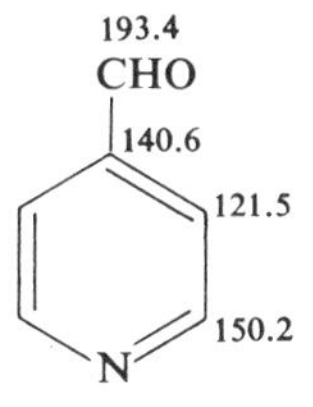

Ref. 14

42 **2-Pyridone**
(DMSO-d_6)

140.8
104.8 119.8
162.3
135.2 N O
H

Ref. 13

43 ***N*-Methyl-2-pyridone**
(DMSO-d_6)

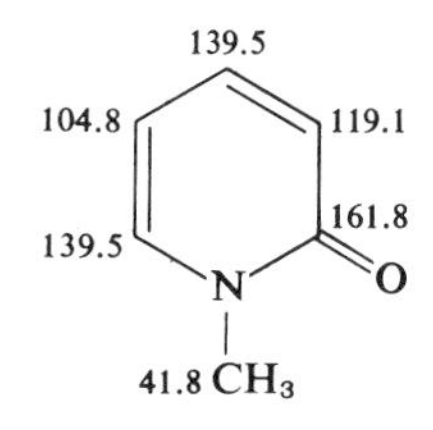

Ref. 13

44 **4-Pyridone**
(DMSO-d_6)

O
175.7
115.9
139.8
N
H

Ref. 13

45 ***N*-Methyl-4-pyridone**
(DMSO-d_6)

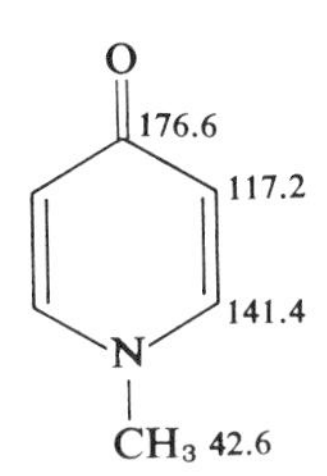

Ref. 13

46 **2,3-Trimethylenepyridine**

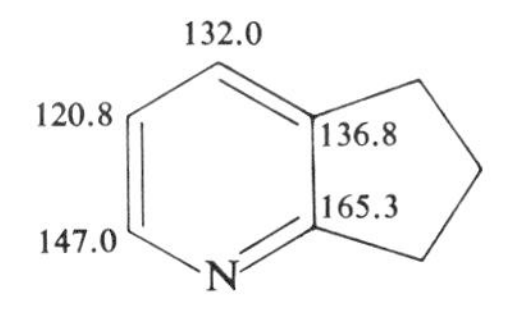

Ref. 15

47 **5,6,7,8-Tetrahydroquinoline**

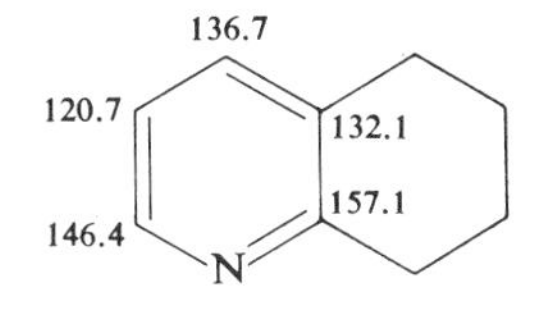

Ref. 15

48 2,3:5,6-Di(trimethylene)pyridine

128.4
134.1
163.2
N

Ref. 15

49 1,2,3,4,5,6,7,8-Octahydroacridine

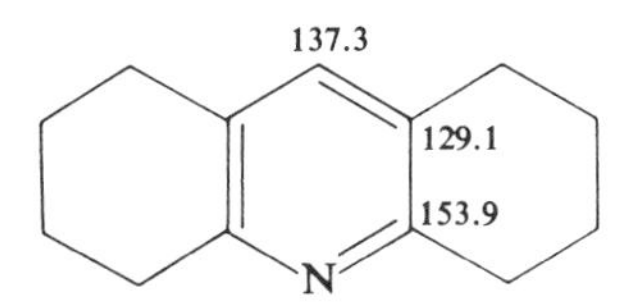

Ref. 15

50 1,2,3,4,7,8,9,10-Octahydrophenanthridine

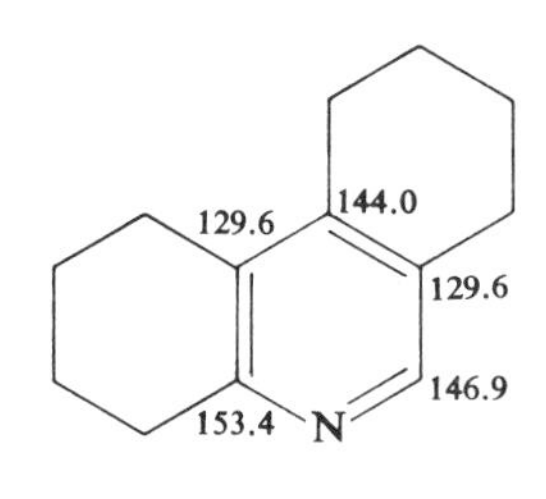

Ref. 15

51 2,4,6-Triphenylpyridinium salt
($TFA–CD_2Cl_2$, 4:1)

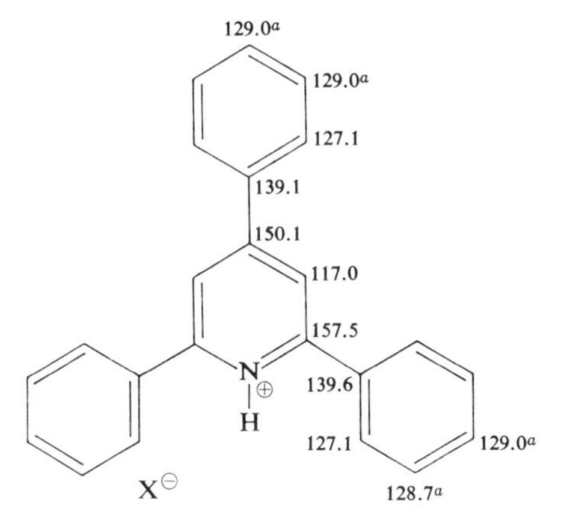

Ref. 12

52 2-Styrylpyridine

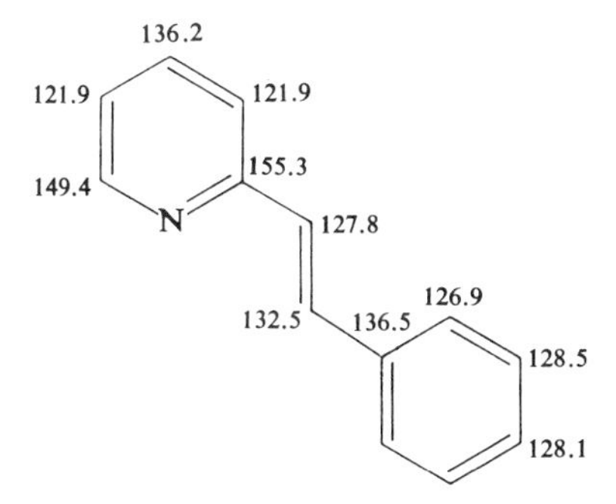

Ref. 16

53 3-Styrylpyridine

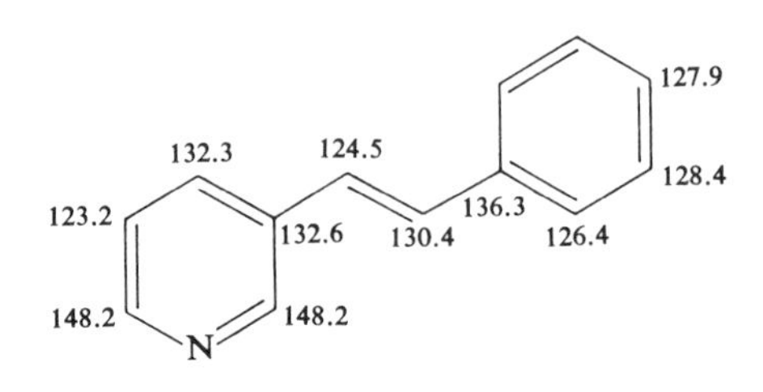

Ref. 16

54 4-Styrylpyridine

128.5
128.6
126.8
135.9
133.5
125.7
144.5
120.7
120.7
149.7
149.7
N

Ref. 16

55 **1′-Ethynyl-1′,1′-di-α-picolyl-α-picoline**

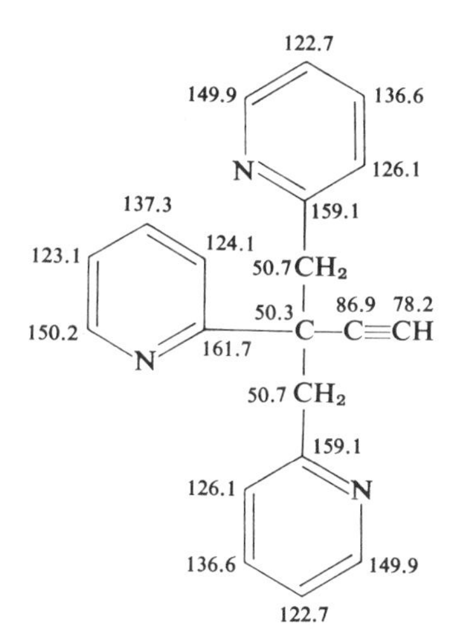

Ref. 17

56 **1′,1′,1′-Tri-α-picolyl-α-picoline**

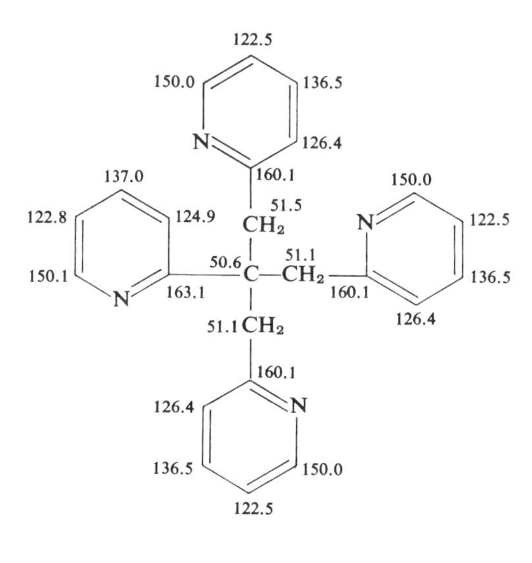

Ref. 17

57 **Nicotinic acid anion**
(D_2O with NaOH)

138.6
133.5 $CO_2^{\ominus}$
124.8
173.8
151.5
150.3
N

Ref. 18

58 **Anabasine**

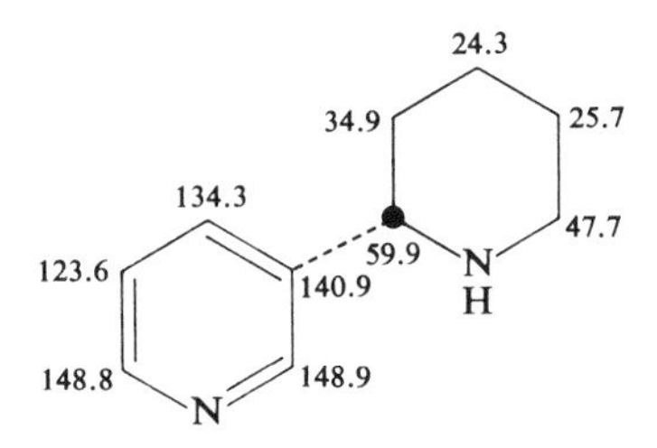

Ref. 18

59 **Nicotine**

35.2
22.6
134.9
57.0
68.9
123.6
N
138.8
148.5
149.5
40.3 CH_3
N

Ref. 18

60 Nornicotine

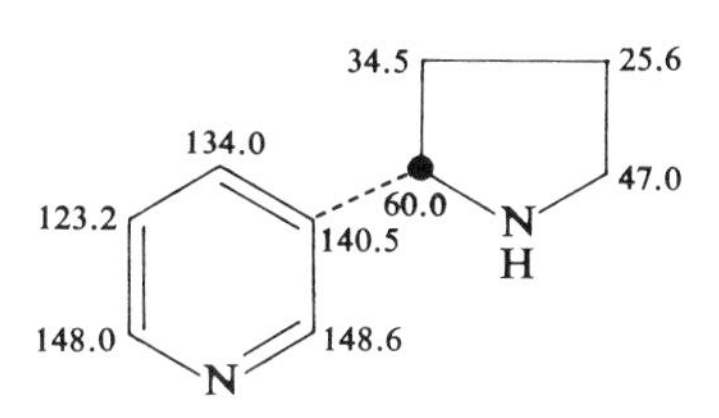

Ref. 18

61 Anatabine perchlorate
(D_2O)

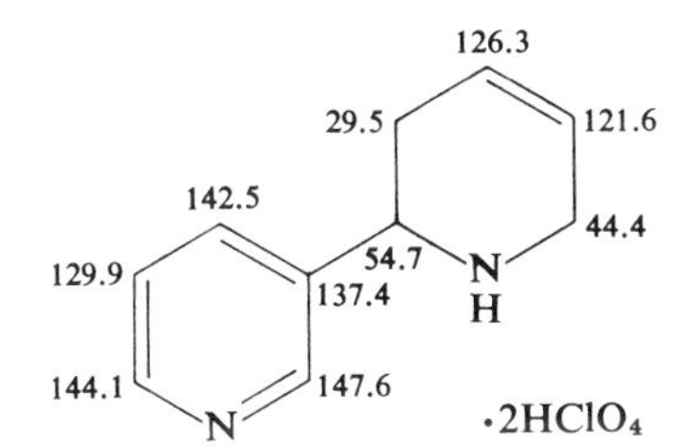

Ref. 18

62 *N,N*-Diethyl-2-pyridinecarboxamide

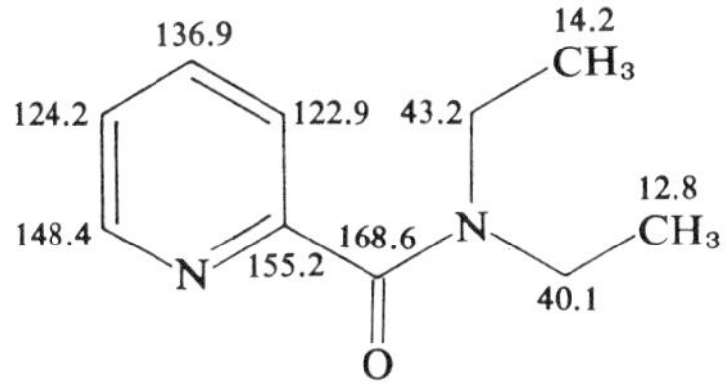

136.9
14.2
CH_3
124.2
122.9
43.2
12.8
148.4
168.6
N
CH_3
N
155.2
40.1
O

Ref. 19

63 ***N,N*-Diethylnicotinamide**

O
134.1 39.6[a]
13.2
123.4 168.6 N CH_3
133.1
150.2 147.2 43.4[a]
N CH_3
13.2

Ref. 19

64 ***N,N*-Diethyl-4-pyridinecarboxamide**

12.7
CH_3
39.3
14.1
O N CH_3
168.6 43.1
144.8
120.7 120.7
150.1 150.1
N

Ref. 19

5. PIPERIDINES[4,20]

65 Piperidine

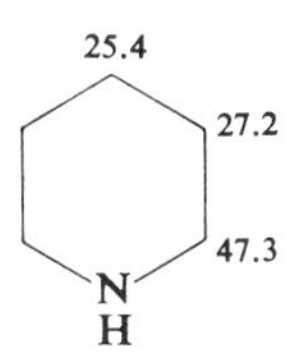

Ref. 21;
see also
Refs. 22, 23

66 *N*-Methylpiperidine

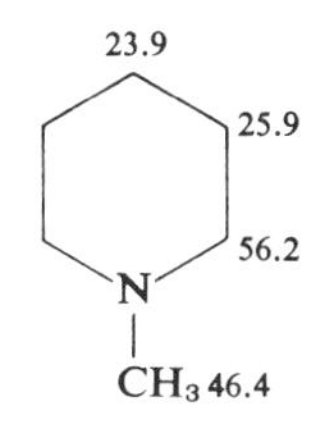

Ref. 21;
see also
Refs. 22, 24

67 α-Pipecoline (2-methylpiperidine)

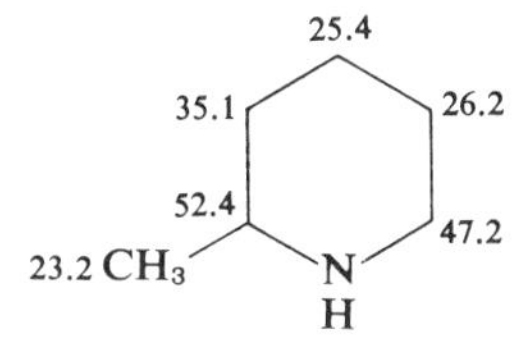

Ref. 21;
see also
Refs. 22, 23

68 **β-Pipecoline**
(C_6D_6)

35.0 20.0
CH_3
28.0 33.3
47.9 56.0
N
H

Ref. 23;
see also Ref. 22

69 **γ-Pipecoline**
(C_6D_6)

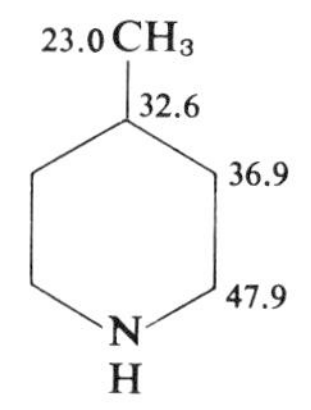

Ref. 23;
see also Ref. 22

70 ***N*-Methyl-α-pipecoline**

24.8
34.7 26.2
58.9 56.8
CH_3
20.2
N
CH_3 42.9

Ref. 21;
see also
Refs. 22, 24

71 ***N*-Methyl-β-pipecoline**

32.5
19.6
CH_3
25.3
30.8
55.7
64.0
N
CH_3
46.4

Ref. 25

72 ***N*-Methyl-γ-pipecoline**

(C_6D_6)

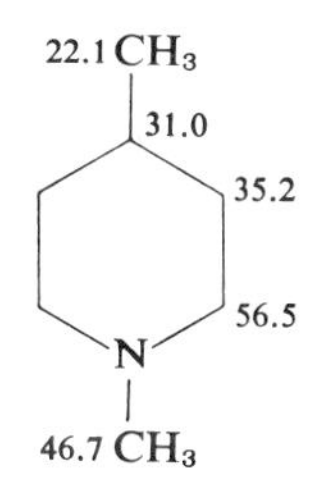

Ref. 23

73 ***cis*-2,6-Dimethylpiperidine**

(C_6D_6)

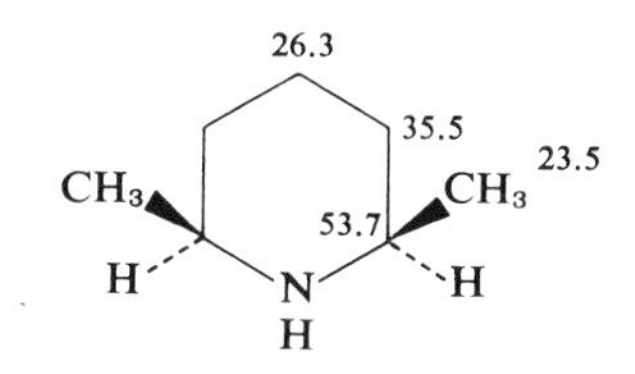

Ref. 23;
see also Ref. 22

74 ***trans*-2,6-Dimethylpiperidine**
(C_6D_6)

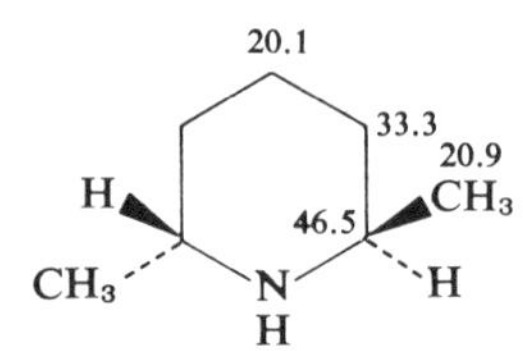

Ref. 23

75 **3,3-Dimethylpiperidine**

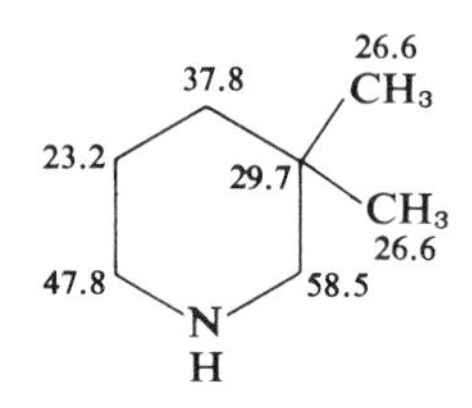

Ref. 24

76 ***cis*-3,5-Dimethylpiperidine**
(C_6D_6)

43.7 19.9
CH_3 CH_3
33.0
H H
55.0
N
H

Ref. 23

77 ***trans*-3,5-Dimethylpiperidine**
(C_6D_6)

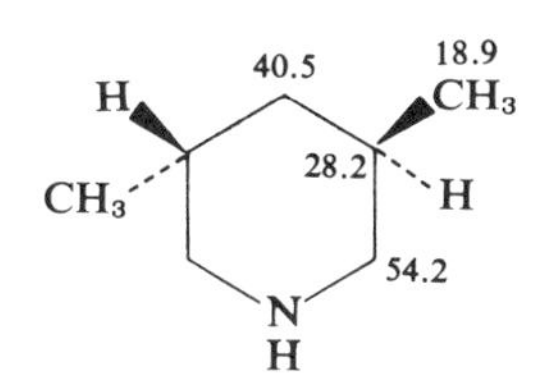

Ref. 23

78 **1,2,6-Trimethylpiperidine**

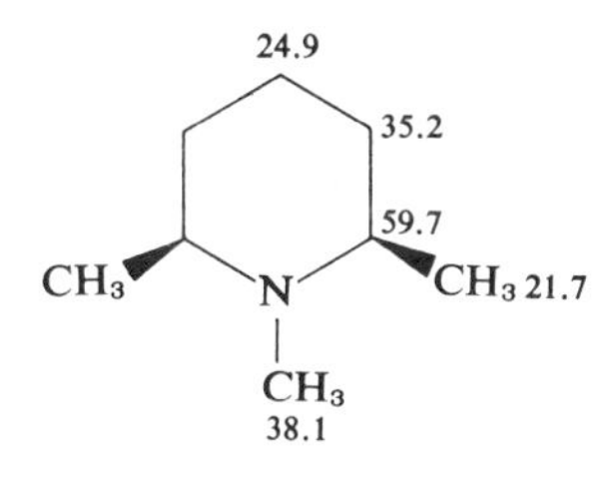

Ref. 26

79 **2,2,6,6-Tetramethylpiperidine**
(neat, values adjusted from C_6H_{12} reference)

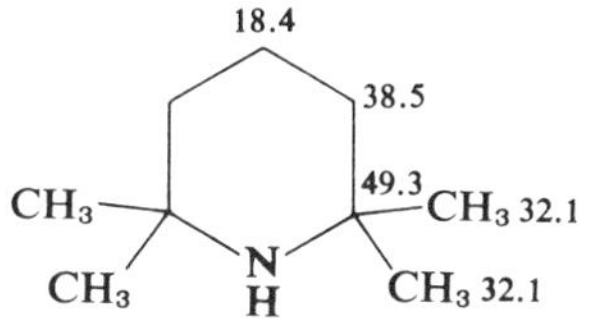

18.4
38.5
49.3
CH3
CH3 32.1
CH3
N
H
CH3 32.1

Ref. 22

80 1-Methyl-3-carbomethyoxypiperidine

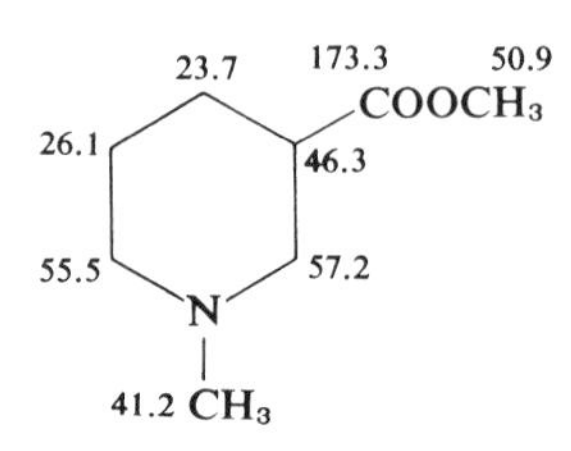

Ref. 24

81 *N*-Ethylpiperidine

(neat, values adjusted from C_6H_{12} reference)

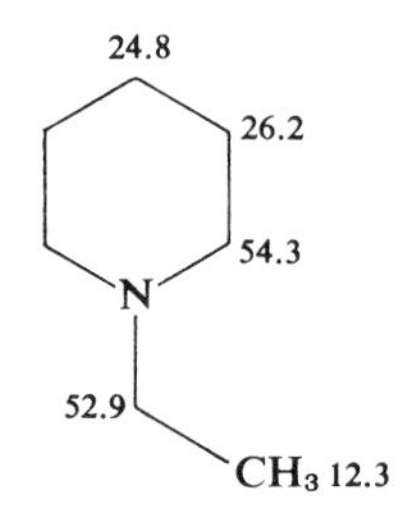

Ref. 22

82 2-Ethylpiperidine

(neat, values adjusted from C_6H_{12} reference)

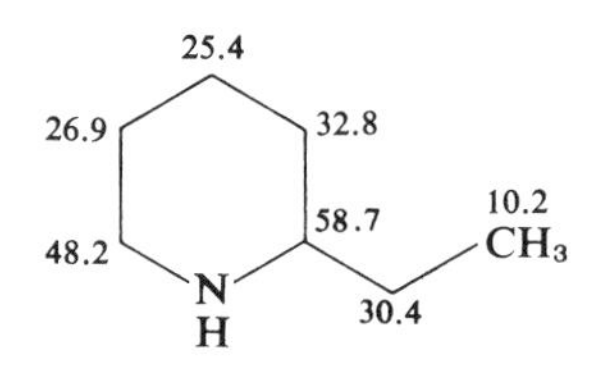

Ref. 22

83 1-Ethyl-3-hydroxypiperidine

33.2
OH
23.0
66.2
52.9[a]
60.7
N
52.3[a]
11.7
CH_3

Ref. 26

84 *N*-(2-Hydroxyethyl)piperidine
(neat, values adjusted from C_6H_{12} reference)

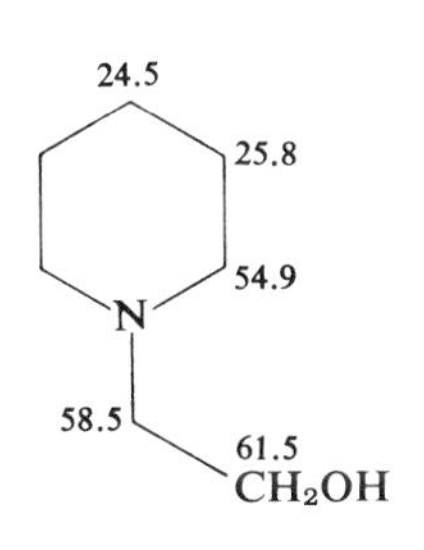

Ref. 22

85 Coniine

24.6
32.4
26.1
18.1
56.1
46.5
CH_3
13.1
39.1
N
H

Ref. 21

86 1,3-Diethyl-3-hydroxypiperidine

32.2 OH 34.1
7.0
21.7 CH3
69.0
51.9 62.8
N
53.1
CH3 12.0

Ref. 27

87 1,3-Diethyl-3,4-epoxypiperideine

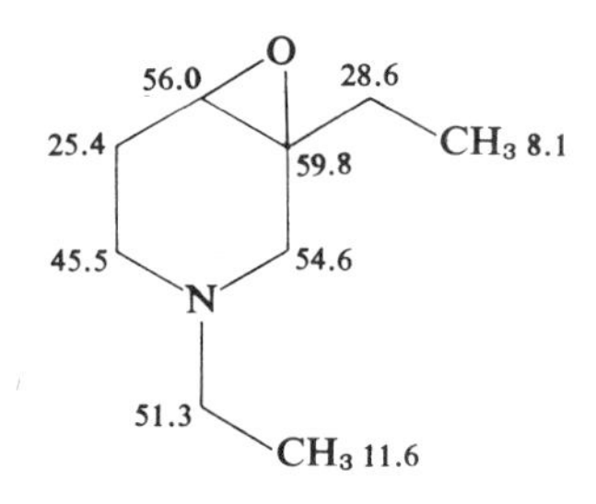

Ref. 27

88 1-Methyl-4-phenylpiperidine

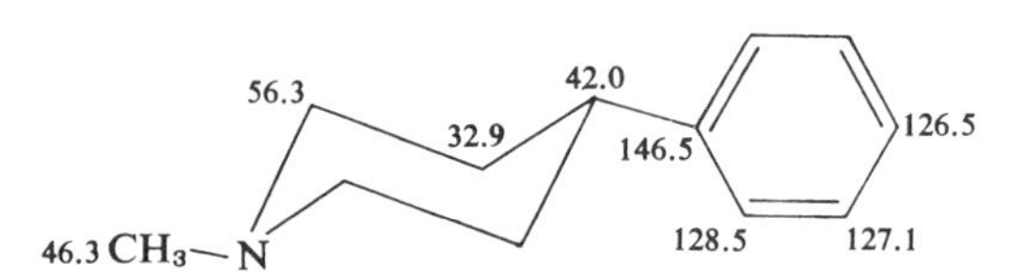

Ref. 28

89 1-Methyl-4-phenylpiperidin-4-ol

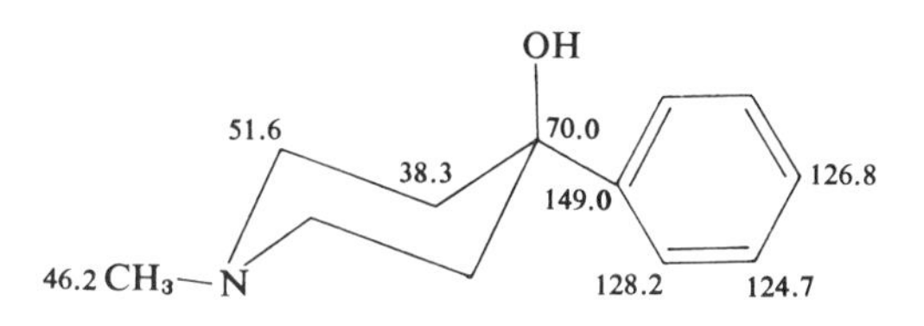

Ref. 28

90 1,2-Dimethyl-4-phenylpiperidin-4-ol (α isomer)

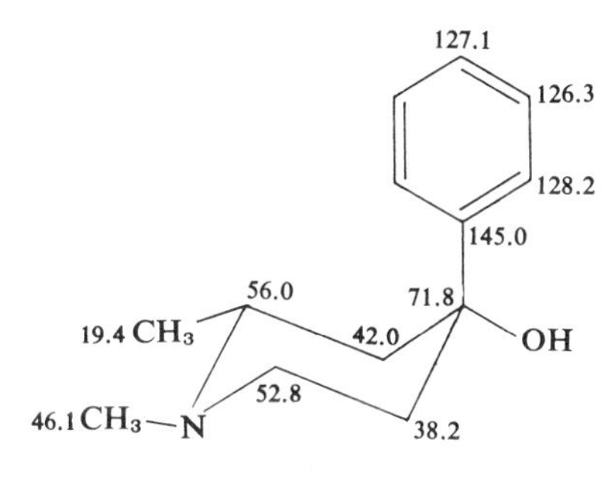

Ref. 28

91 1,2-Dimethyl-4-phenylpiperidin-4-ol (β isomer)

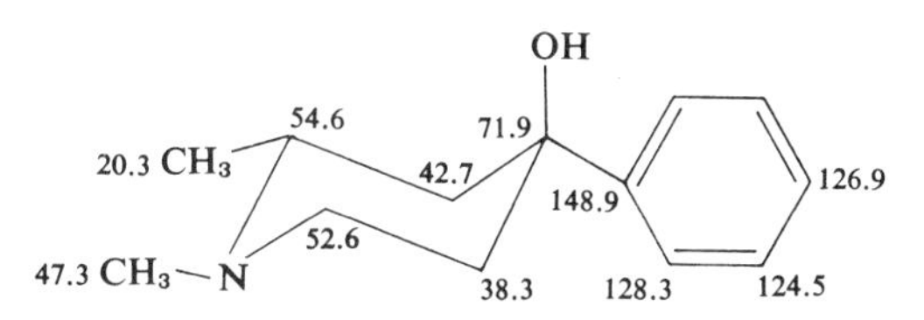

Ref. 28

92 1,3-Dimethyl-4-phenylpiperidin-4-ol (α isomer)

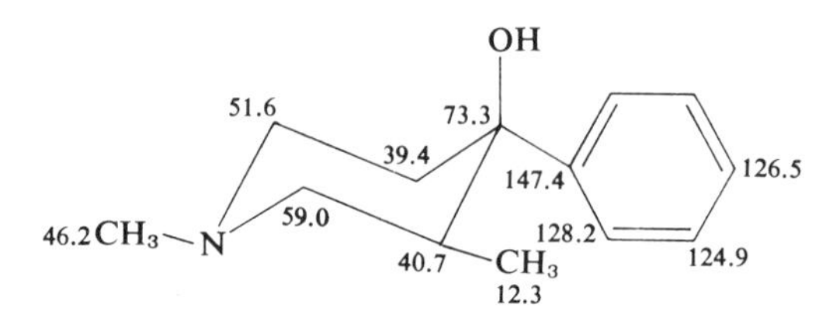

Ref. 28

93 1,3-Dimethyl-4-phenylpiperidin-4-ol (β isomer)

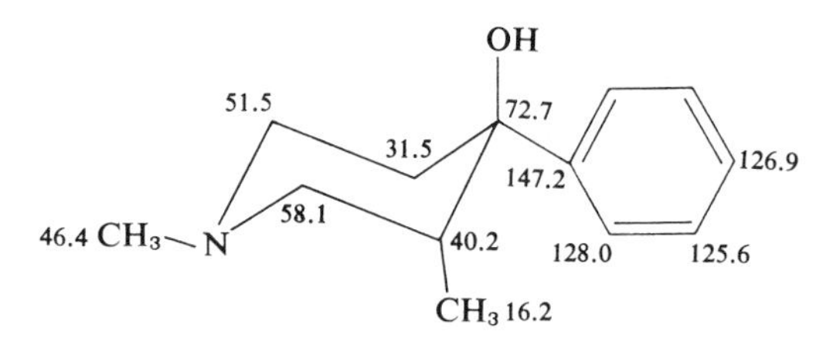

Ref. 28

94 1-Methyl-Δ^3-piperideine
(CCl_4)

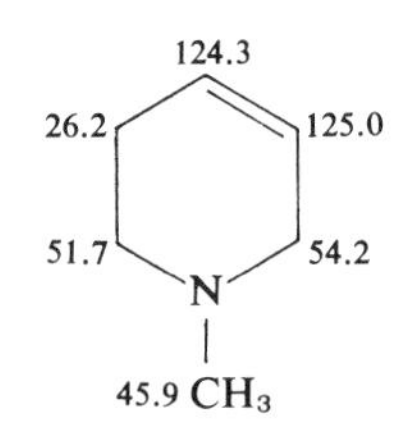

Ref. 24;
see also Ref. 18

95 **2-Methyl-Δ[1]-piperideine**

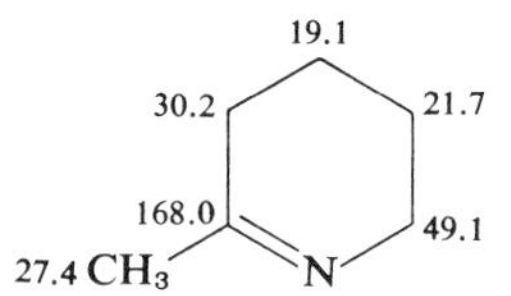

Ref. 29

96 **1,6-Dimethyl-Δ[3]-piperideine**

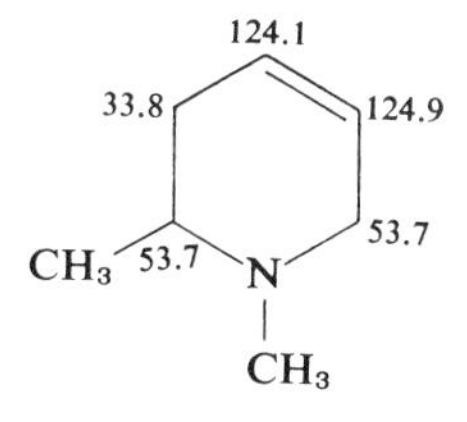

Ref. 18

97 **1,2-Dimethyl-Δ[3]-piperideine**

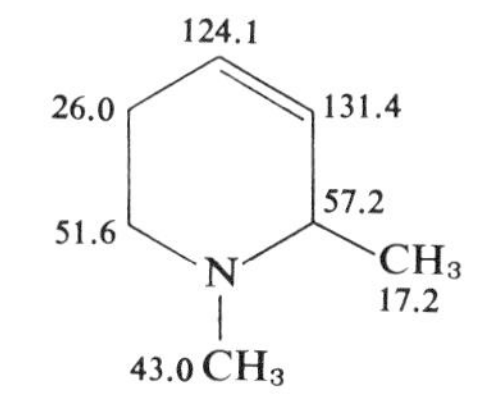

Ref. 24

98 1,2-Dimethyl-Δ⁴-piperideine

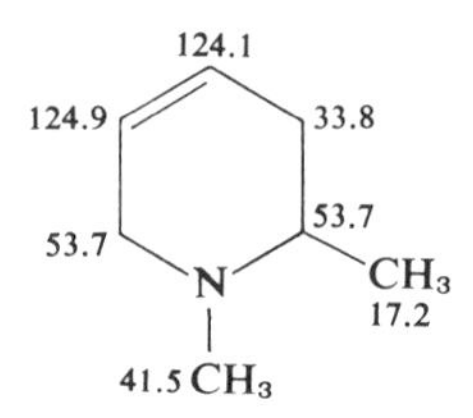

Ref. 24

99 1,3-Diethyl-Δ³-piperideine

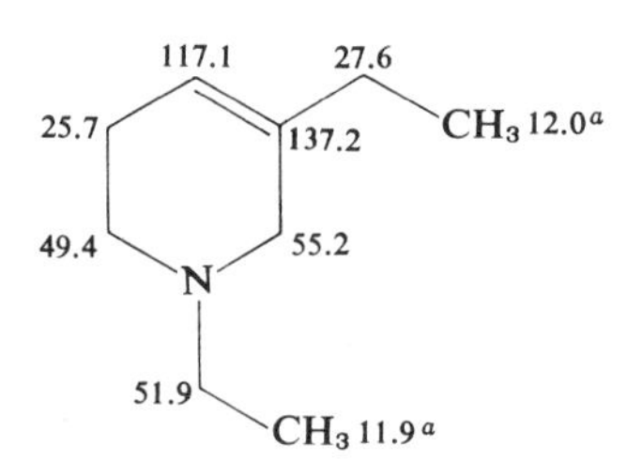

Ref. 27

100 1-Methyl-1,4,5,6-tetrahydronicotinonitrile

20.1[a] 122.9
CN
20.7[a]
70.8
46.2
147.0
N
CH_3 41.8

Ref. 29

101 3-Carbomethoxy-Δ²-piperideine

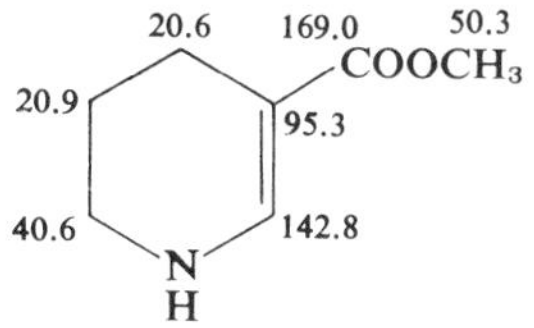

20.6 169.0 50.3
$COOCH_3$
20.9 95.3
40.6 142.8
N
H

Ref. 24

102 Methyl 1-methyl-1,4,5,6-tetrahydronicotinate

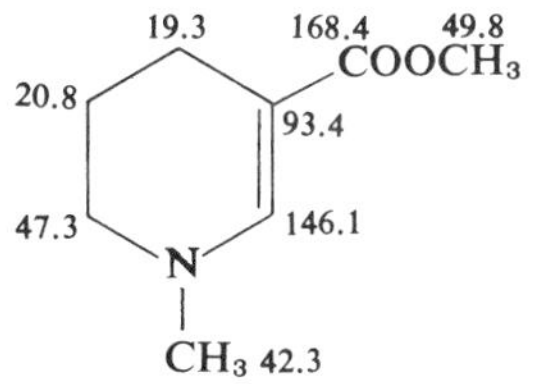

19.3 168.4 49.8
$COOCH_3$
20.8 93.4
47.3 146.1
N
CH_3 42.3

Ref. 29;
see also Ref. 24

103 Arecoline

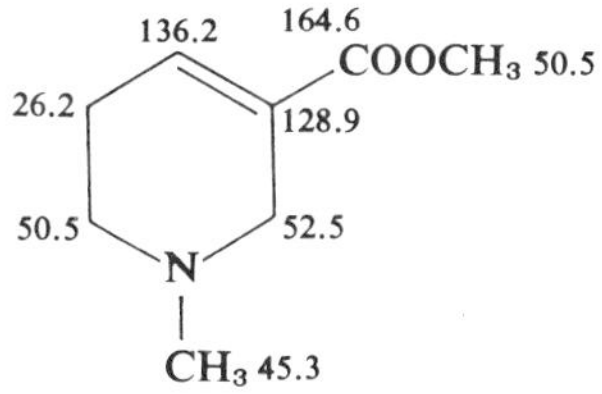

136.2 164.6
$COOCH_3$ 50.5
26.2 128.9
50.5 52.5
N
CH_3 45.3

Ref. 21;
see also
Refs. 24, 30

104 **Methyl 1-methyl-1,4-dihydronicotinate**

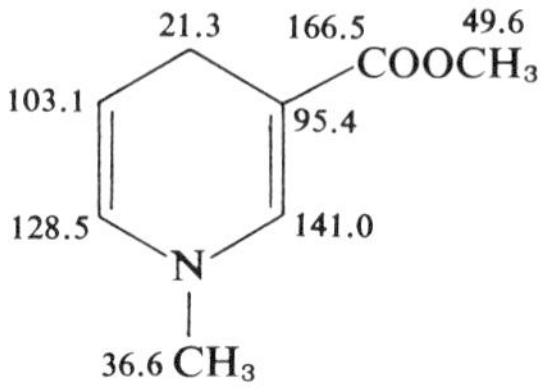

21.3 166.5 49.6
COOCH$_3$
103.1 95.4
128.5 141.0
N
36.6 CH$_3$

Ref. 24

105 **Arenaïne**

($CHCl_3$, values adjusted from CS_2 reference)

20.6
CH$_3$
40.7 65.1 142.5
116.1
29.9 CH$_2$
N NH
63.1 158.5
42.3 179.8 NH
CH$_3$
11.5
O

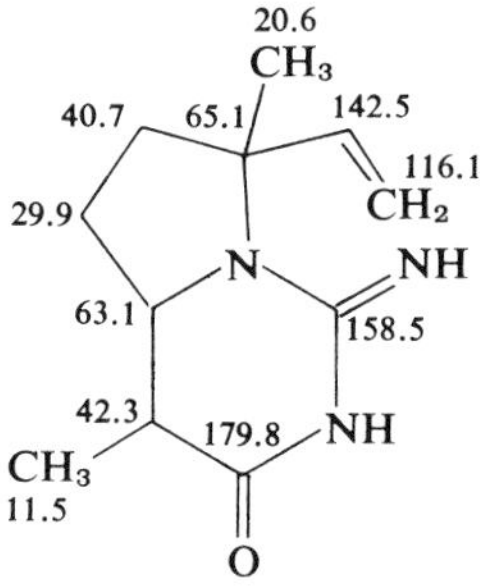

Ref. 31

106

(CCl_4, values adjusted from CS_2 reference)

O
121.0 45.6
130.0
107.9 168.2 N 25.8
148.0
107.4 45.6 24.5
O 148.0 25.8
O
101.2

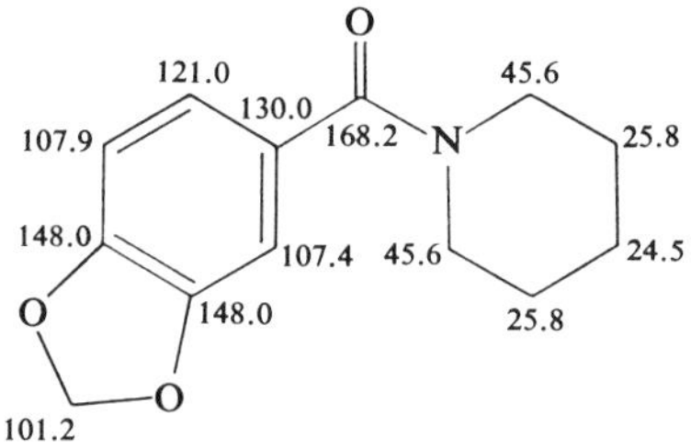

Ref. 32

107

(CCl_4, values adjusted from CS_2 reference)

O
123.1 141.4 163.8 44.8
108.0 129.9 115.5 N 26.0
147.8 106.0 44.8 24.6
O 148.3 26.0
100.6 O

Ref. 32

108 Piperine

($CHCl_3$, values adjusted from CS_2 reference)

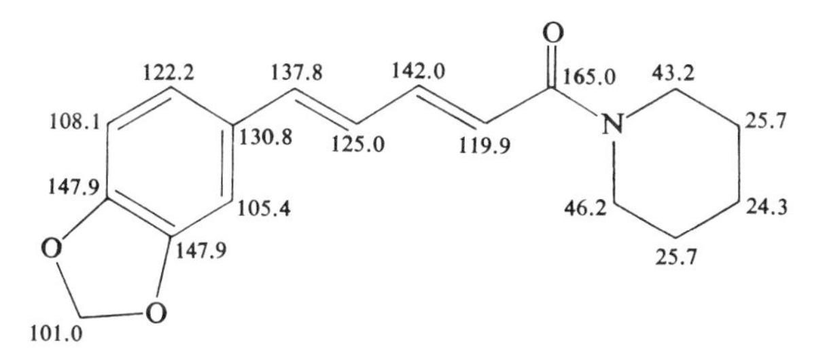

Ref. 32

109 Quinuclidine

20.8
26.8
47.6
N

Ref. 21

110 **Quinuclidine *N*-oxide**

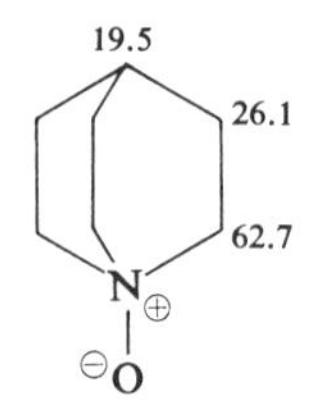

Ref. 21

111 ***syn*-3-Ethylidenequinuclidine**

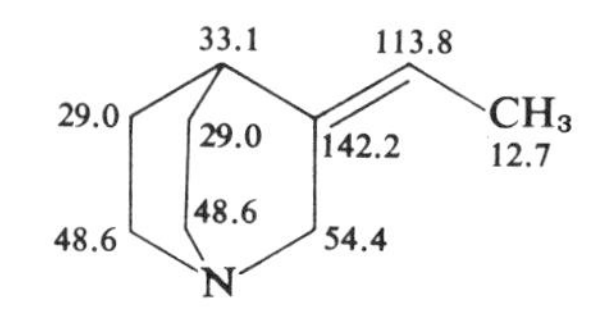

Ref. 33

112 ***anti*-3-Ethylidenequinuclidine**

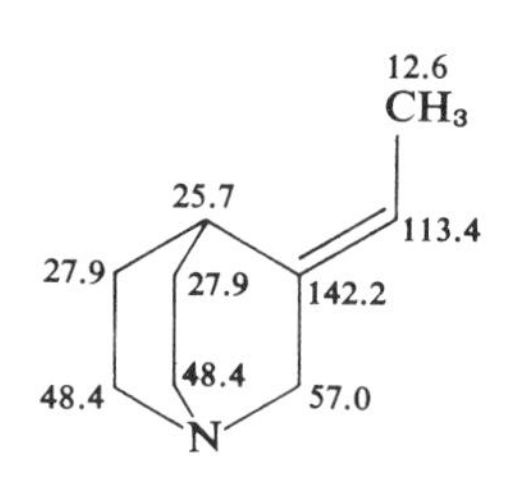

Ref. 33

113 **3-Ethyl-Δ^2-quinuclideine**

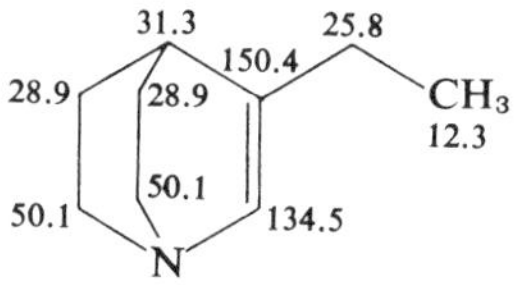

Ref. 33

114 **3-Quinuclidone**

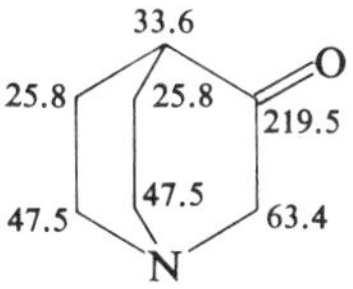

Ref. 33

115 **2-Azabicyclo[2,2,2]octan-3-one**

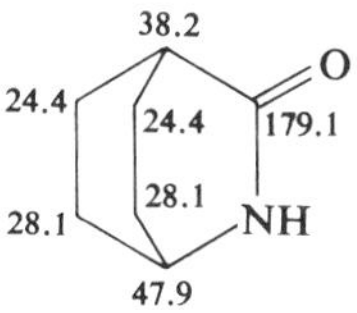

Ref. 33

116 **2-Azabicyclo[2,2,2]octane**

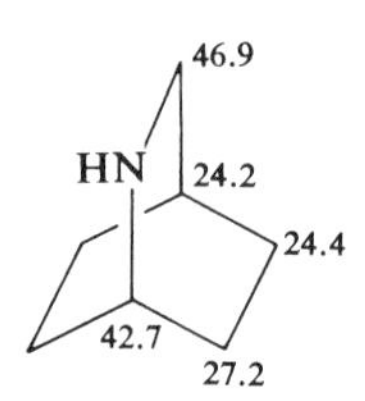

Ref. 34

117 ***N*-Methyl-2-azabicyclo[2,2,2]octane**

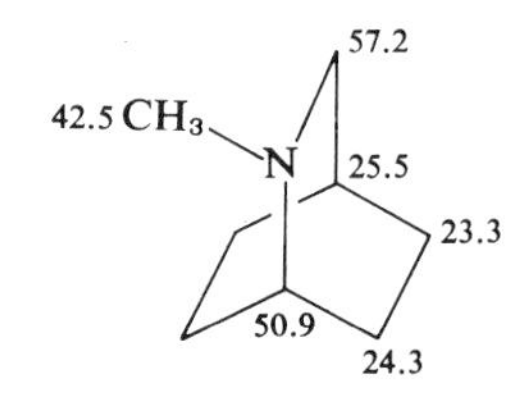

Ref. 34

118 **Dioscorine**

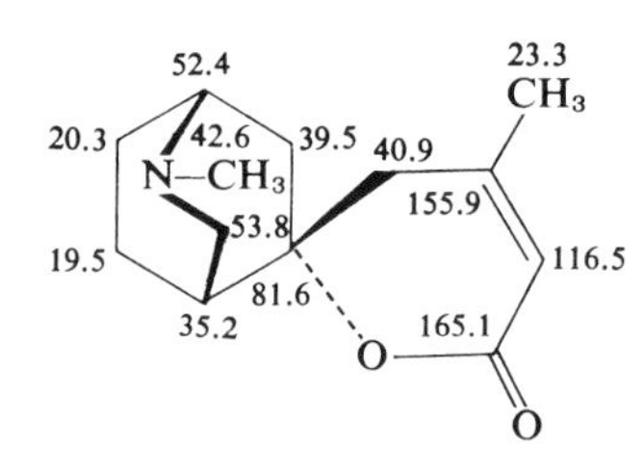

Ref. 35

119 2-Piperidone

22.4
21.0 41.9
31.3 NH
O

Ref. 26

120 1-Methyl-4-piperidone
($CHCl_3$)

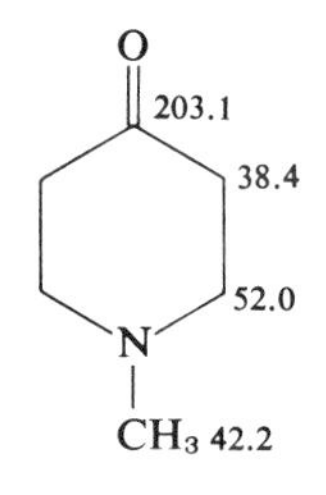

Ref. 36;
see also Ref. 37

121 1,2-Dimethyl-4-piperidone
($CHCl_3$–DMSO)

O
202.5
37.3 44.6
58.7
51.8 N CH_3 17.3
CH_3 ~39.5

Ref. 36

122 **1,3-Dimethyl-4-piperidone**
($CHCl_3$)

O
203.8
CH_3 10.9
37.4
40.7
53.1
60.4
N
CH_3 43.1

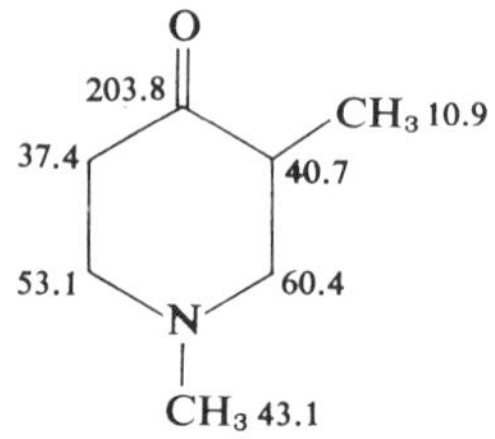

Ref. 36

123 **1,2,5-Trimethyl-4-piperidone**
($CHCl_3$)

O
203.3
10.7 CH_3 41.5
45.5
60.9
N
59.8
CH_3 17.9
CH_3 39.9

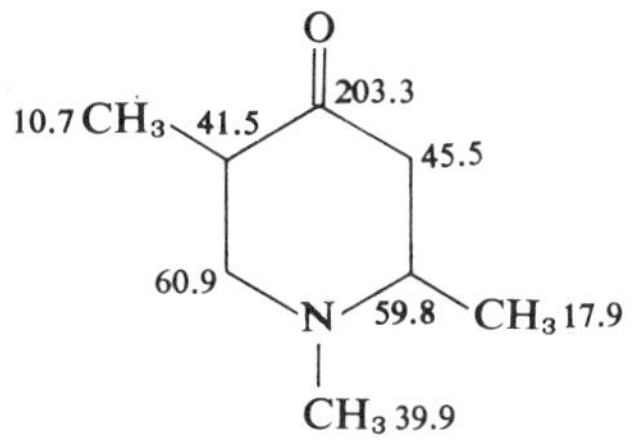

Ref. 36

6. SIMPLE ENAMINES

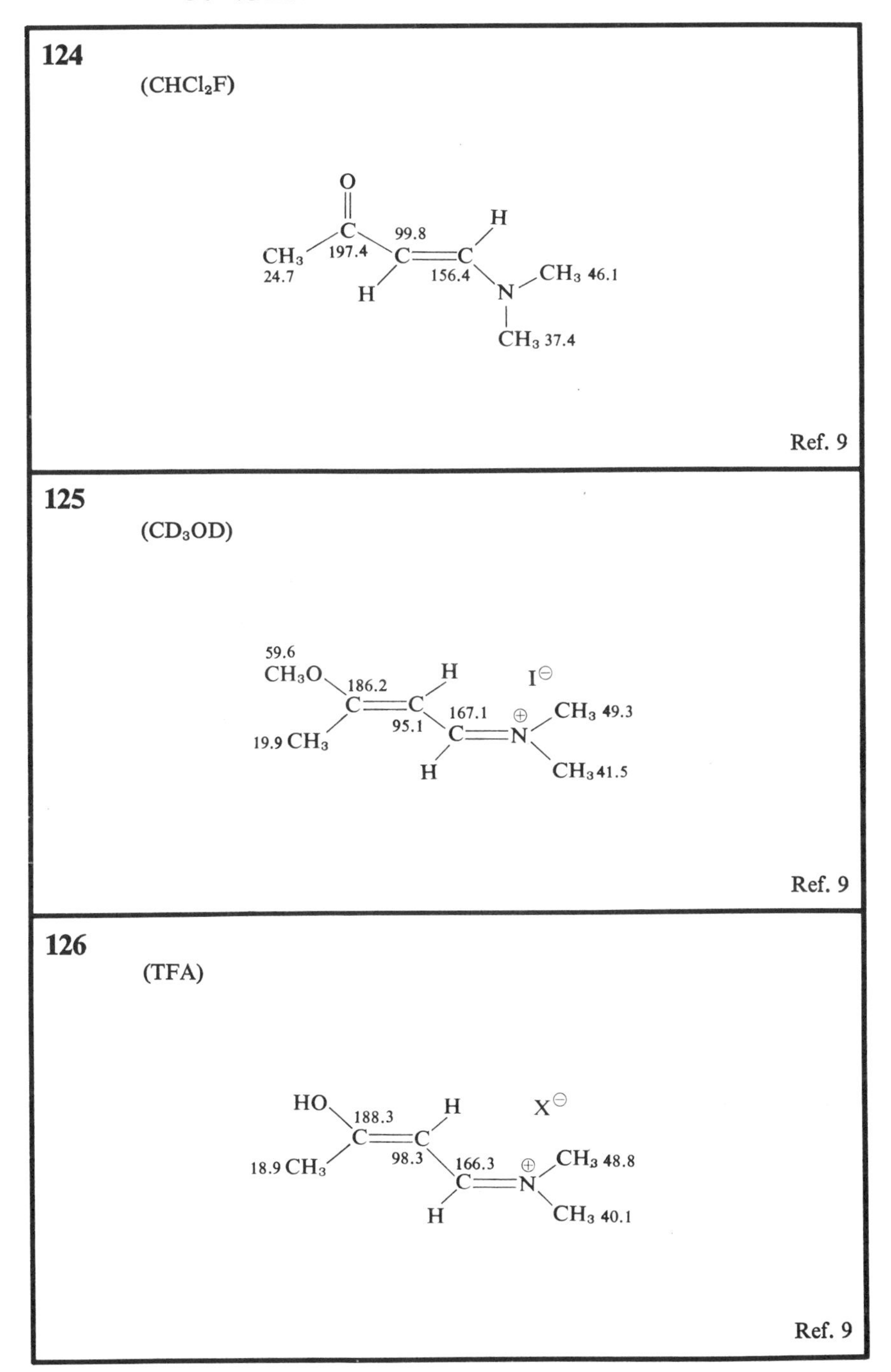

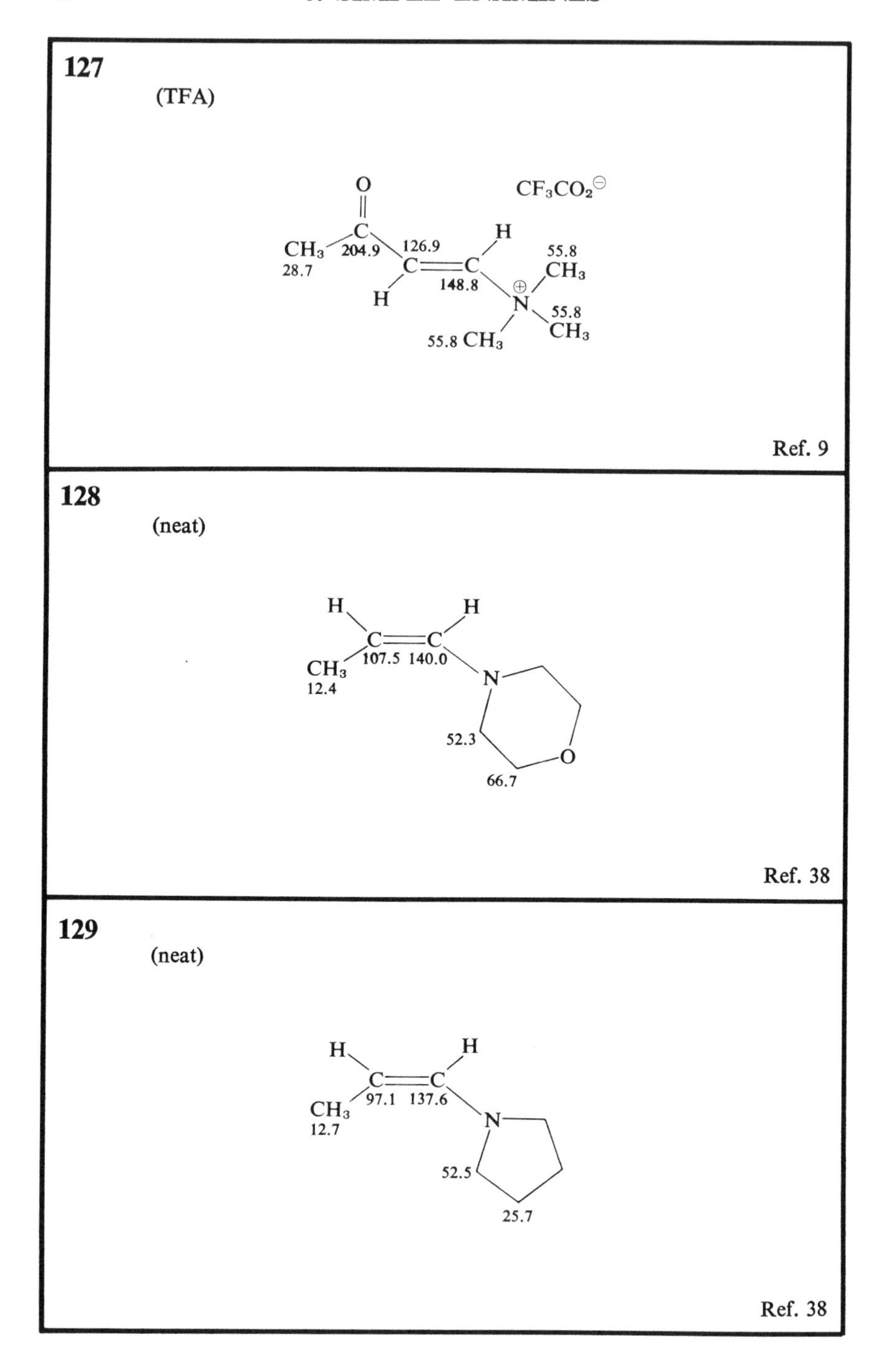
127
(TFA)
O
C
CH_3
204.9
28.7
126.9
C
H
H
C
148.8
$CF_3CO_2^{\ominus}$
55.8
CH_3
N
55.8
CH_3
55.8 CH_3
Ref. 9
128
(neat)
H
H
C
C
CH_3
107.5
140.0
12.4
N
52.3
O
66.7
Ref. 38
129
(neat)
H
H
C
C
CH_3
97.1
137.6
12.7
N
52.5
25.7
Ref. 38

130

(neat)

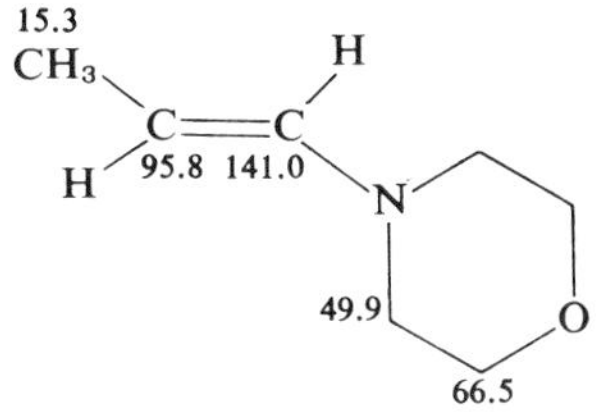

15.3
CH_3
H
C=C
H
95.8 141.0
N
49.9
O
66.5

Ref. 38

131

(neat)

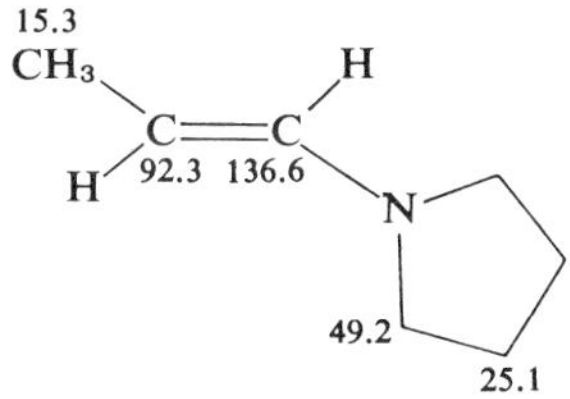

15.3
CH_3
H
C=C
H
92.3 136.6
N
49.2
25.1

Ref. 38

132

(neat)

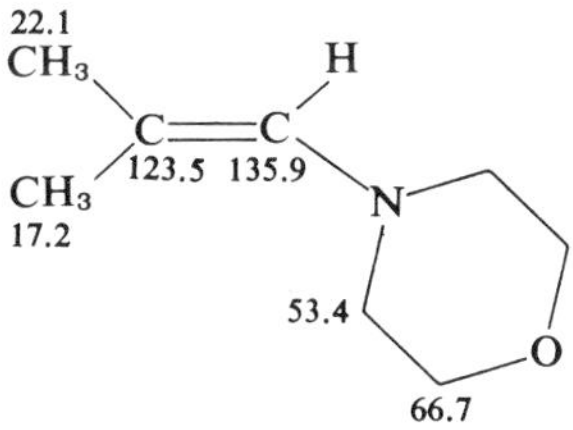

22.1
CH_3
H
C=C
123.5 135.9
CH_3
17.2
N
53.4
O
66.7

Ref. 38

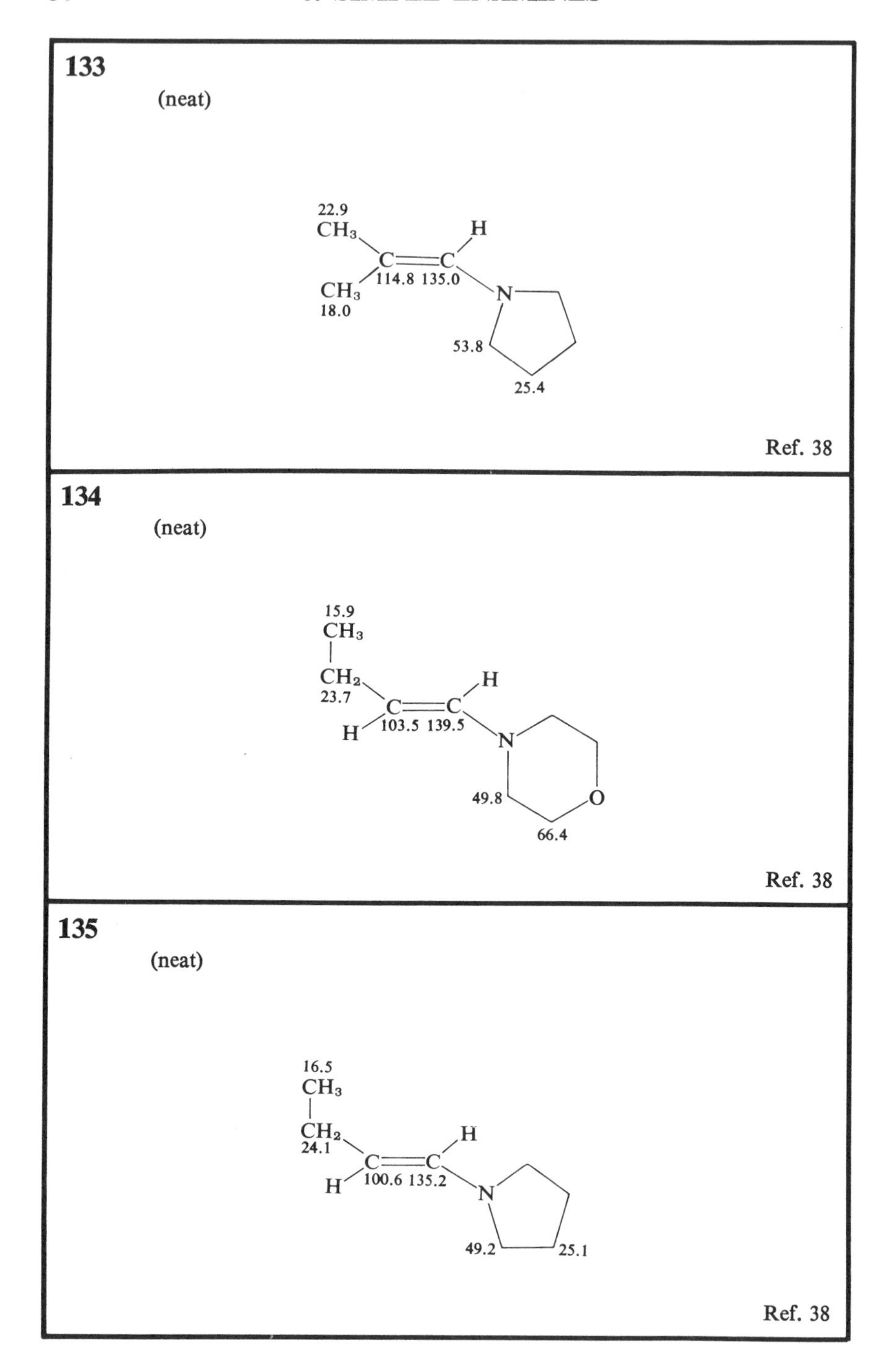
133
(neat)
22.9
CH3
H
C
C
114.8 135.0
N
CH3
18.0
53.8
25.4
Ref. 38
134
(neat)
15.9
CH3
CH2
H
23.7
C
C
103.5 139.5
H
N
49.8
O
66.4
Ref. 38
135
(neat)
16.5
CH3
CH2
H
24.1
C
C
100.6 135.2
H
N
49.2
25.1
Ref. 38

136 ***N*-Cyclohexenylpiperidine**

(C_2Cl_4)

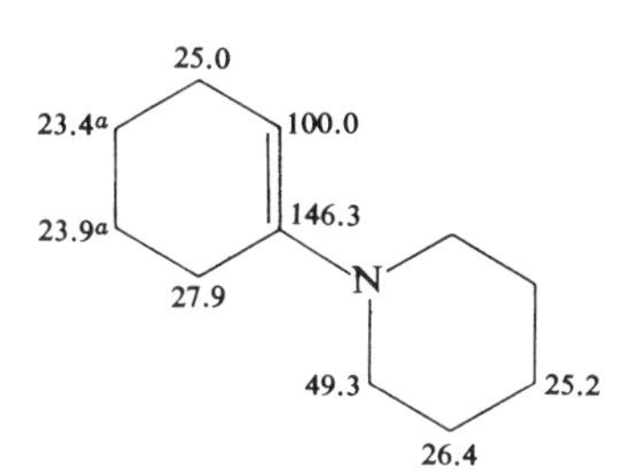

Ref. 39

137 ***N*-Cyclohexenylpyrrolidine**

(C_2Cl_4)

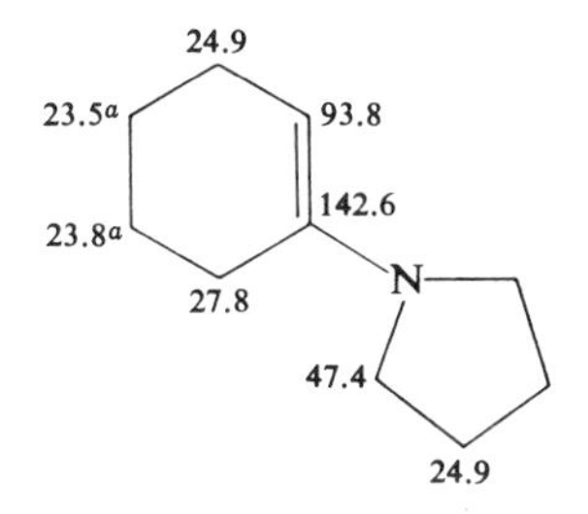

Ref. 39

138

(C_2Cl_4)

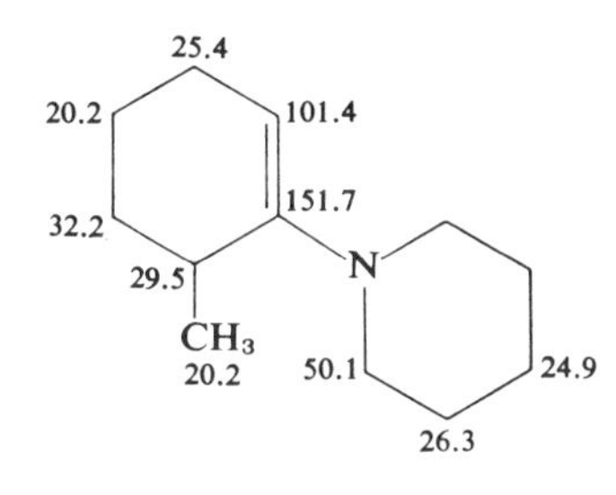

Ref. 39

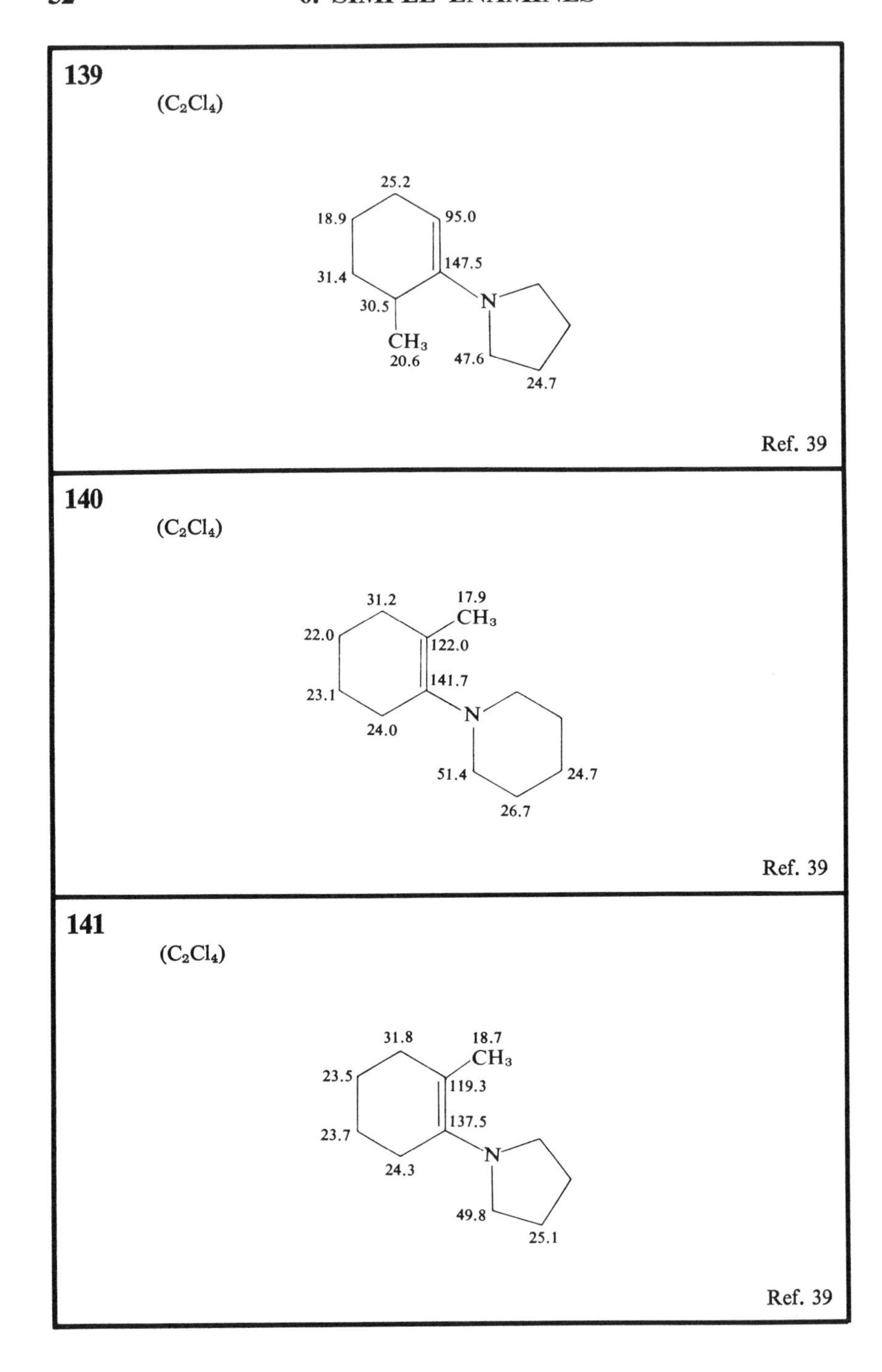
139
(C_2Cl_4)
25.2
18.9
95.0
147.5
31.4
30.5
N
CH3
20.6
47.6
24.7
Ref. 39
140
(C_2Cl_4)
31.2
17.9
CH3
22.0
122.0
141.7
23.1
N
24.0
51.4
24.7
26.7
Ref. 39
141
(C_2Cl_4)
31.8
18.7
CH3
23.5
119.3
137.5
23.7
N
24.3
49.8
25.1
Ref. 39

142 ***N*-Cyclopentenylpiperidine**
(C_2Cl_4)

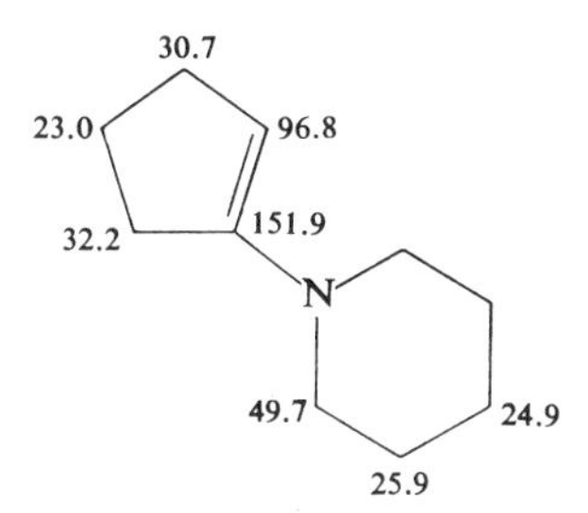

Ref. 39

143 ***N*-Cyclopentenylpyrrolidine**
(C_2Cl_4)

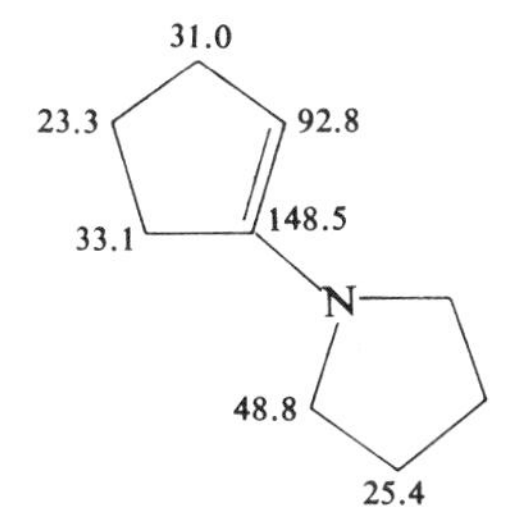

Ref. 39

144

O
196.8
35.8
99.5
164.6
22.3
27.1
N
47.6
24.3
25.6

Ref. 39

145

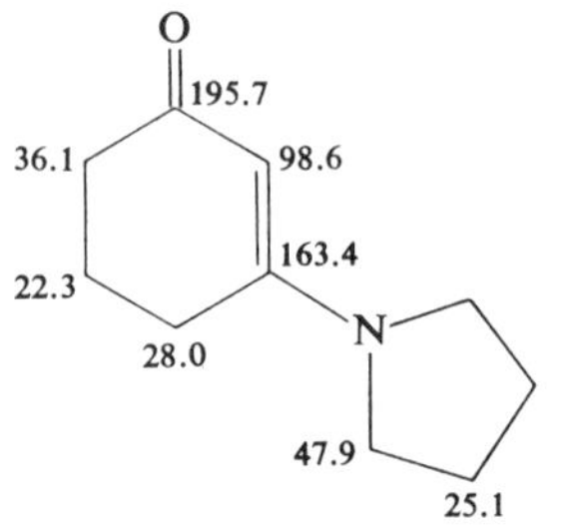

O
195.7
36.1
98.6
163.4
22.3
N
28.0
47.9
25.1

Ref. 39

146 1,1-Diphenyl-2-azadeca-1,3Z,5Z,7E-tetraene-9-one

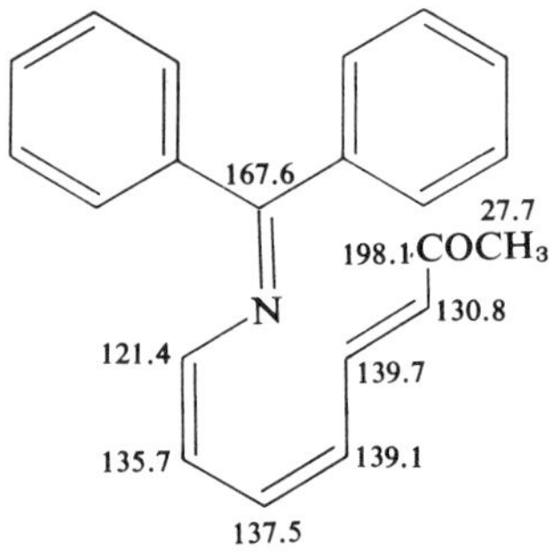

167.6
27.7
198.1 $COCH_3$
N
130.8
121.4
139.7
135.7
139.1
137.5

Ref. 40

7. TROPANES[138]

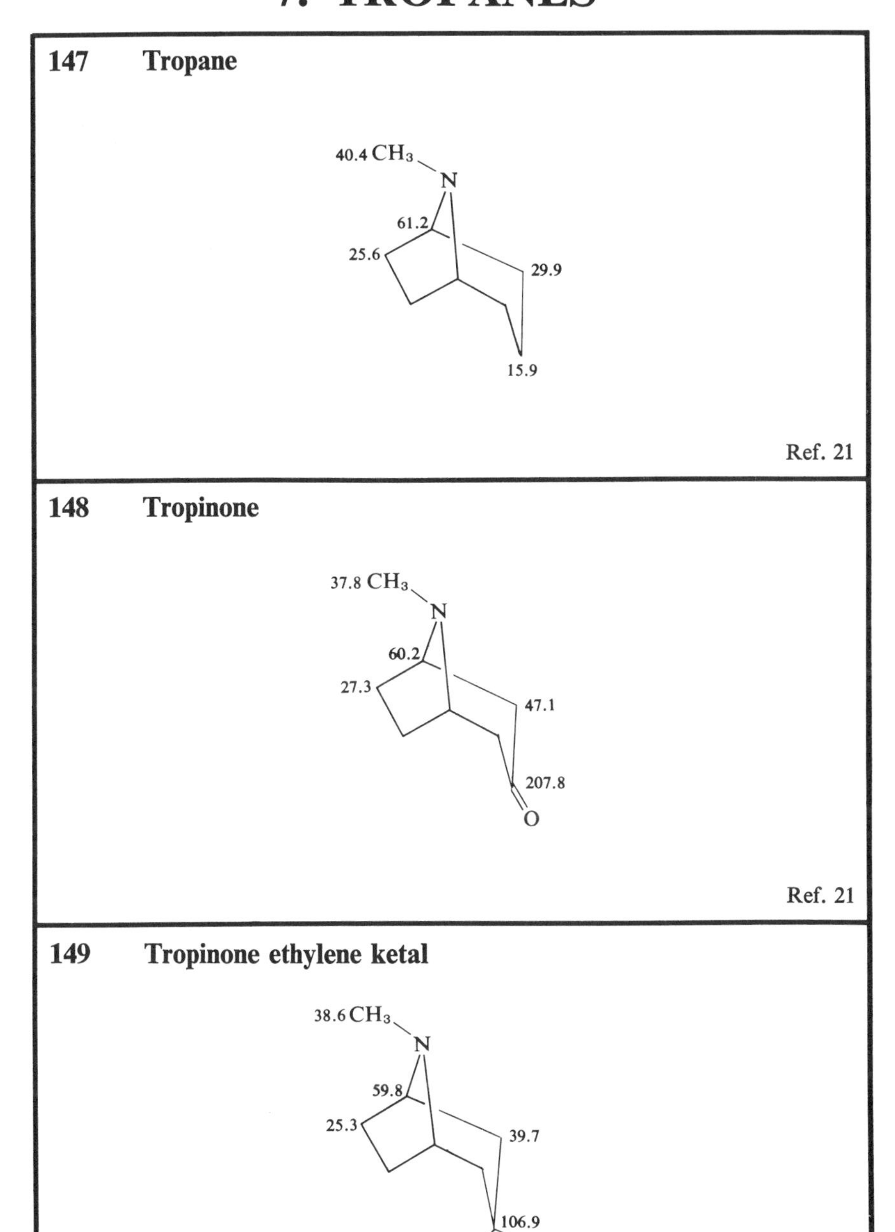

150 **(3R)-Hydroxytropane**

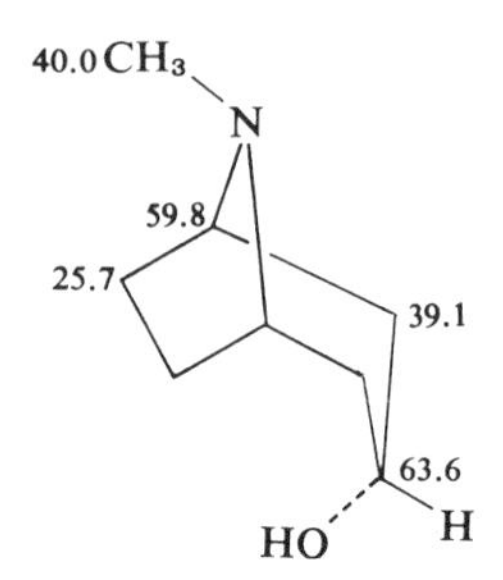

Ref. 21

151 **Pseudotropine**

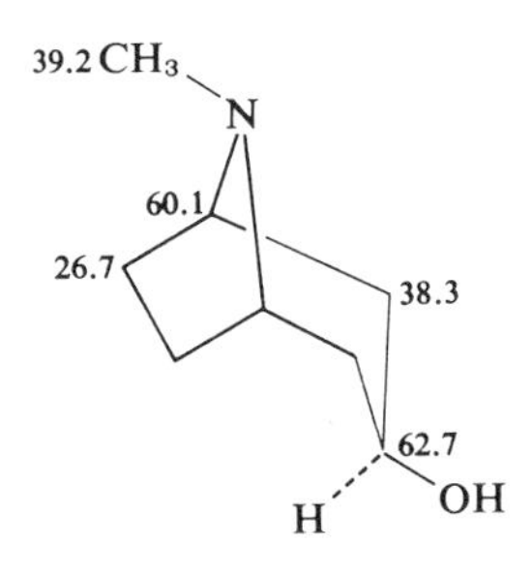

Ref. 21

152 **(3R)-Hydroxytropane benzoate**

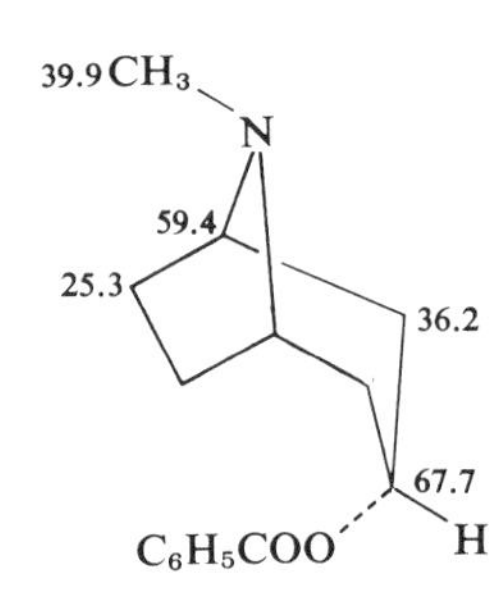

Ref. 21

153 Tropacocaine

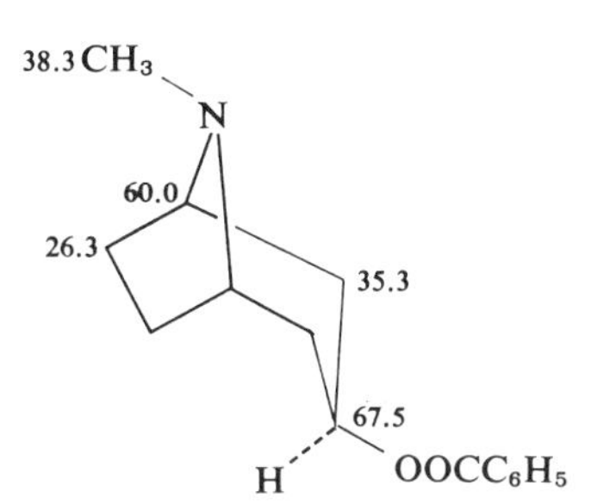

Ref. 21

154 Cocaine

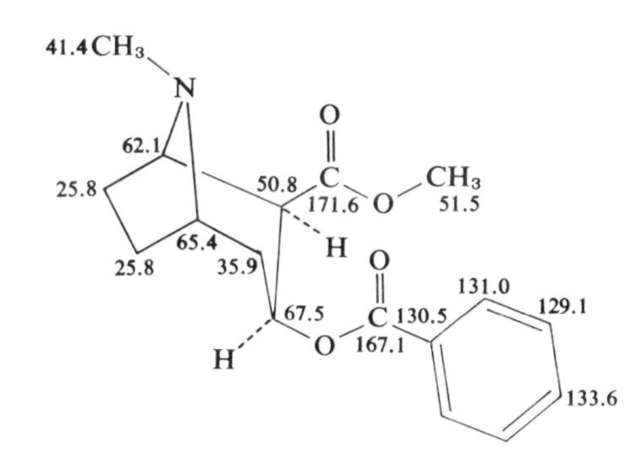

Ref. 41

155 Atropine

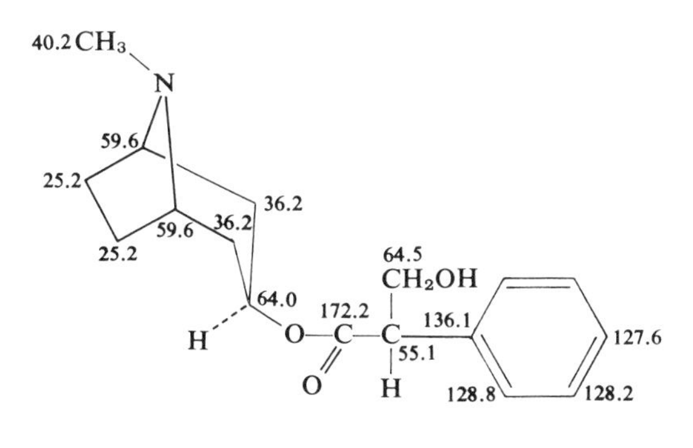

Ref. 42;
see also Ref. 41

156 Tropidine

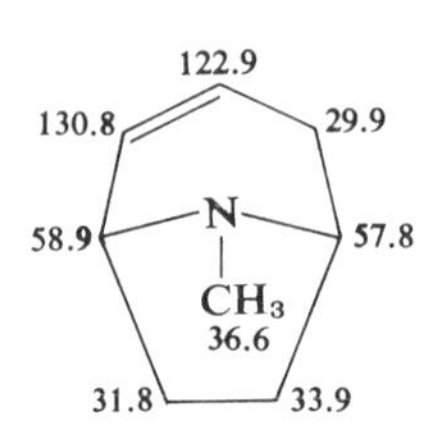

Ref. 21

157 Scopolamine

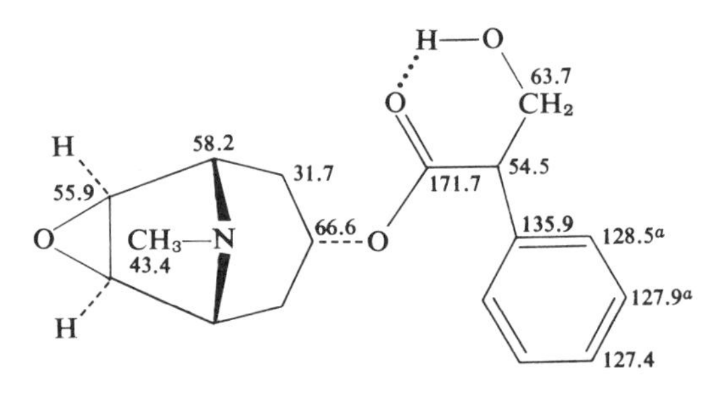

Ref. 21

158 Scopolamine *N*-oxide

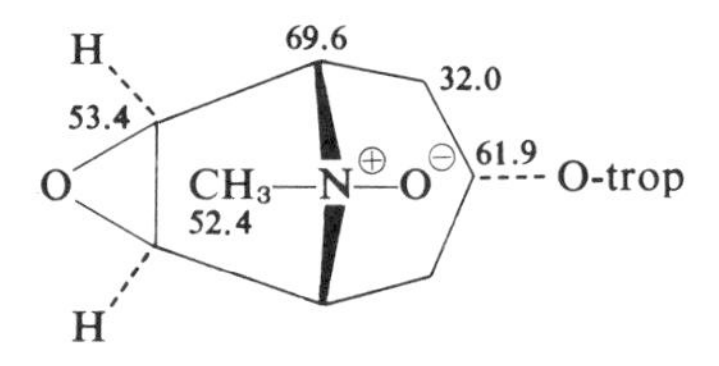

Ref. 21

8. QUINOLIZIDINES[43,136,139]

Simple Quinolizidines

159 Quinolizidine

33.2
62.9
24.4
N
25.6
56.4

Ref. 21;
see also
Refs. 26, 44

160 Indolizidine

30.7
64.1
30.1
24.2
20.3
25.1
N
53.9
52.7

Ref. 21

161

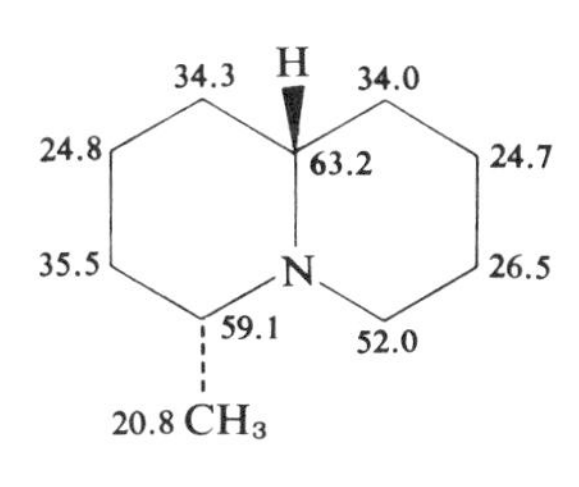

Ref. 44

162

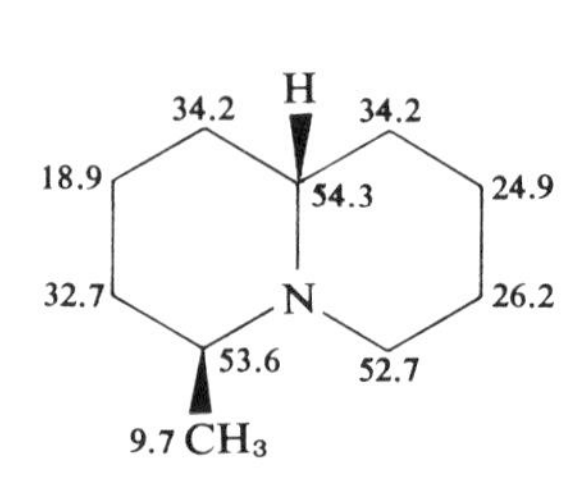

Ref. 44

163 Quinolizidine methiodide
(CD_3OD)

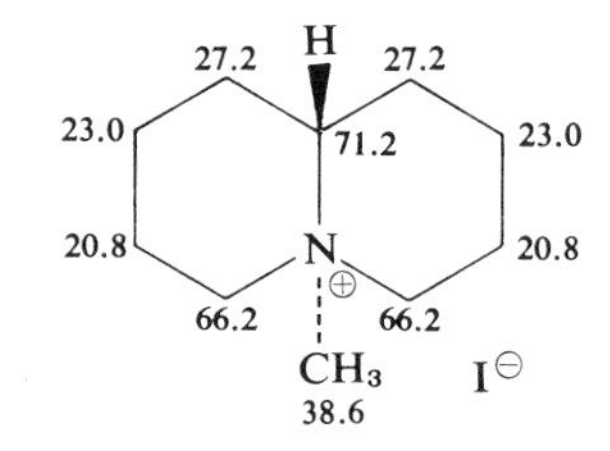

Ref. 44

164 *trans*-10-Methylquinolizidine *N*-metho salt
(TFA-*d*)

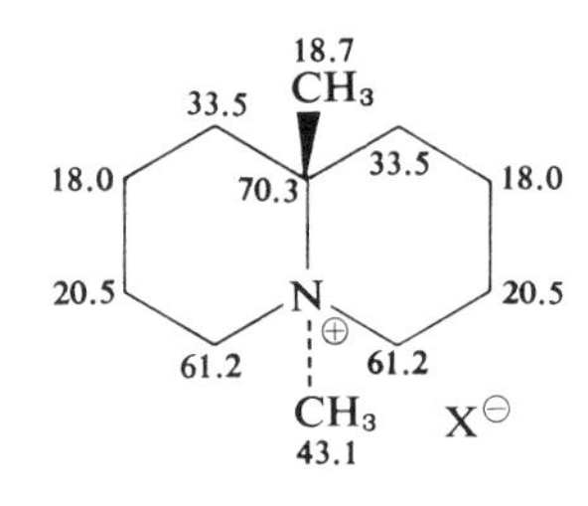

Ref. 45

165 ***cis*-10-Methylquinolizidine *N*-metho salt**
(TFA-*d*)

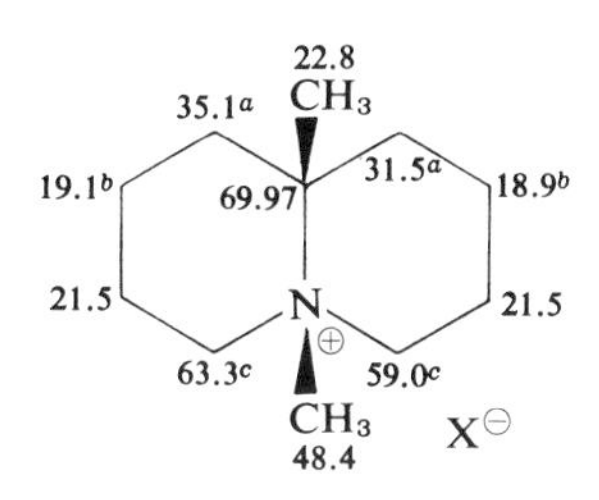

Ref. 45

166

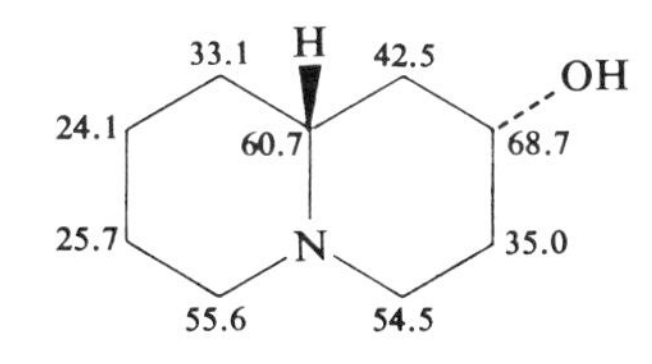

Ref. 26

167

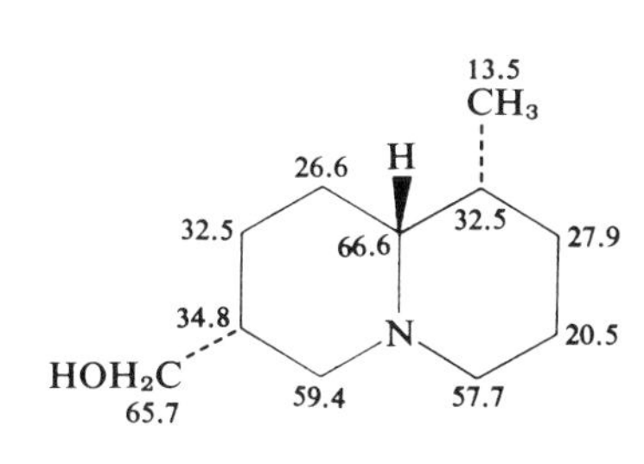

Ref. 26

168 Lupinine

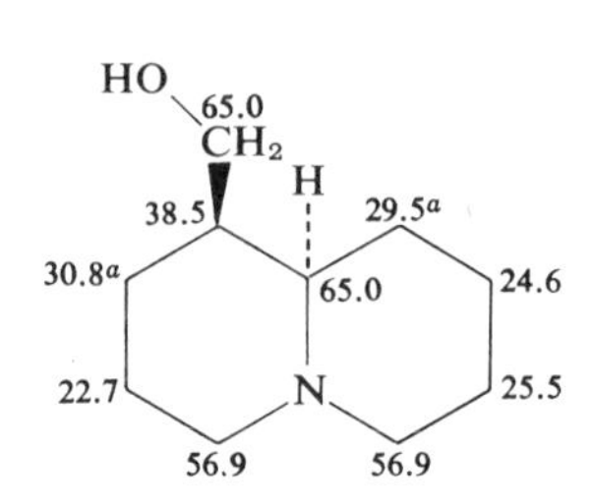

Ref. 29;
see also Ref. 26

169 Epilupinine

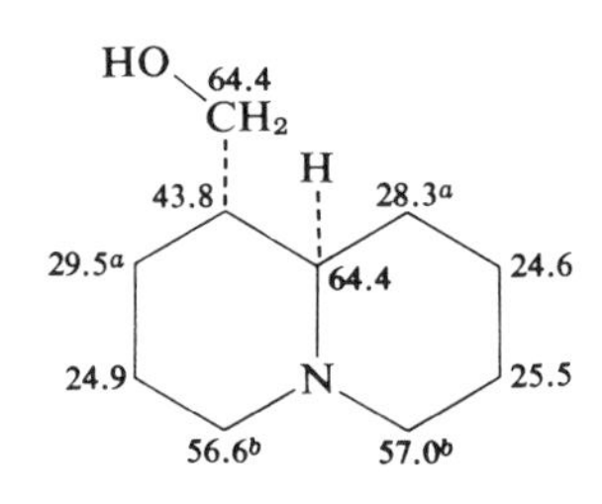

Ref. 29;
see also Ref. 26

170

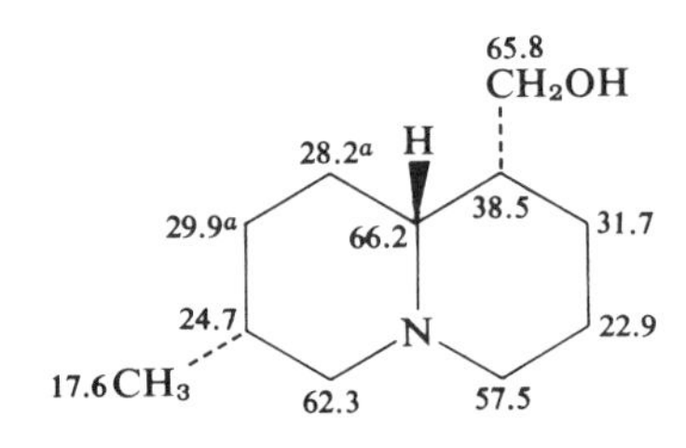

Ref. 26

171

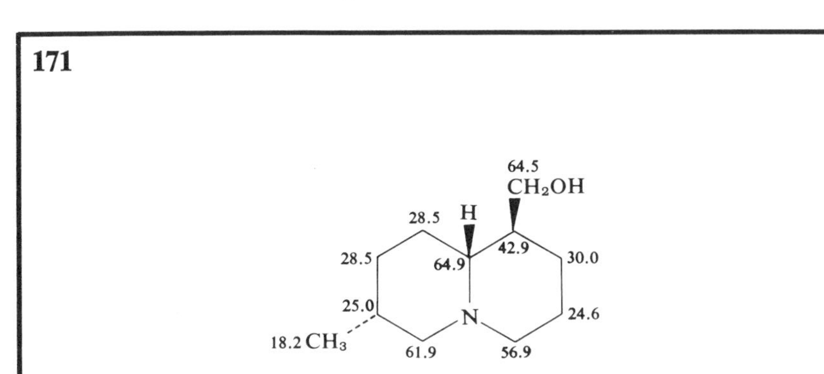

Ref. 26

172

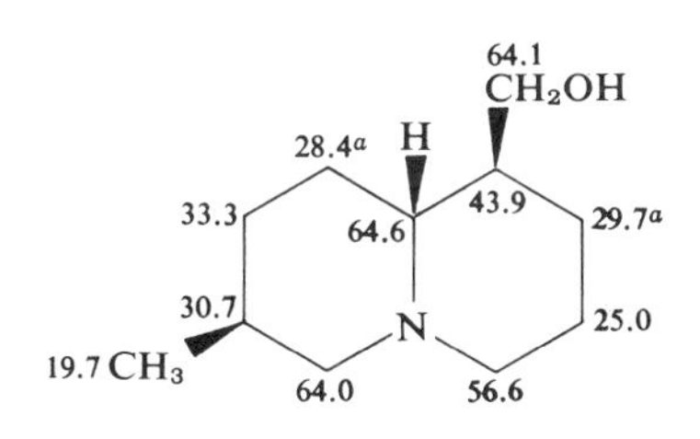

Ref. 26

173

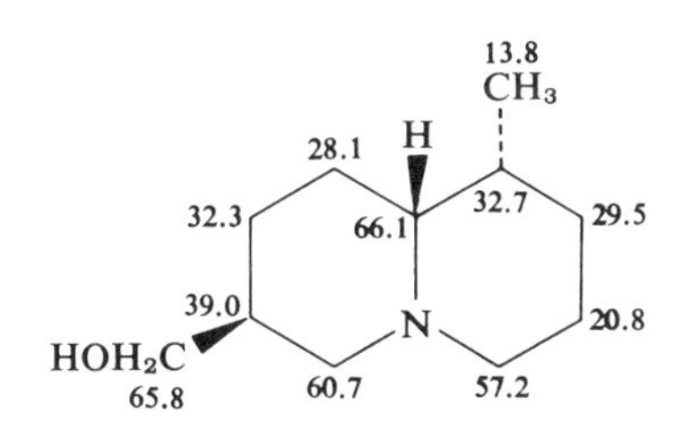

Ref. 26

174 4-Quinolizidone

Ref. 26

175

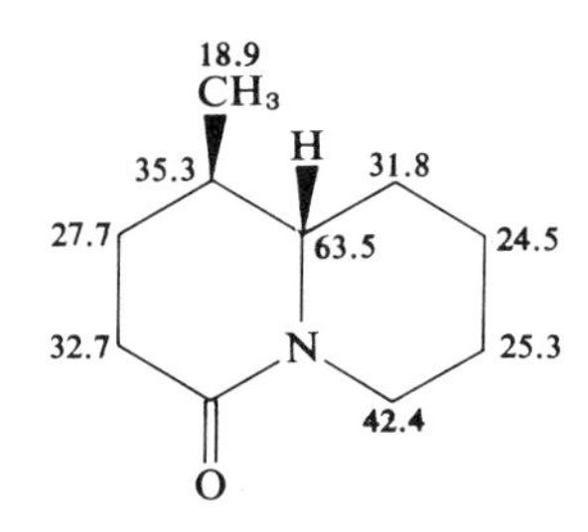

Ref. 26

176

Ref. 26

b. Julolidines

177 Julolidine

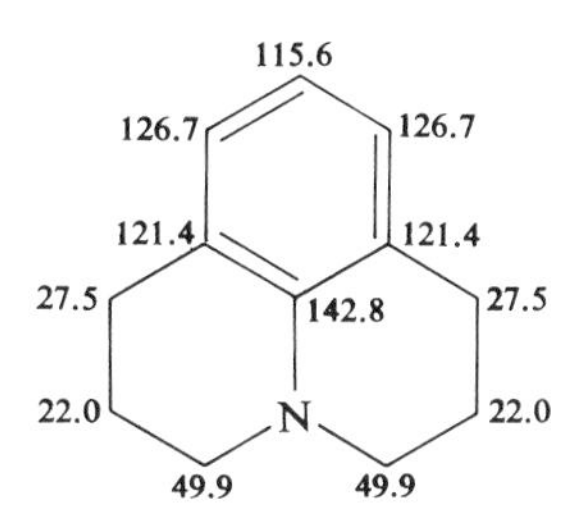

Ref. 29

178 Hexahydrojulolidine isomer

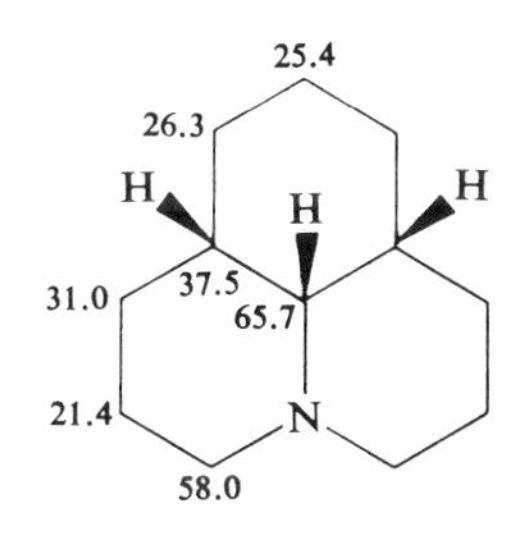

Ref. 26

179 Hexahydrojulolidine isomer

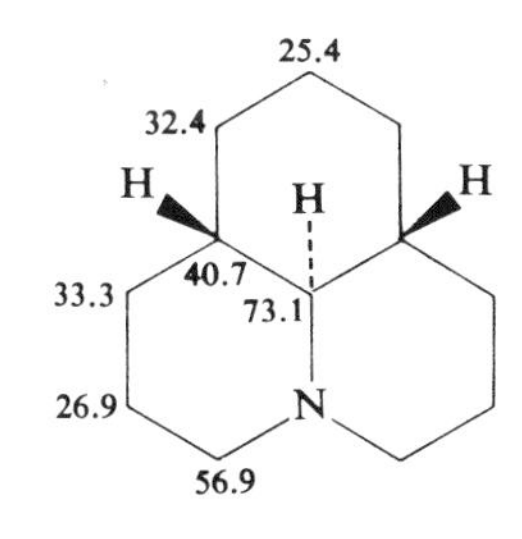

Ref. 26

180 9-Aza-8-methyl-7a,10,10a,10b-tetrahydrojulolidine

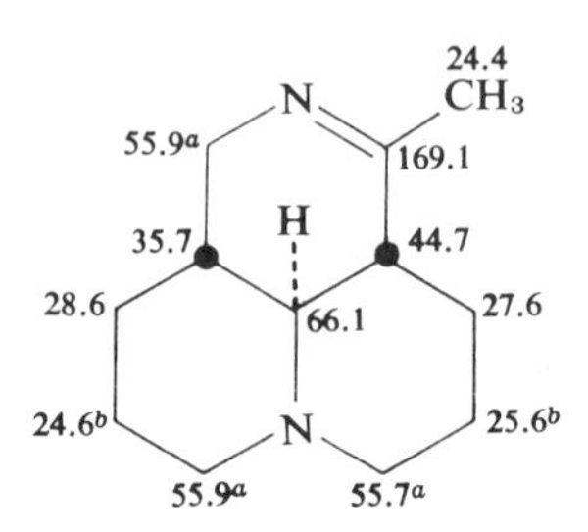

Ref. 29

181 7a,8,9,10,10a,10b-Hexahydro-8-julolidone

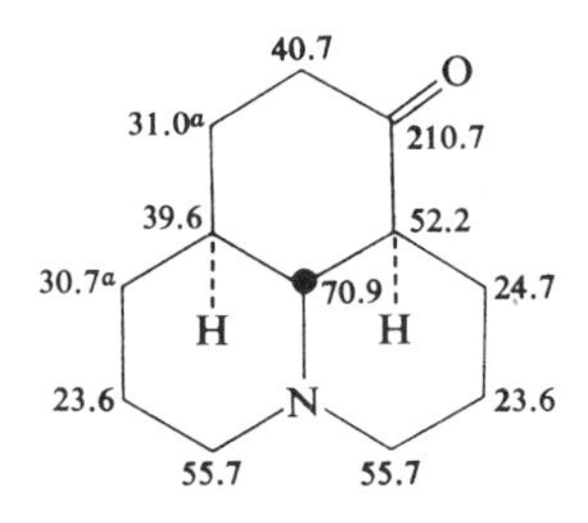

Ref. 29

182 7a,8,10a,10b-Tetrahydro-8-julolidone

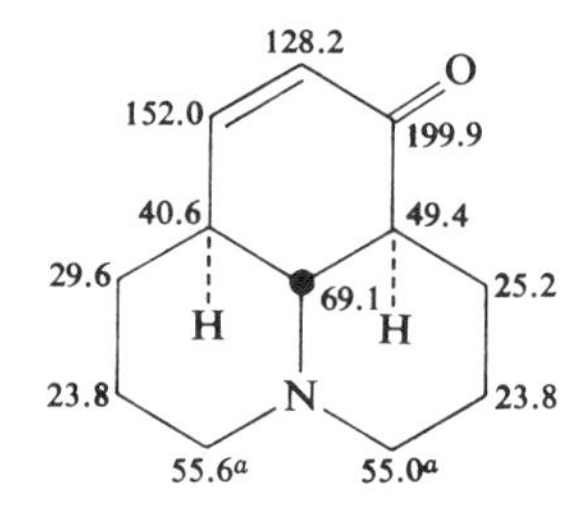

Ref. 29

183 **10-Acetonyl-7a,8,9,10,10a,10b-hexahydro-8-julolidone**

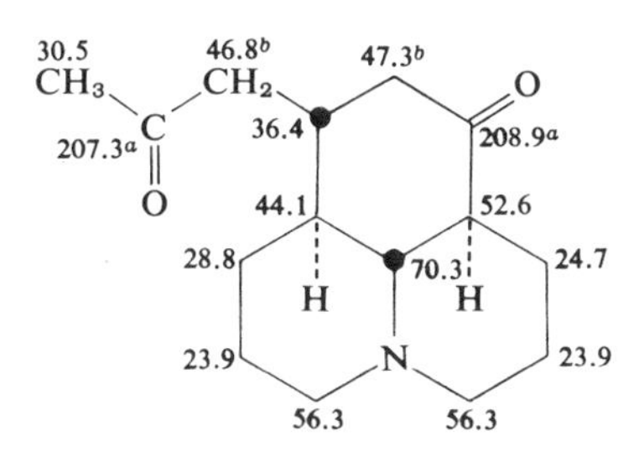

Ref. 29

184 **Methyl *trans,trans*-7a,8,10a,10b-tetrahydro-8-julolidone-10b-carboxylate**

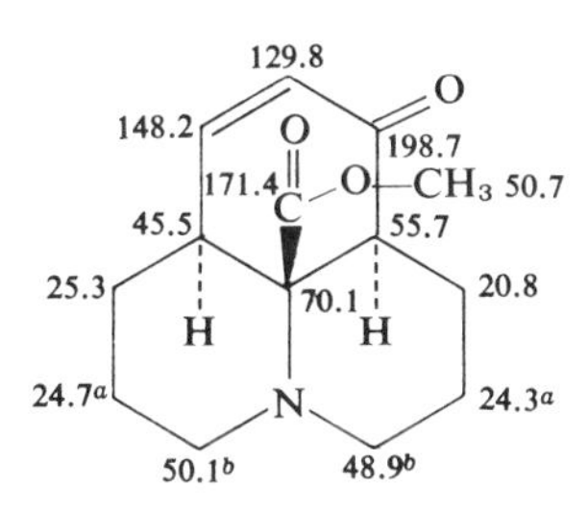

Ref. 29

185 **Methyl *cis,cis*-7a,8,10a,10b-tetrahydro-8-julolidone-10b-carboxylate**

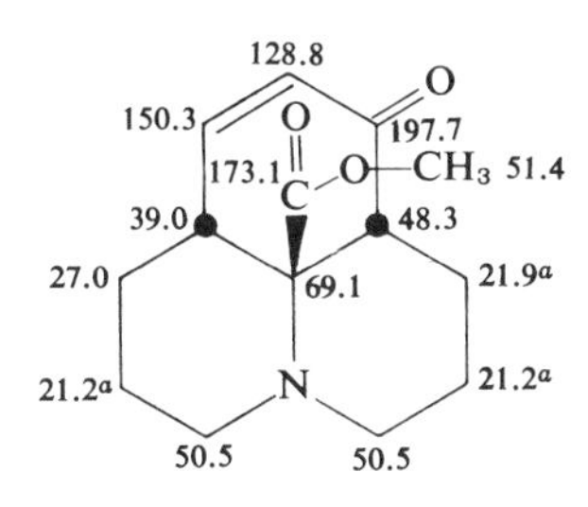

Ref. 29

186 **Methyl *cis,trans*-7a,8,10a,10b-tetrahydro-8-julolidone-10b-carboxylate**

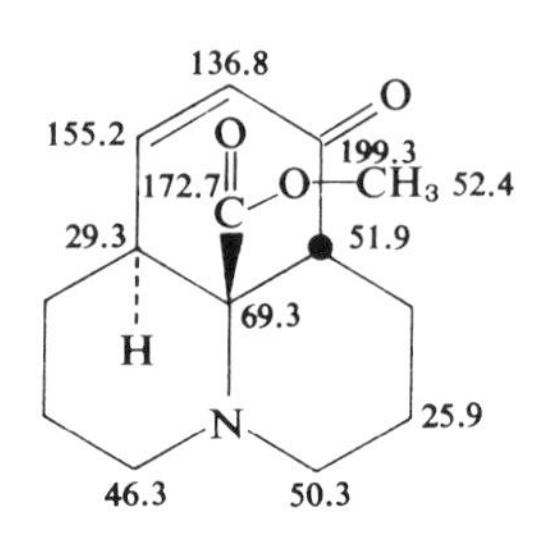

Ref. 29

187 **Methyl *cis,cis*-10-methoxy-7a,8,10a,10b-tetrahydro-8-julolidone-10b-carboxylate**

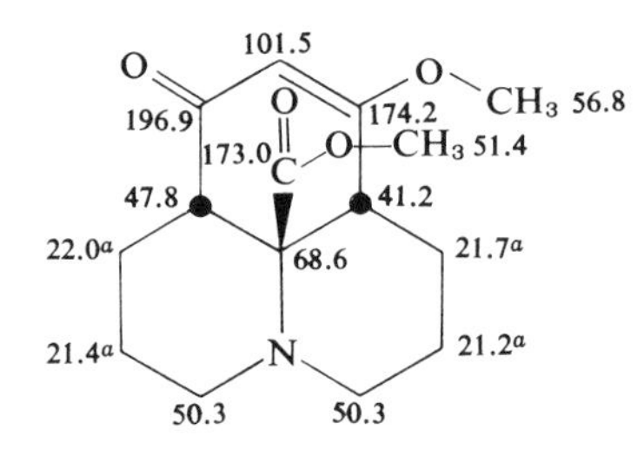

Ref. 29

188 **Methyl *trans,trans*-10-methoxy-7a,8,10a,10b-tetrahydro-8-julolidone-10b-carboxylate**

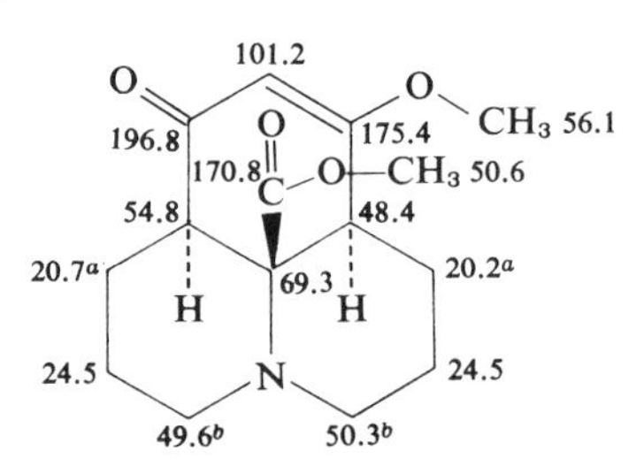

Ref. 29

189 **Methyl *trans,cis*-10-methoxy,7a,8,10a,10b-tetrahydro-8-julolidone-10b-carboxylate**

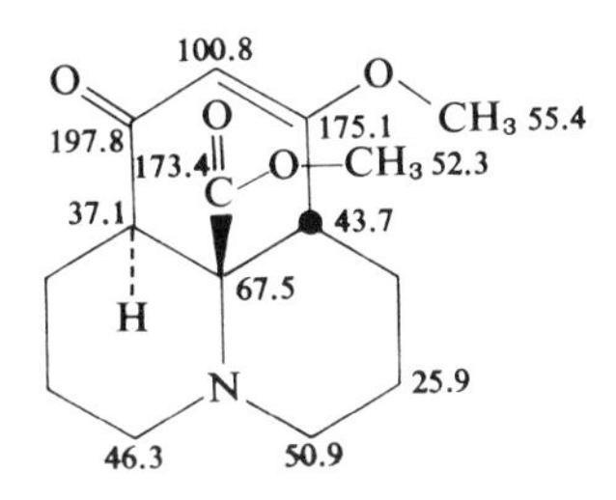

Ref. 29

190 **Methyl *trans,trans*-7a,8,9,10,10a,10b-hexahydro-8,10-julolidone-10b-carboxylate**

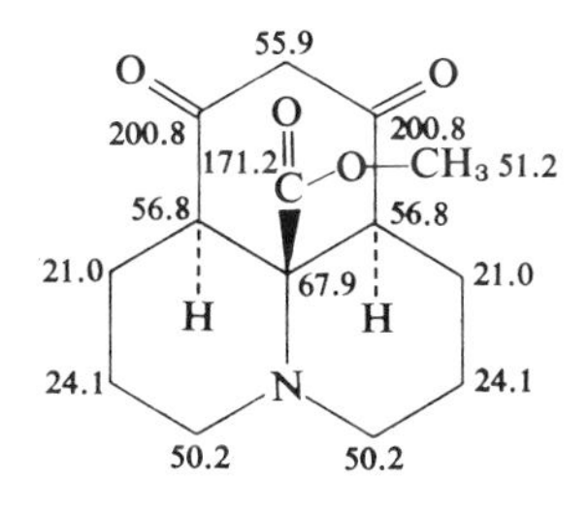

Ref. 29

191 **Carboethoxyisosophoramine**

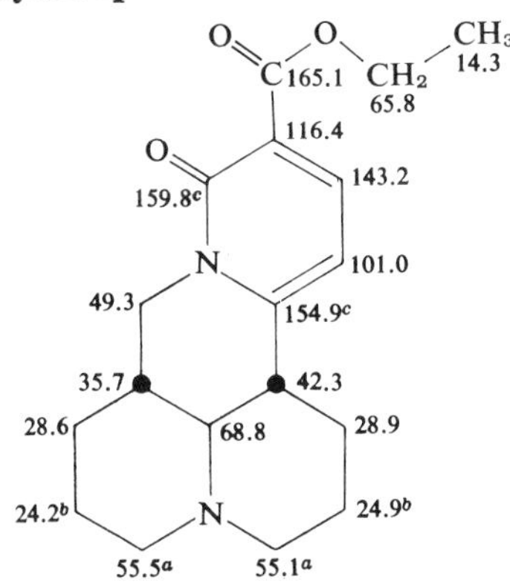

Ref. 29

c. Matrines

192 Matridine

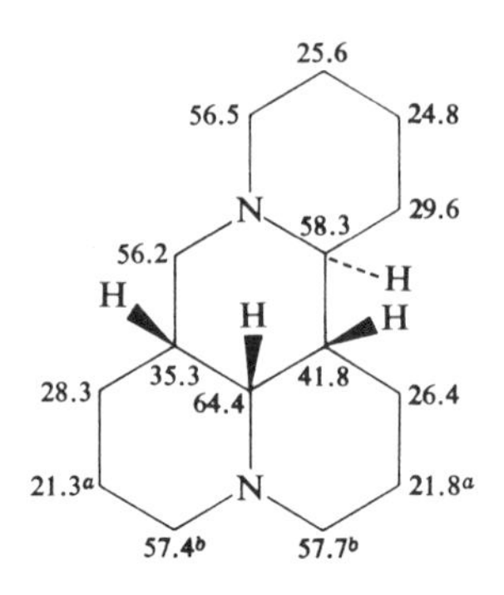

Ref. 26

193 Matridine isomer

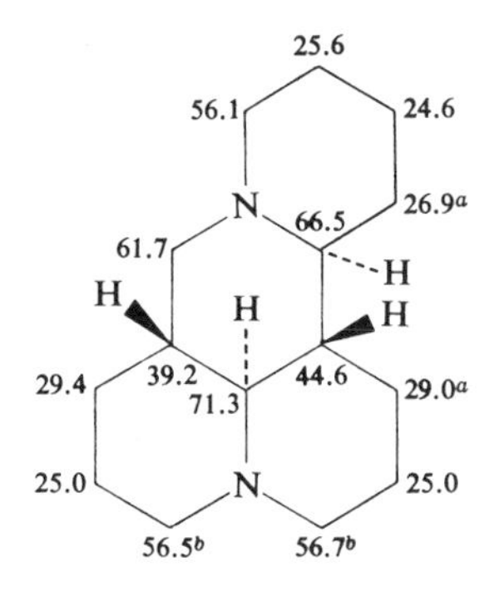

Ref. 26

194 Matrine

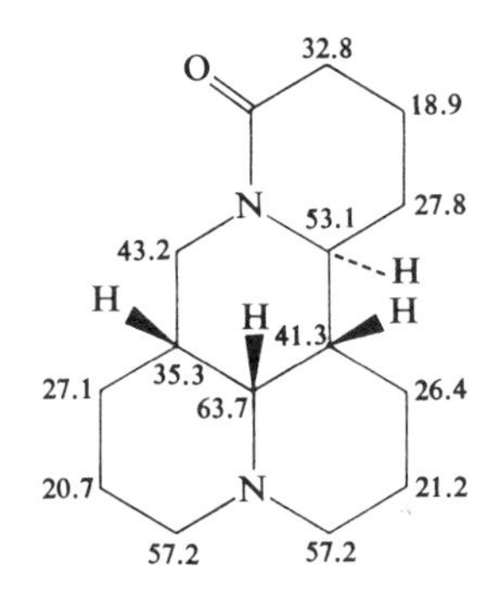

Ref. 26

195 Leontine

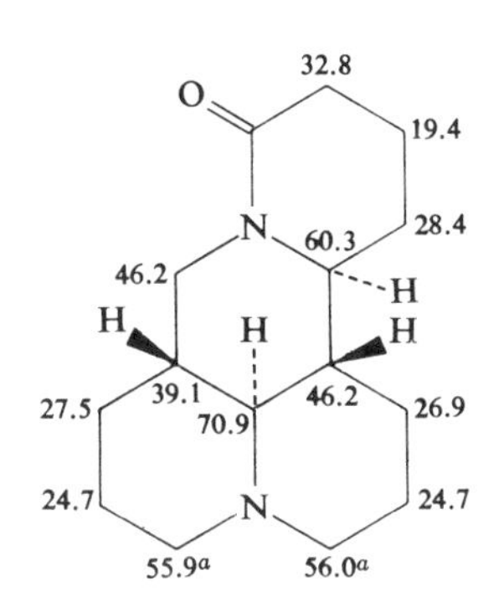

Ref. 26

196 7a-β-Hydroxymatrine

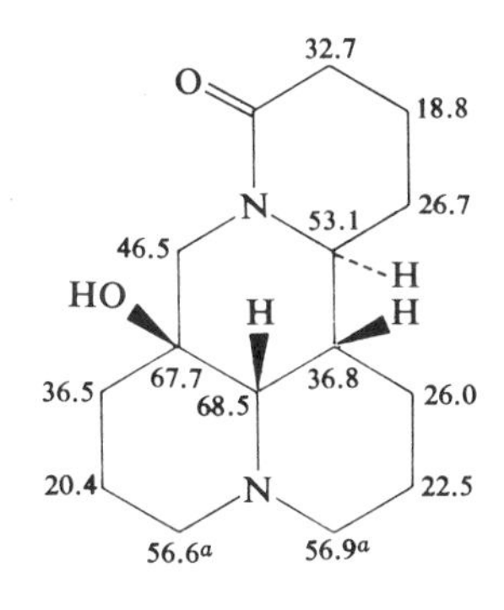

Ref. 26

d. Sparteines

197 Sparteine

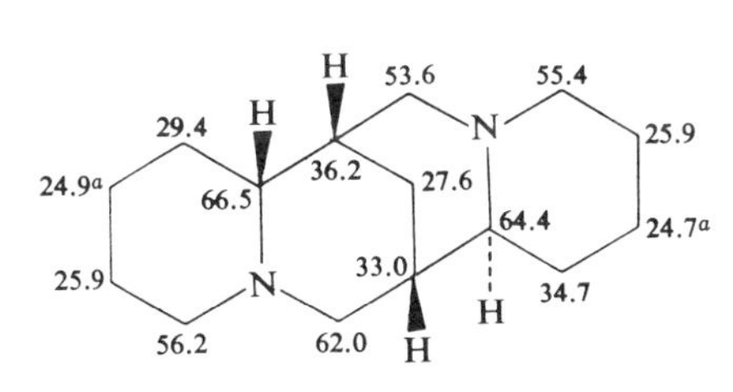

Ref. 26

198 **α-Isosparteine**

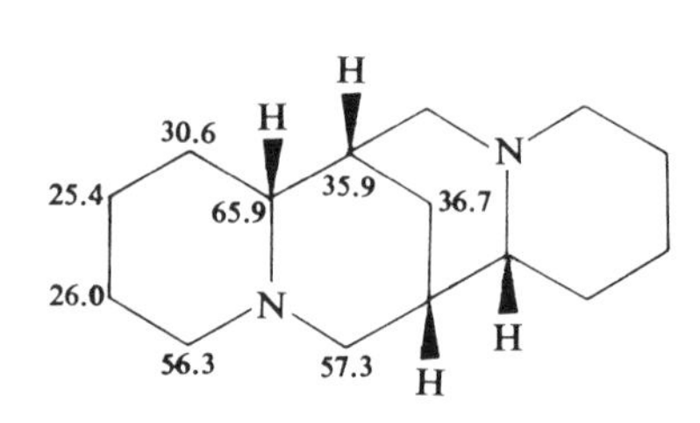

Ref. 26

199 **7-Hydroxy-α-isosparteine**

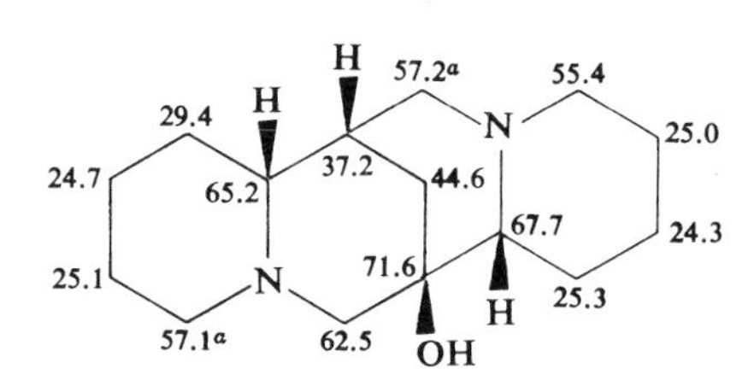

Ref. 26

200 **8-Hydroxysparteine**

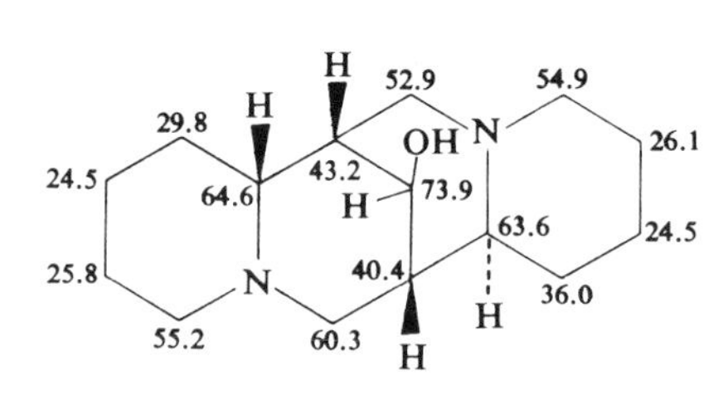

Ref. 26

201 12α-Hydroxysparteine (retamine)

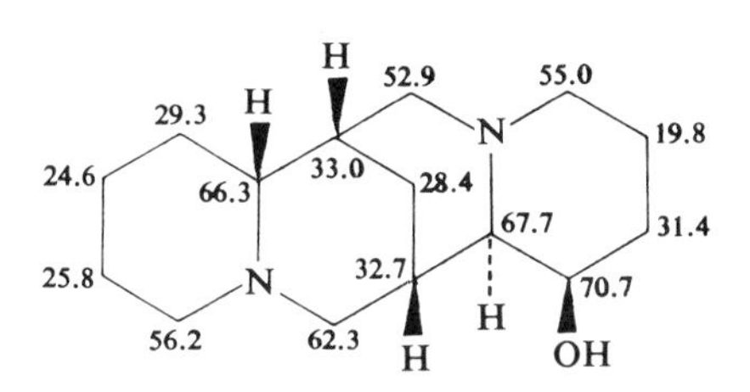

Ref. 26

202 13α-Hydroxysparteine

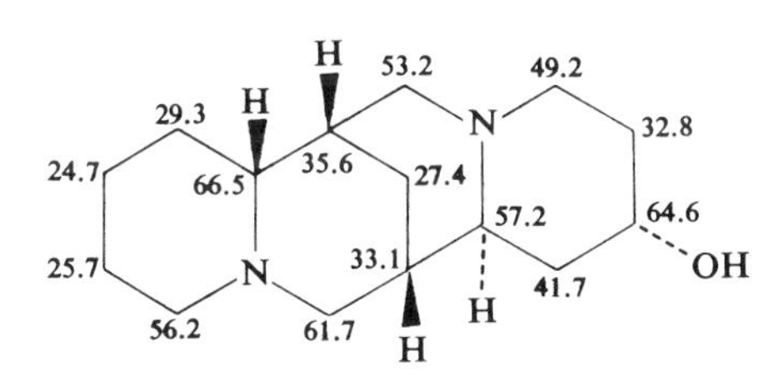

Ref. 26

203 13α-Acetoxysparteine

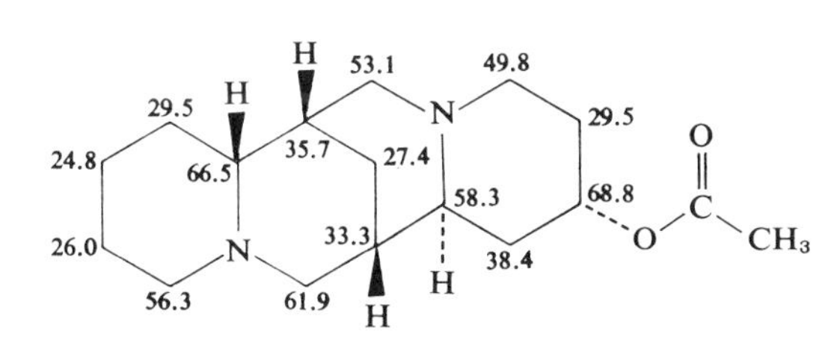

Ref. 26

204 8-Oxosparteine

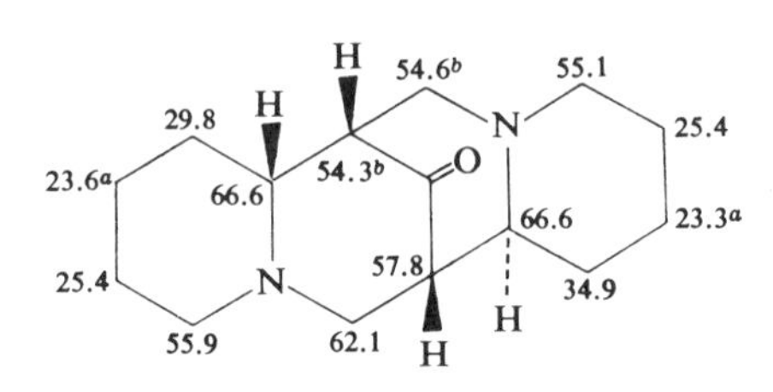

Ref. 26

205 17-Oxosparteine

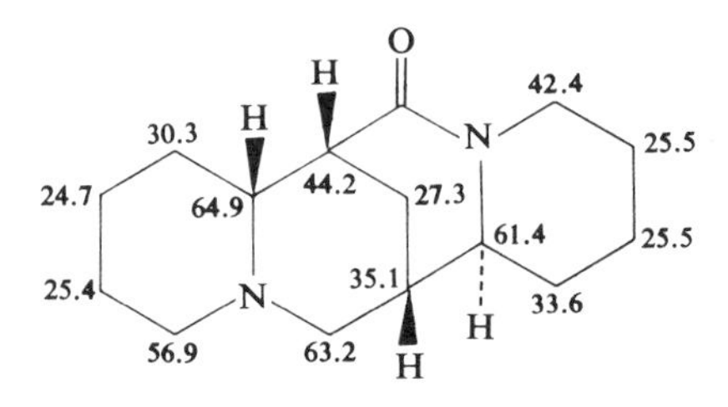

Ref. 26

206 17-Oxo-α-isosparteine

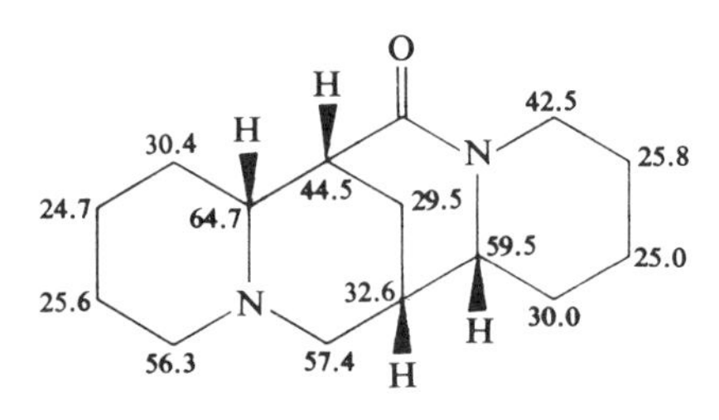

Ref. 26

207 13,17-Dioxosparteine

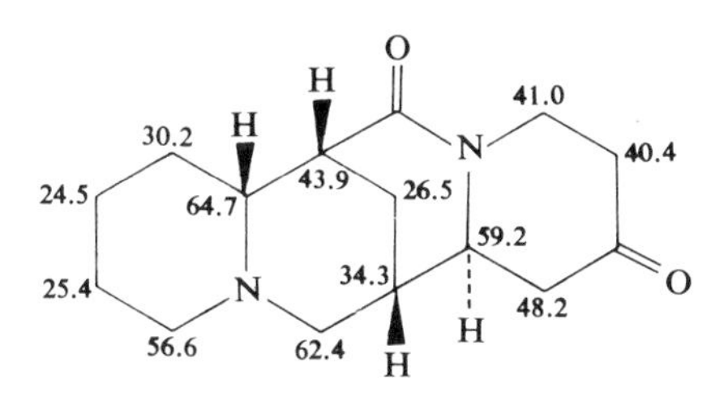

Ref. 26

208 6,17-Dioxo-α-isosparteine

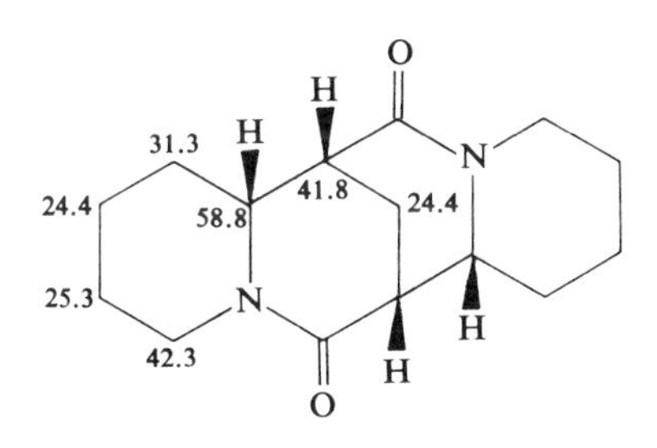

Ref. 26

209 4-Oxosparteine (lupanine)

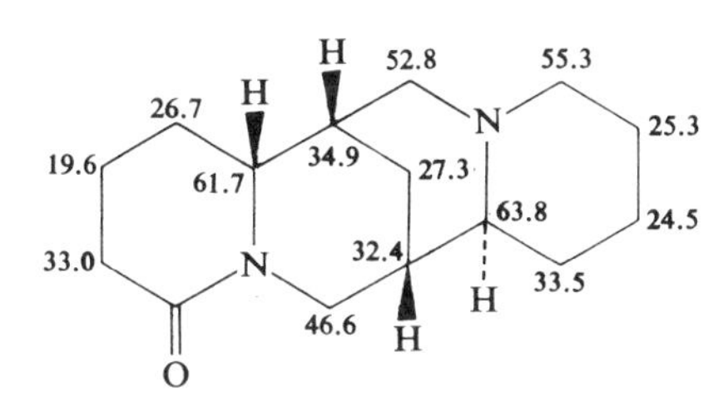

Ref. 26

210 **4-Oxo-α-isosparteine (isolupanine)**

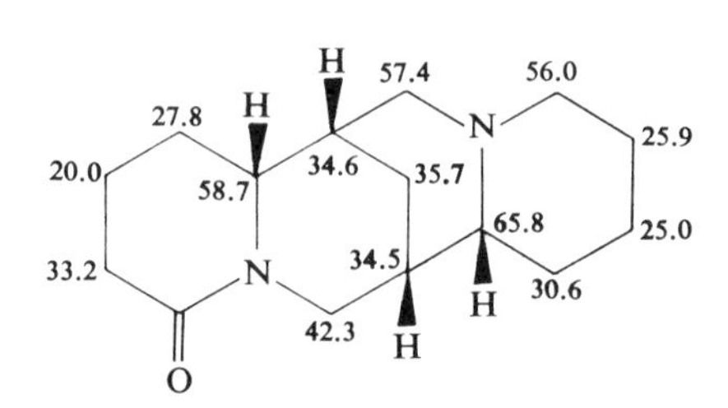

Ref. 26

211 **13α-Hydroxylupanine**

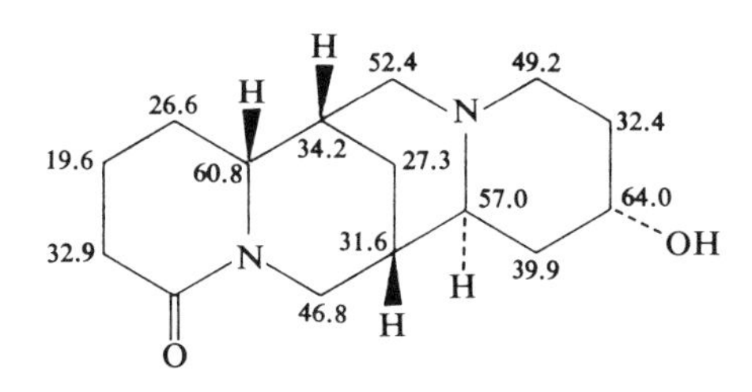

Ref. 26

212 **13β-Hydroxylupanine**

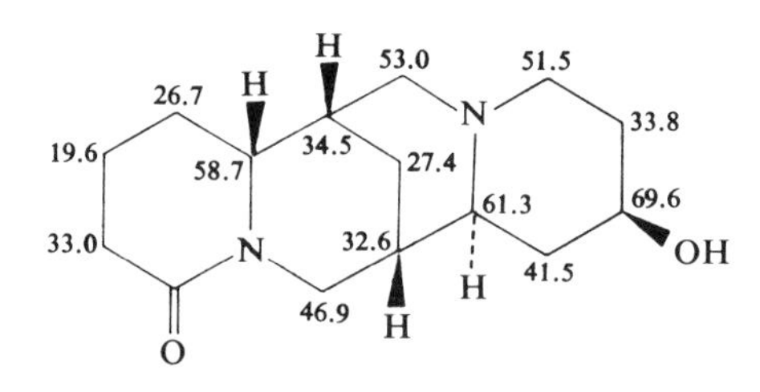

Ref. 26

213 **13β-Hydroxyisolupanine**

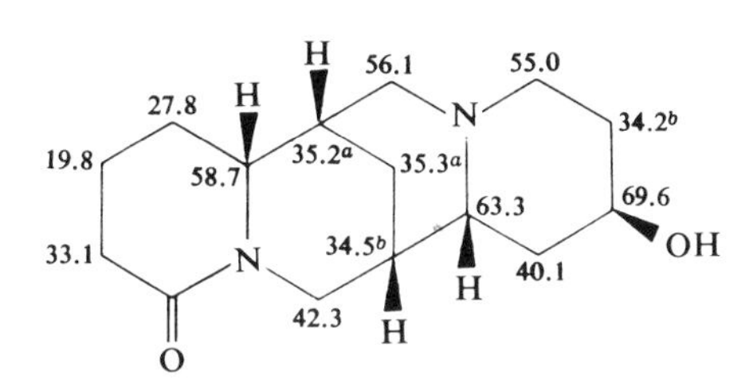

Ref. 26

214 **Cytisine**

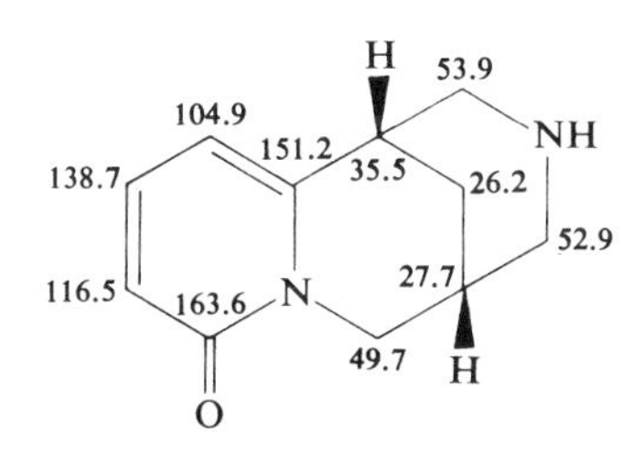

Ref. 26

215 **Anagyrine**

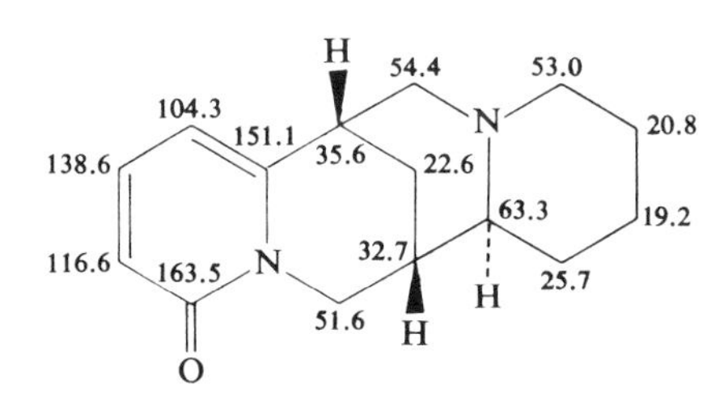

Ref. 26

e. Nuphar Alkaloids

216 Nupharidine

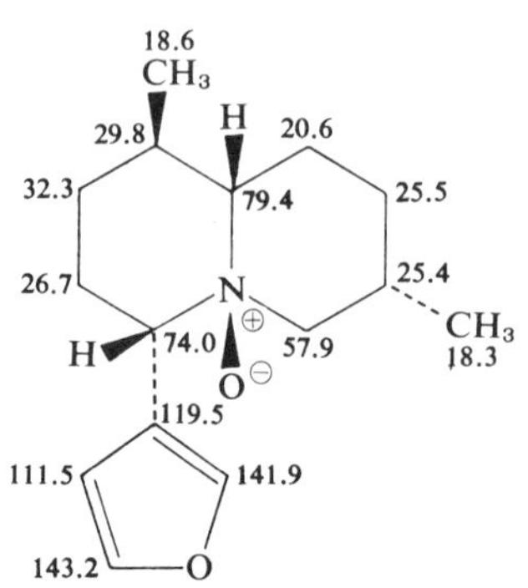

Ref. 46

217 7-Epinupharidine

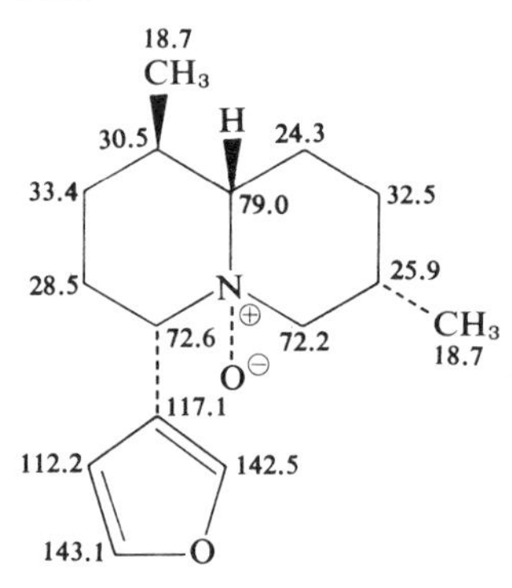

Ref. 46

218 Deoxynupharidine

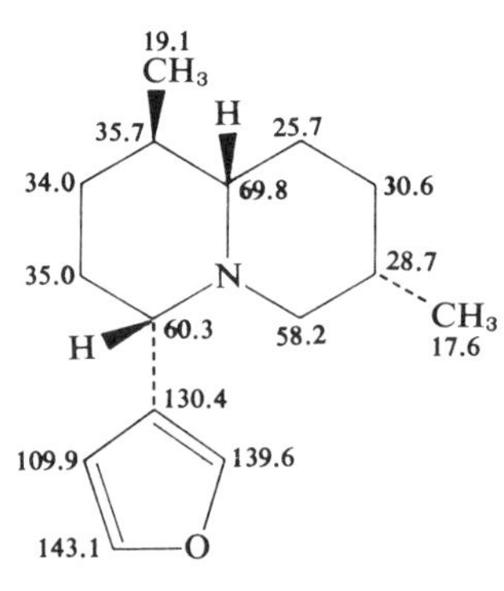

Ref. 46

219 7-Epideoxynupharidine

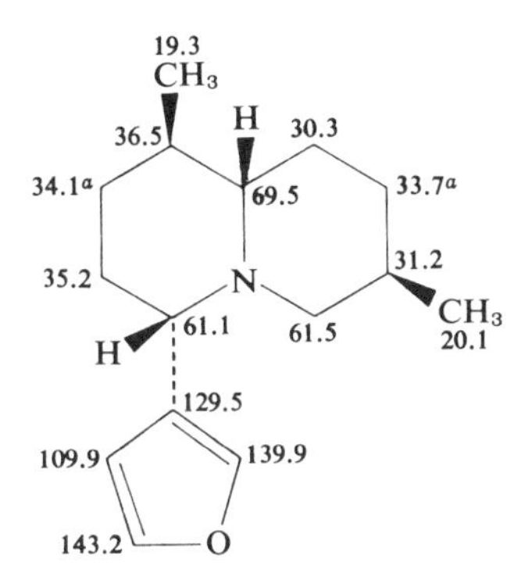

Ref. 46

220 Nupharolutine

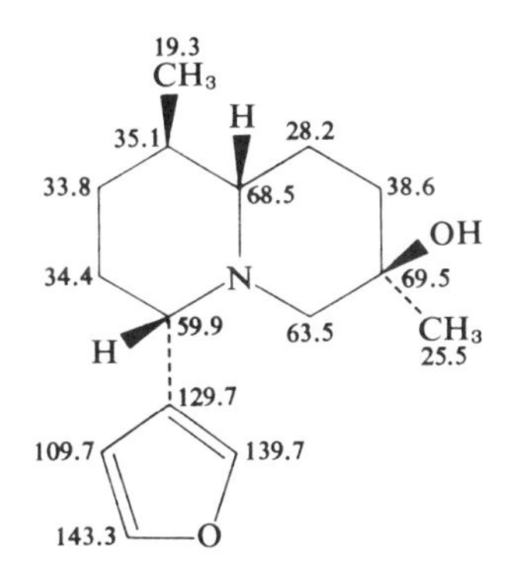

Ref. 46

221 7-Epinupharolutine

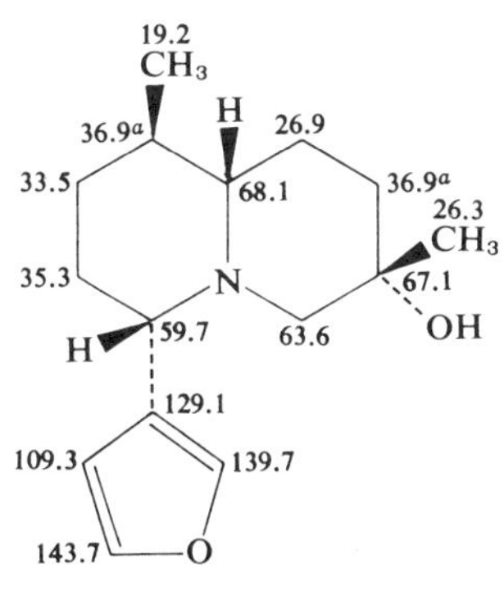

Ref. 46

222 Nupharamine

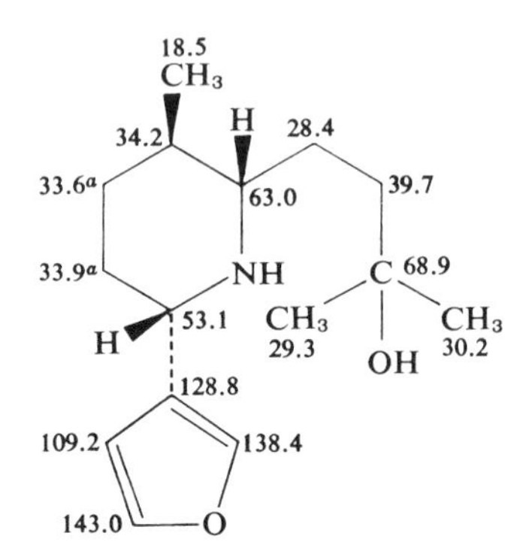

Ref. 47

223 Deoxynupharamine

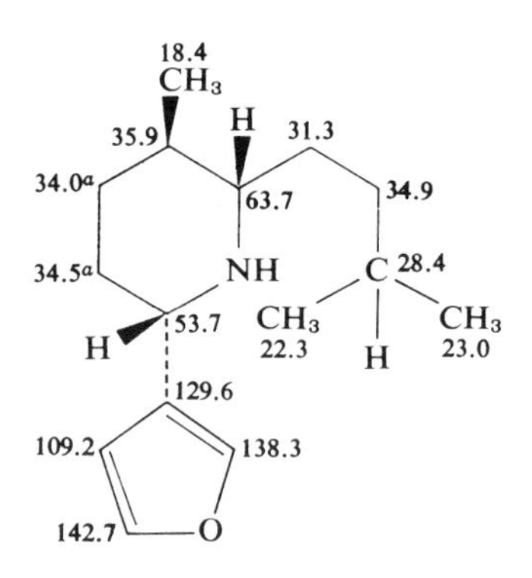

Ref. 47

224 Anhydronupharamine

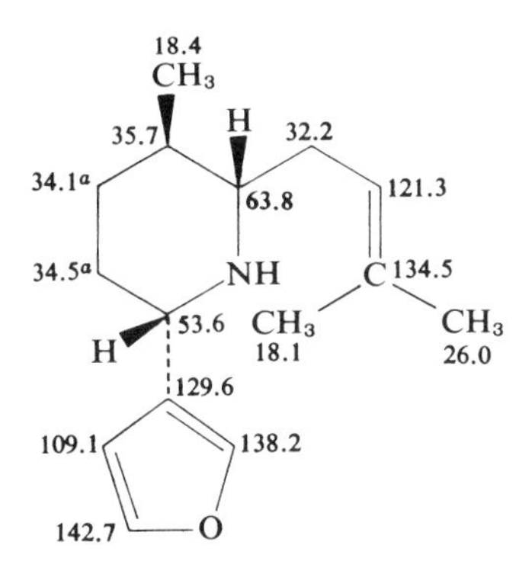

Ref. 47

225 Nuphamine

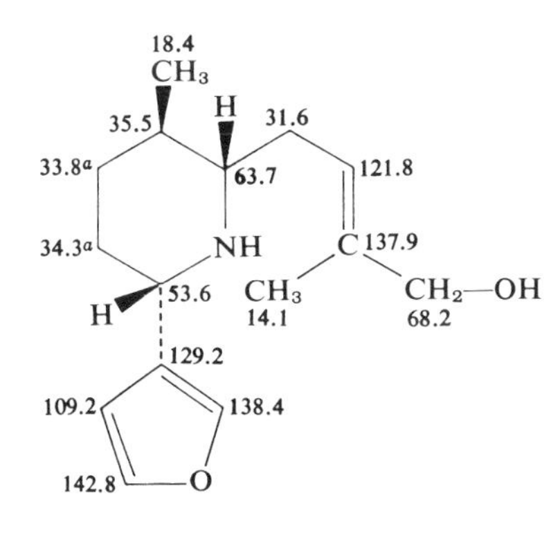

Ref. 47

f. Lycopodium Alkaloids

226 Lycopodine

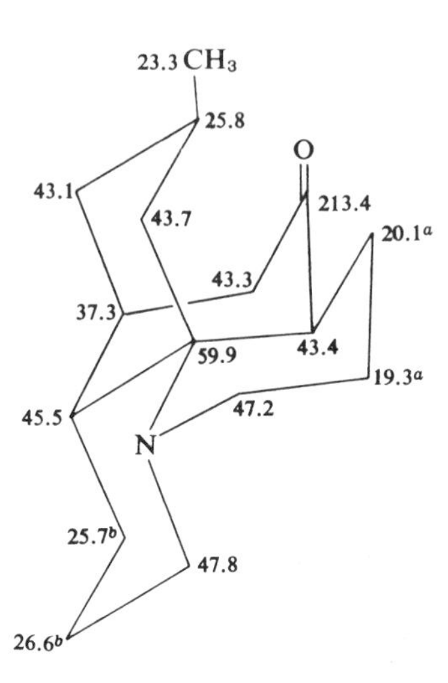

Ref. 48

227 Dihydrolycopodine

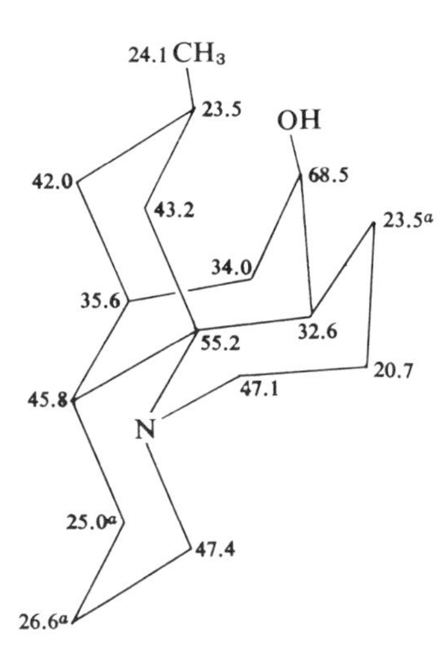

Ref. 48

228 Epidihydrolycopodine

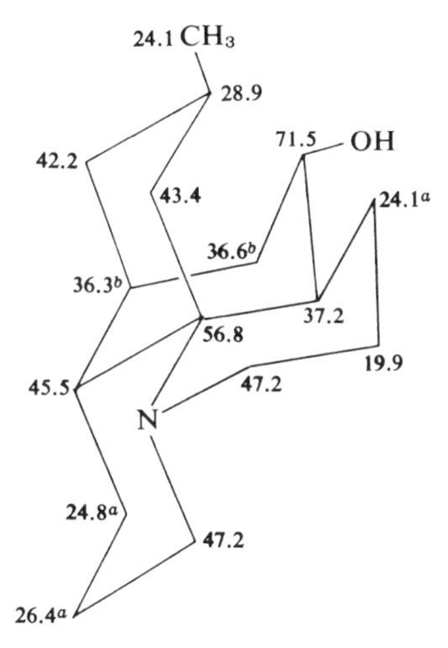

Ref. 48

229 α-Lofoline

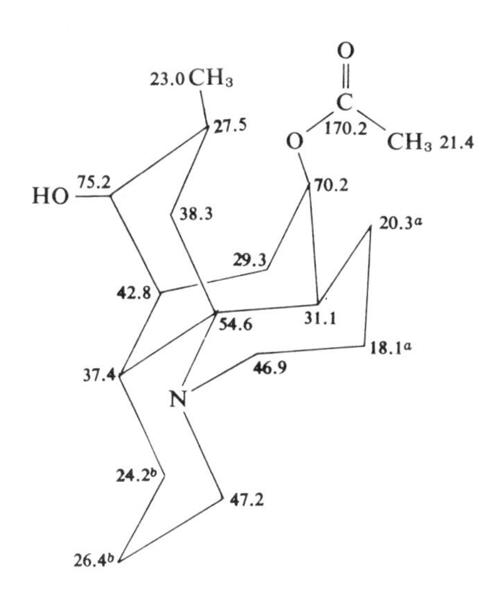

Ref. 48

230 Clavolonine

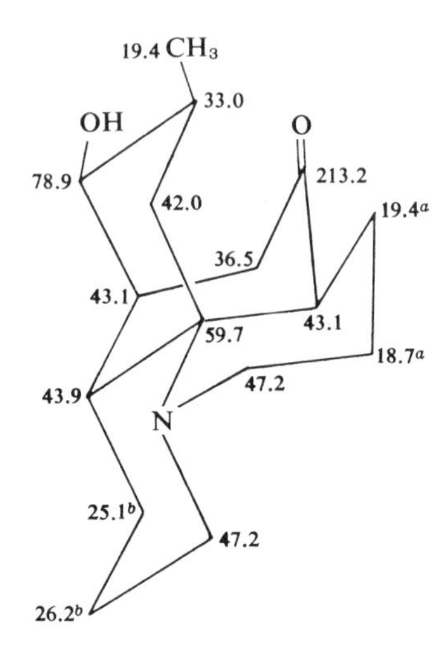

Ref. 48

231 Flabelliformine

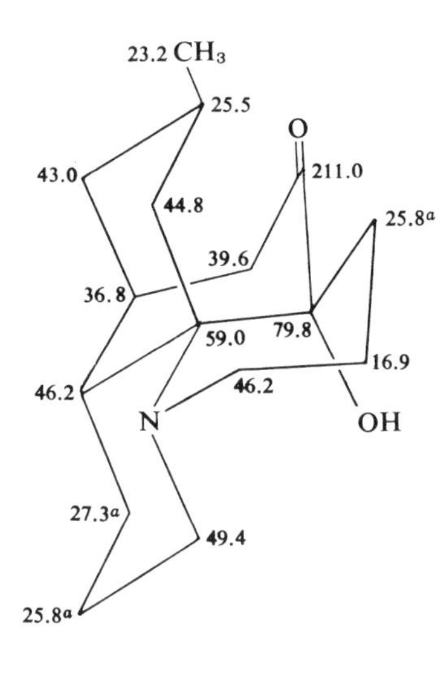

Ref. 48

232 Lycodoline

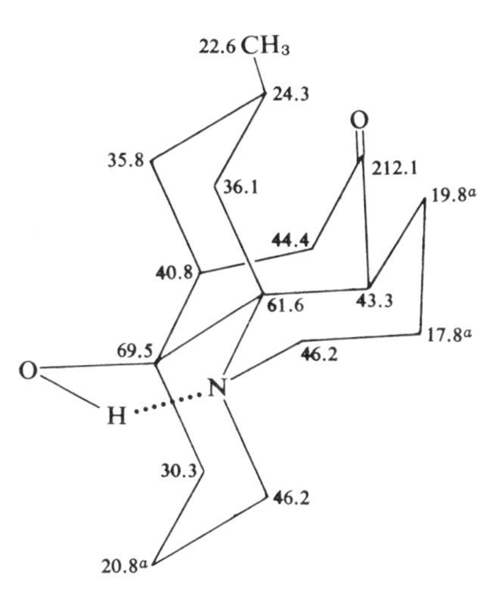

Ref. 48

233 Alkaloid L-23

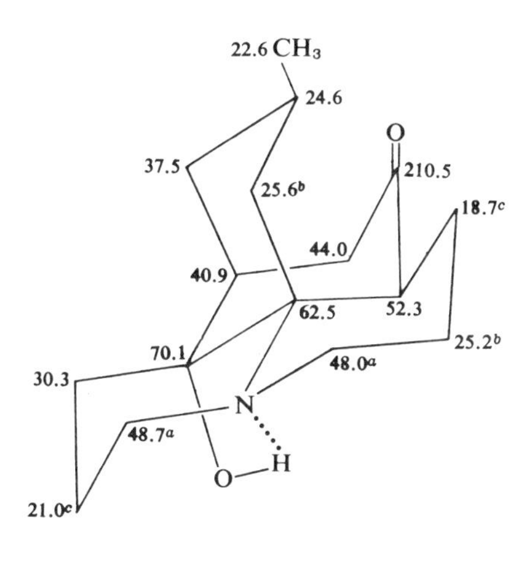

Ref. 48

9. QUINOLINES[49]

Simple Quinolines[137]

234 Quinoline

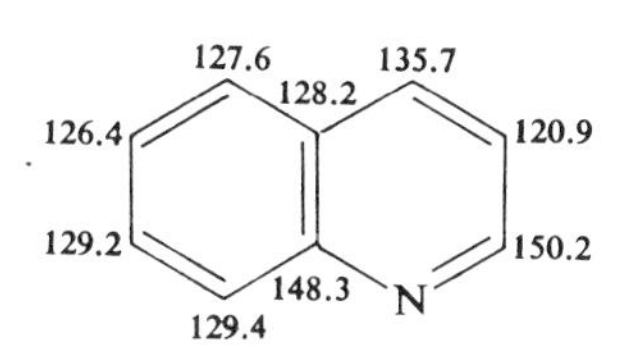

Ref. 50;
see also Ref. 18

235 2-Methylquinoline

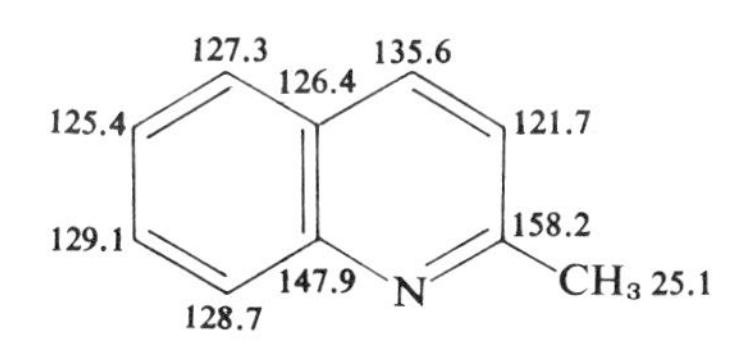

Ref. 50

236 3-Methylquinoline

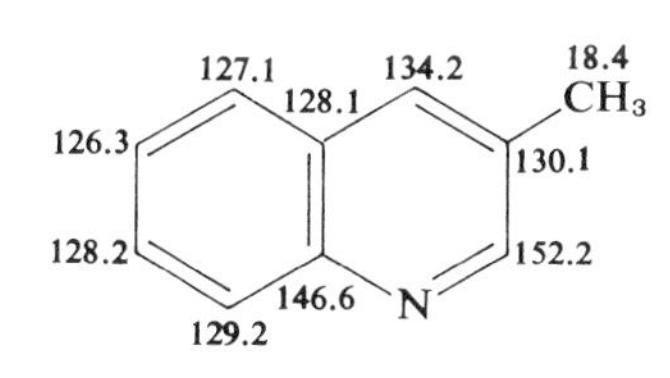

Ref. 50

237 4-Methylquinoline

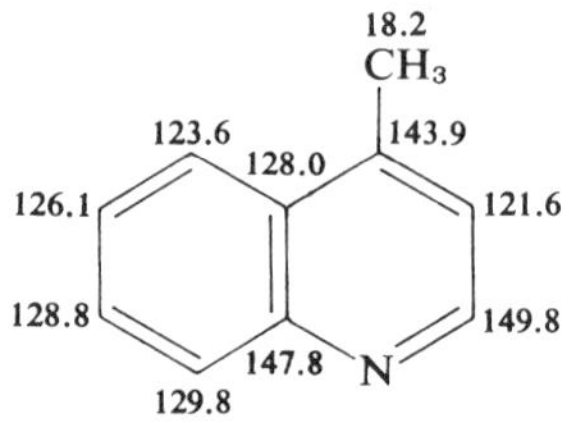

18.2
CH_3
123.6
143.9
128.0
126.1
121.6
128.8
149.8
147.8
N
129.8

Ref. 50

238 6-Methylquinoline

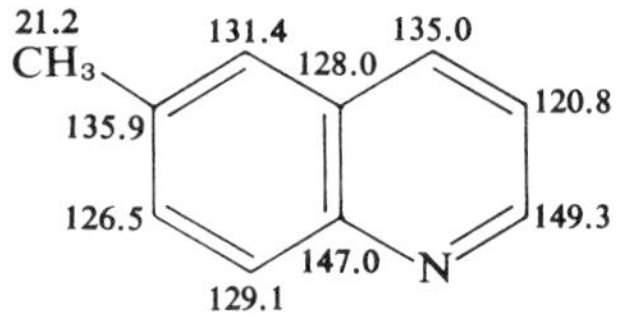

21.2
CH_3
131.4
135.0
128.0
120.8
135.9
126.5
149.3
147.0
N
129.1

Ref. 50

239 8-Methylquinoline

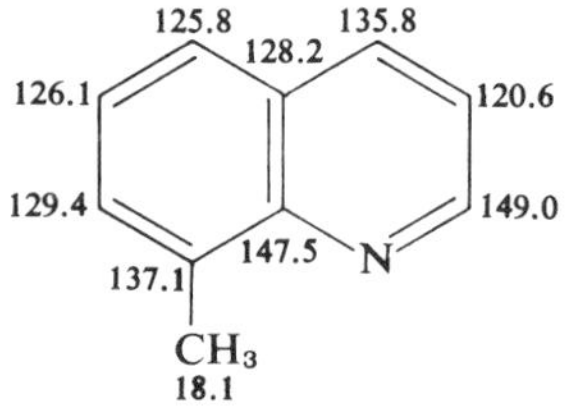

125.8
135.8
128.2
126.1
120.6
129.4
149.0
147.5
N
137.1
CH_3
18.1

Ref. 50

240 5,8-Dimethylquinoline

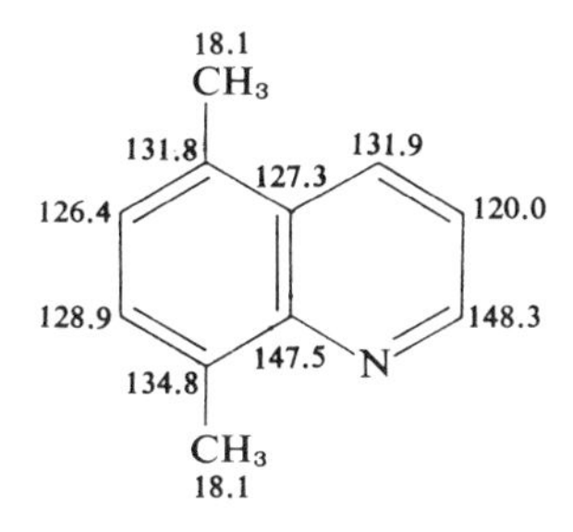

Ref. 51

241 6,8-Dimethylquinoline

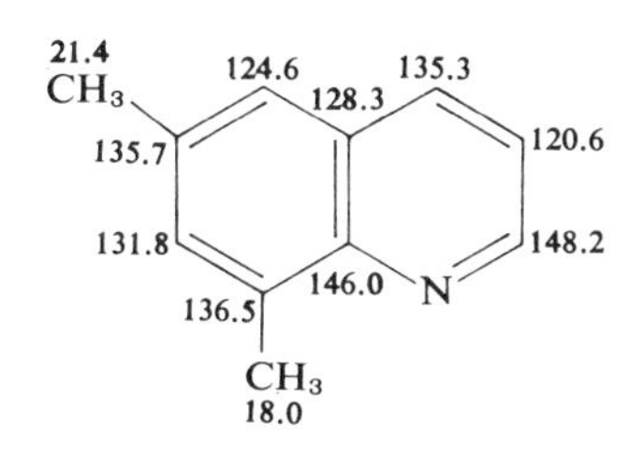

Ref. 51

242 7,8-Dimethylquinoline

124.7 135.8
126.5
129.3 119.6
134.1 149.9
CH_3 147.3 N
20.5 136.9
CH_3
13.3

Ref. 51

243 **6-Methoxyquinoline**

55.2
CH_3O
105.1
134.5
129.3
121.2
157.7
122.1
147.8
144.5
N
130.8

Ref. 51

244 **3-Aminoquinoline**

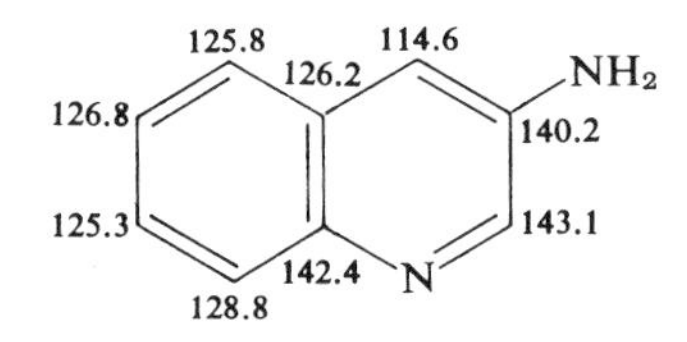

Ref. 18

245 **5-Aminoquinoline**
(acetone-d_6)

NH_2
145.5
131.0
119.1
109.2
119.8
130.8
150.6
150.3
N
118.6

Ref. 52

246 3-Nitroquinoline

133.6 132.4
126.2 NO_2
129.0 141.6
130.0 144.2
150.3 N
130.1

Ref. 18

Quinolones

247 2-Quinolone
($DMSO\text{-}d_6$–$CDCl_3$, 9:1)

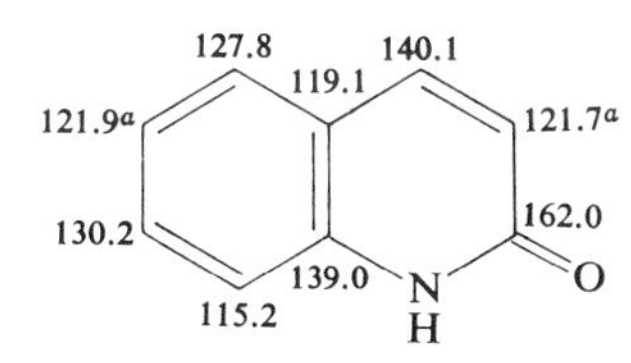

Ref. 53

248 4-Methyl-2-quinolone
($DMSO\text{-}d_6$–$CDCl_3$, 9:1)

18.4
CH_3
124.5 147.7
119.6
121.5 120.9
161.6
130.1
138.7 N O
115.4 H

Ref. 53

249 **6-Methyl-2-quinolone**
(DMSO-d_6–$CDCl_3$, 9:1)

20.3
CH_3
127.3
139.9
119.1
121.7
130.6
161.9
131.4
136.8
N
H
O
115.0

Ref. 54

250 **8-Methyl-2-quinolone**
(DMSO-d_6–$CDCl_3$, 9:1)

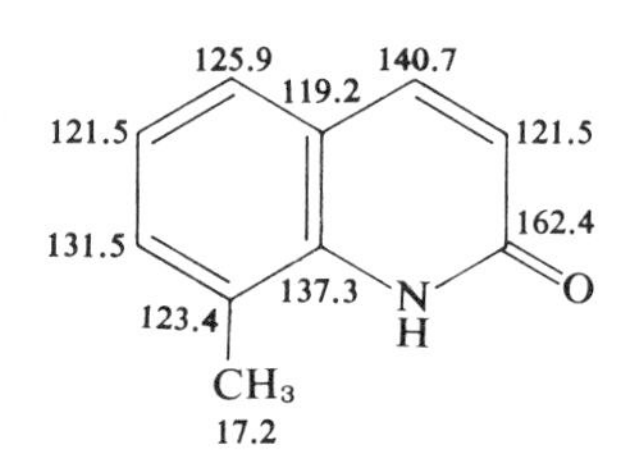

Ref. 54

251 **1,4-Dimethyl-2-quinolone**

18.8
CH_3
125.1
146.3
121.3
121.9
121.1
162.0
130.4
139.8
N
O
114.4
CH_3
29.1

Ref. 55

252 **4,6-Dimethyl-2-quinolone**
(DMSO-d_6–$CDCl_3$, 9:1)

18.5 CH_3
20.6 CH_3
124.1
119.5
147.4
130.4
120.8
161.5
131.3
136.6
N H
O
115.4

Ref. 53

253 **4,7-Dimethyl-2-quinolone**
(DMSO-d_6–$CDCl_3$, 9:1)

18.4 CH_3
124.4
117.6
147.7
123.0
119.8
140.2
161.9
CH_3 21.2
138.8
N H
O
115.2

Ref. 53

254 **4,8-Dimethyl-2-quinolone**
(DMSO-d_6–$CDCl_3$, 9:1)

18.7 CH_3
122.5
119.7
148.1
121.2
120.6
131.4
161.8
137.0
N H
O
123.5
CH_3 17.3

Ref. 53

255 **6-Ethyl-4-methyl-2-quinolone**
(DMSO-d_6–$CDCl_3$, 9:1)

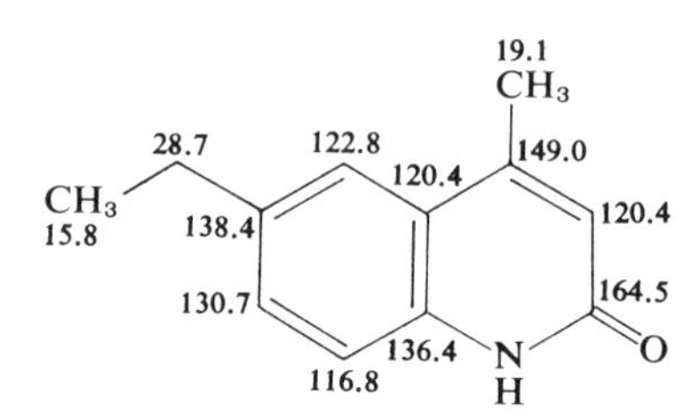

Ref. 54

256 **4,5,7-Trimethyl-2-quinolone**
(DMSO-d_6–$CDCl_3$, 9:1)

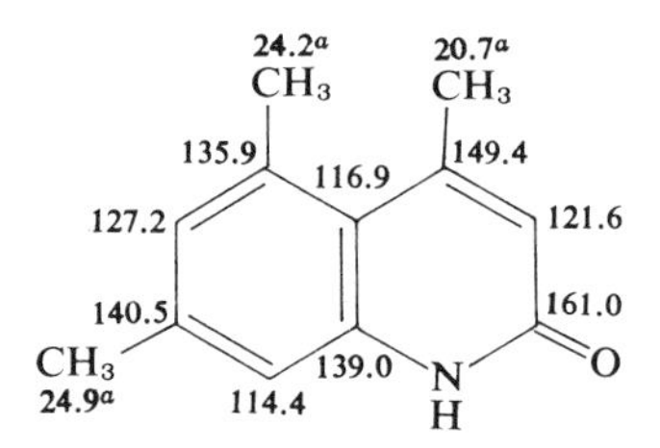

Ref. 53

257 **4,6,7-Trimethyl-2-quinolone**
(DMSO-d_6–$CDCl_3$, 9:1)

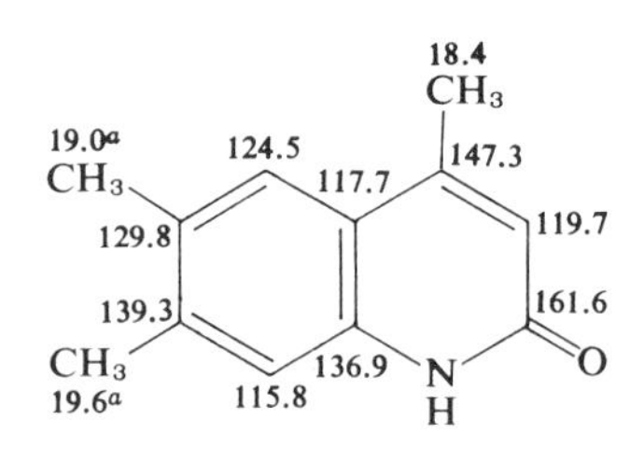

Ref. 53

258 **4,6,8-Trimethyl-2-quinolone**
(DMSO-d_6–$CDCl_3$, 9:1)

18.8
CH_3
20.4 CH_3
122.1
117.6
147.9
130.1
120.6
132.7
161.7
135.0
N
O
H
123.4
CH_3
17.2

Ref. 53

259 **8-Methoxy-4-methyl-2-quinolone**

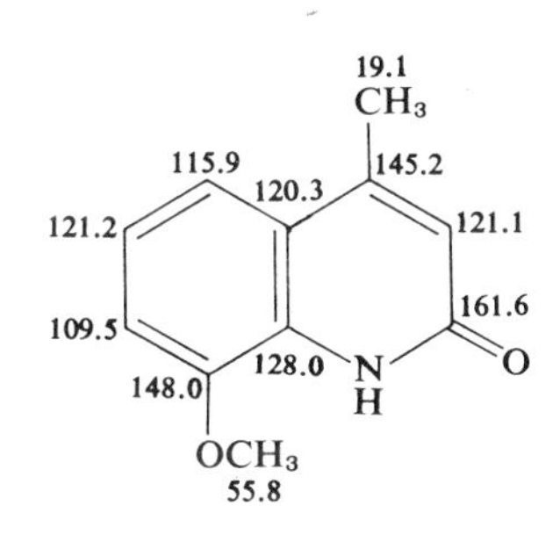

Ref. 55

260 **1,4-Dimethyl-8-methoxy-2-quinolone**

19.5
CH_3
117.5
145.8
123.4
122.3
121.3
163.1
113.7
131.3
N
O
148.7
OCH_3 CH_3
56.8 35.3

Ref. 55

261 **6-Methyl-2H,4H-oxazolo[5,4,3-ij]quinolin-4-one**

17.9
CH_3
114.8
148.0
118.1
123.6
122.0
158.2
107.4
O
131.0
N
150.4
O
CH_2
85.7

Ref. 55

262 **4-Acetoxymethyl-1-methyl-2-quinolone**

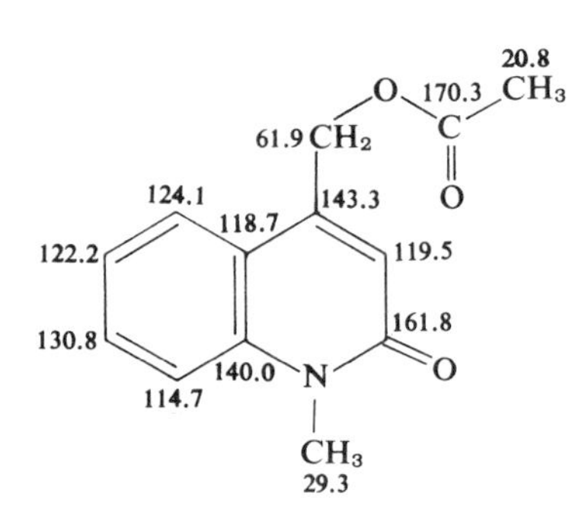

Ref. 55

263 **4-Quinolone**
(DMSO-d_6–$CDCl_3$, 9:1)

O
125.0
177.2
125.9
123.1
108.8
131.5
139.5
140.1
N
118.4
H

Ref. 54

264 2-Methyl-4-quinolone
($DMSO-d_6$–$CDCl_3$, 9:1)

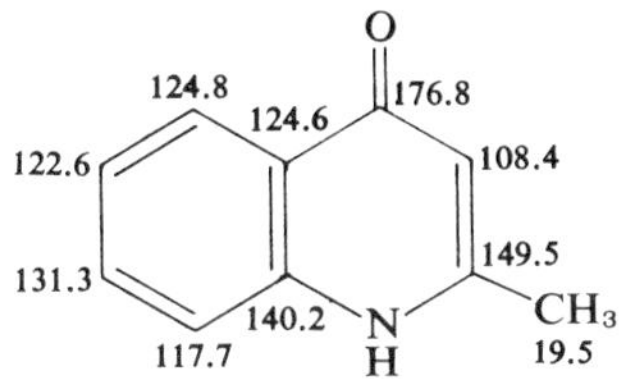

Ref. 54

265 2,5-Dimethyl-4-quinolone
($DMSO-d_6$–$CDCl_3$, 9:1)

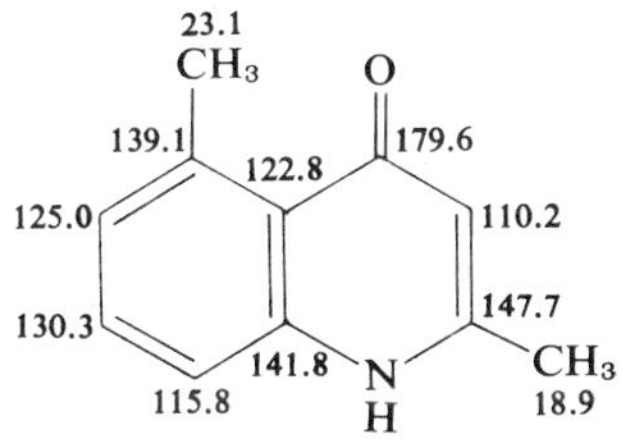

Ref. 54

266 2,6-Dimethyl-4-quinolone
($DMSO-d_6$–$CDCl_3$, 9:1)

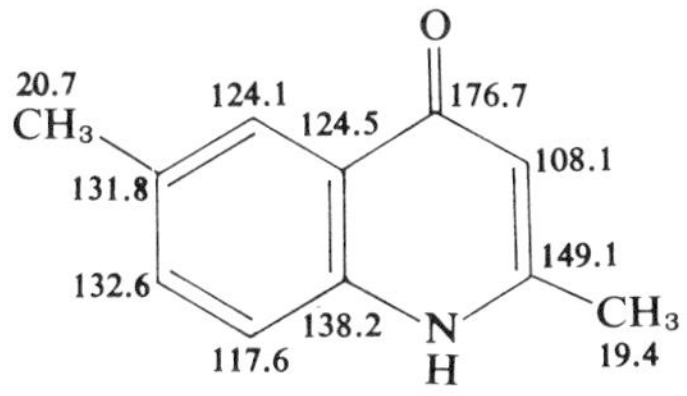

Ref. 54

267 2,8-Dimethyl-4-quinolone
(DMSO-d_6–$CDCl_3$, 9:1)

Ref. 54

268 2,5,8-Trimethyl-4-quinolone
(DMSO-d_6–$CDCl_3$, 9:1)

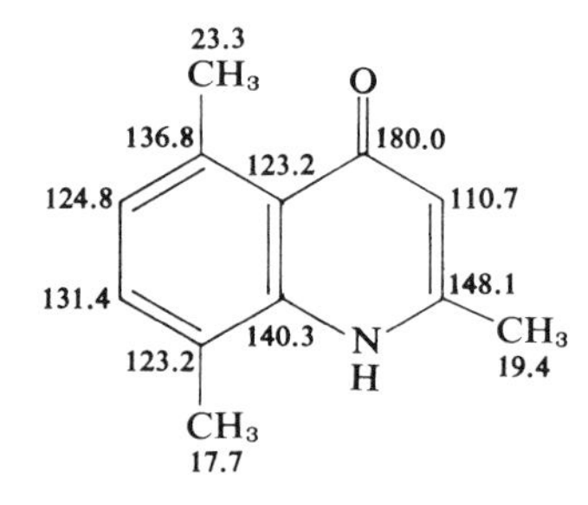

Ref. 54

269 2,6,8-Trimethyl-4-quinolone
(DMSO-d_6–$CDCl_3$, 9:1)

Ref. 54

270 **2,7,8-Trimethyl-4-quinolone**
(DMSO-d_6–$CDCl_3$, 9:1)

O
122.0 177.1
123.2
124.9 108.3
139.3 149.7
CH_3 138.9 N CH_3
20.4 123.2 H 19.8
CH_3
13.1

Ref. 54

c. Reduced Quinolines

271 **1,2,3,4-Tetrahydroquinoline**

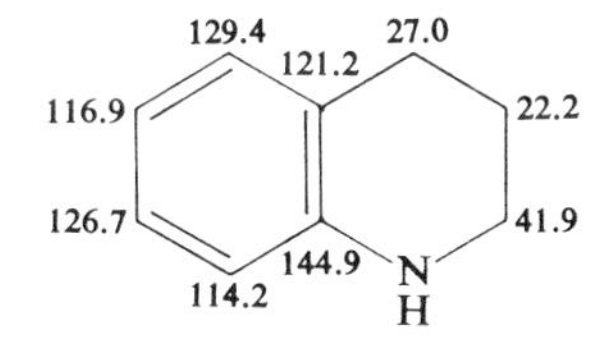

Ref. 84

272 ***N*-Formyl-1,2,3,4-tetrahydroquinoline**

129.6 27.1
128.8
124.4 22.2
127.1 40.2
137.2 N
116.9
161.0
H O

Ref. 84

273 ***N*-Acetyl-1,2,3,4-tetrahydroquinoline**

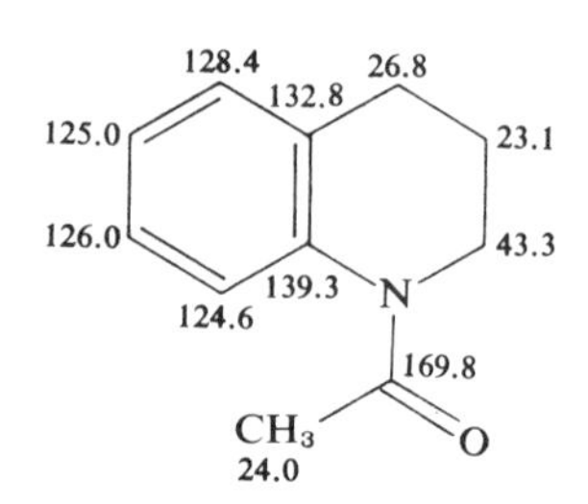

Ref. 84

274 ***cis*-Decahydroquinoline**

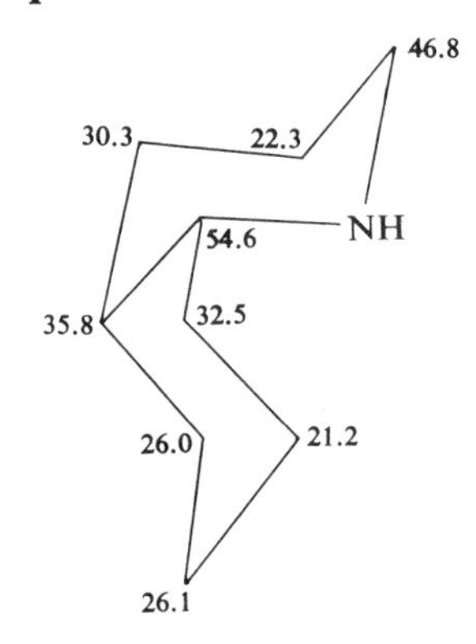

Ref. 23

275 ***trans*-Decahydroquinoline**

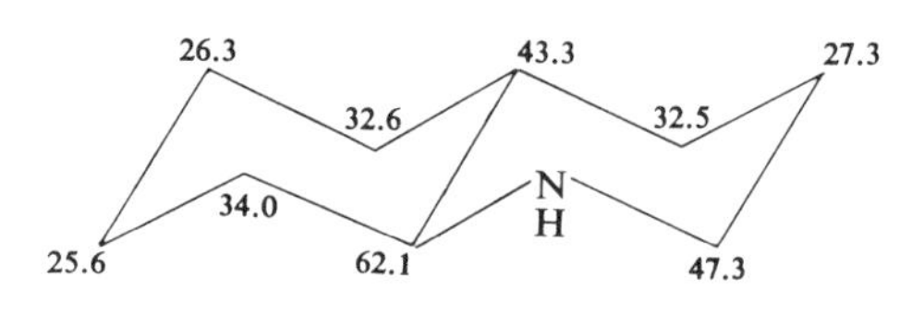

Ref. 57;
see also Ref. 23

276 *N*-Methyl-*trans*-decahydroquinoline

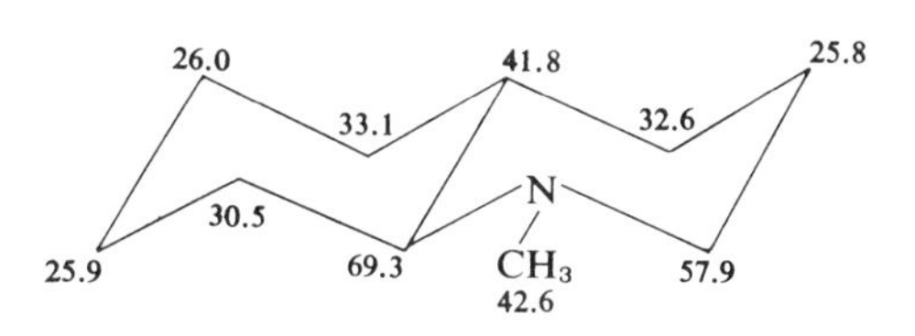

Ref. 57

277 2α-Methyl-*trans*-decahydroquinoline

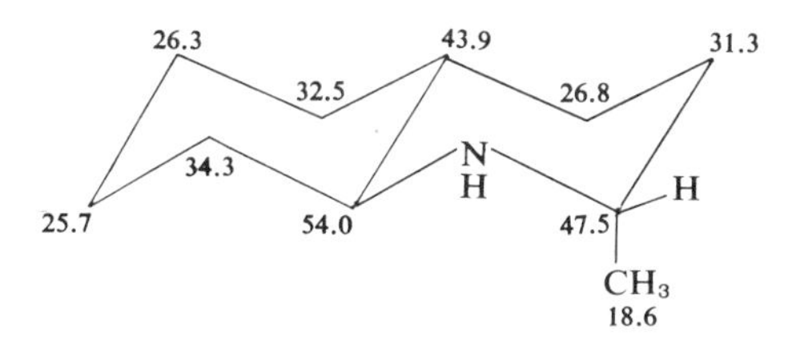

Ref. 57

278 1,2α-Dimethyl-*trans*-decahydroquinoline

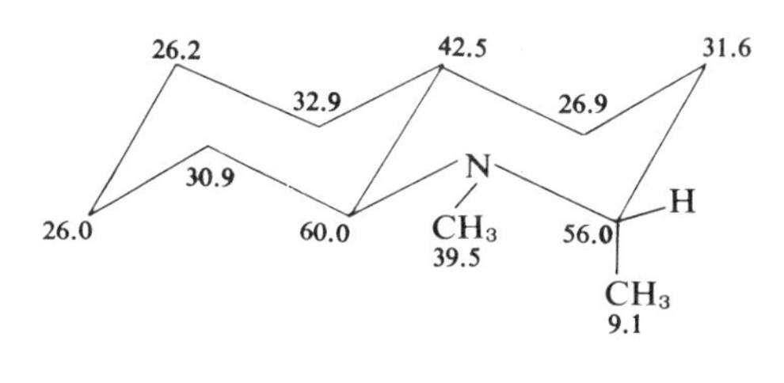

Ref. 57

279 **2β-Methyl-*trans*-decahydroquinoline**

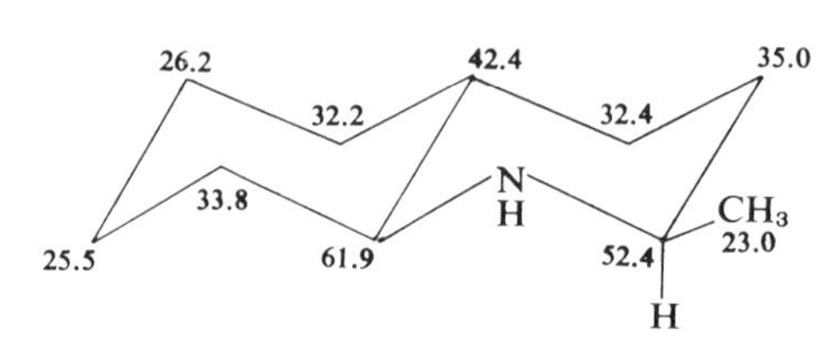

Ref. 57

280 **1,2β-Dimethyl-*trans*-decahydroquinoline**

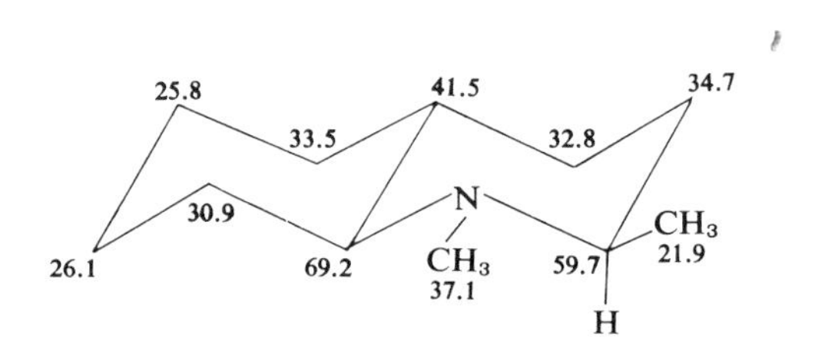

Ref. 57

281 **3α-Methyl-*trans*-decahydroquinoline**

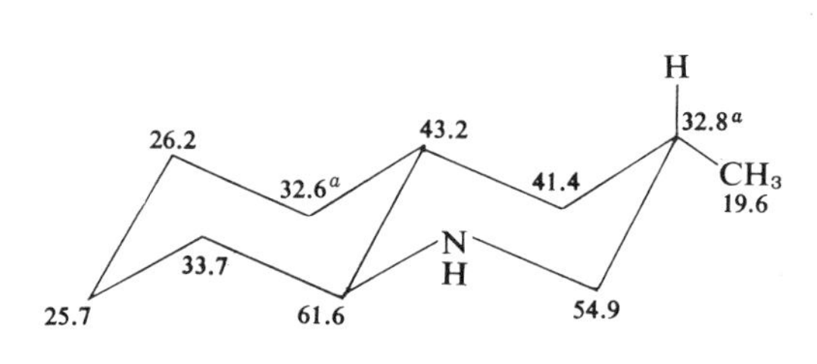

Ref. 57

282 **1,3α-Dimethyl-*trans*-decahydroquinoline**

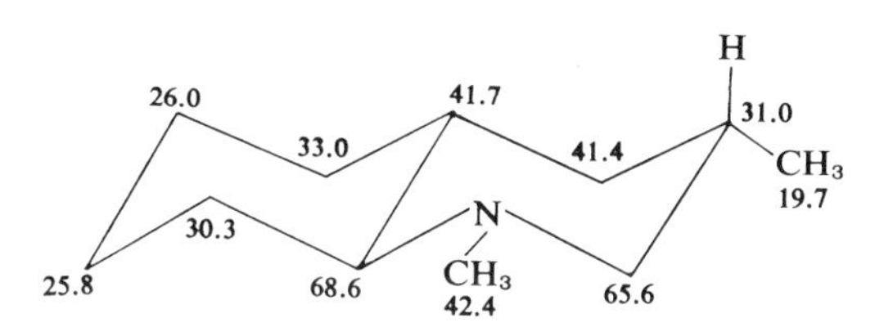

Ref. 57

283 **3β-Methyl-*trans*-decahydroquinoline**

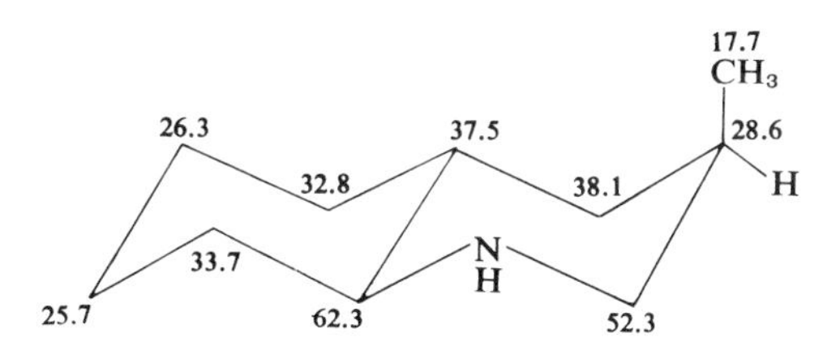

Ref. 57

284 **1,3β-Dimethyl-*trans*-decahydroquinoline**

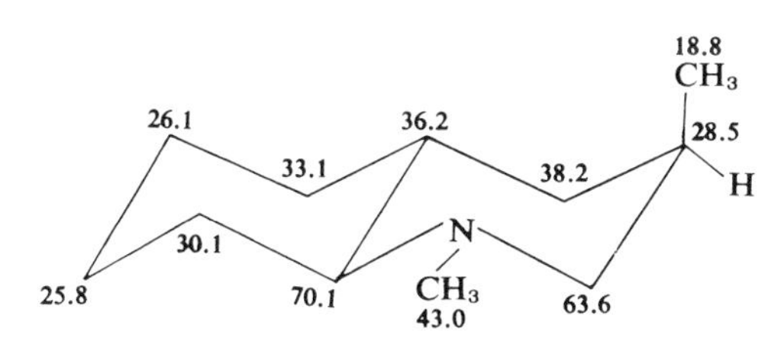

Ref. 57

285 **6α-Methyl-*trans*-decahydroquinoline**

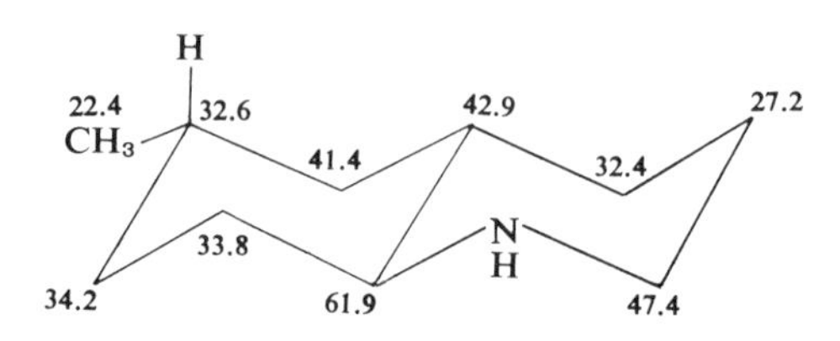

Ref. 57

286 **1,6α-Dimethyl-*trans*-decahydroquinoline**

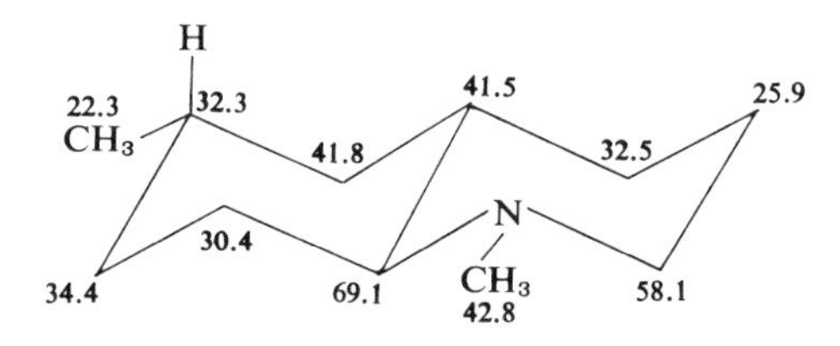

Ref. 57

287 **8α-Methyl-*trans*-decahydroquinoline**

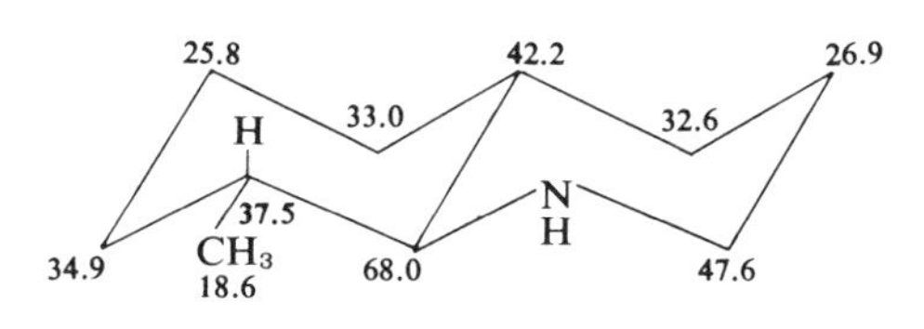

Ref. 57

288 1,8α-Dimethyl-*trans*-decahydroquinoline

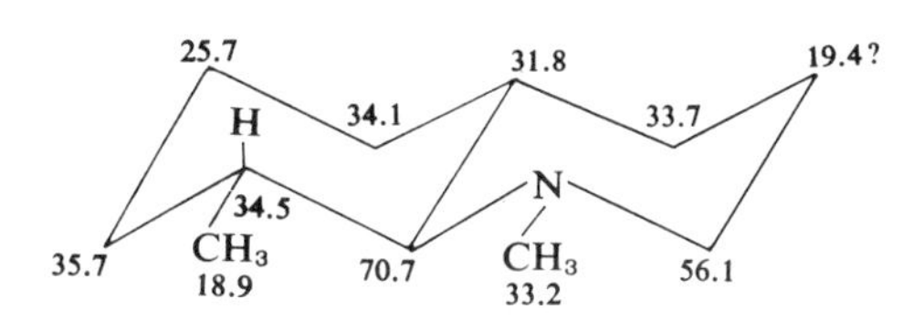

Ref. 57

289 8β-Methyl-*trans*-decahydroquinoline

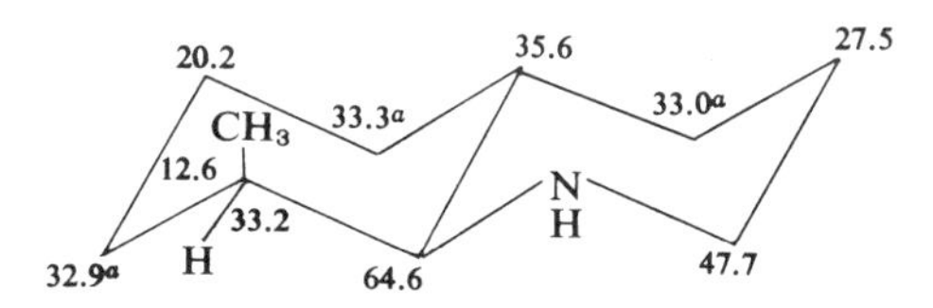

Ref. 57

290 1,8β-Dimethyl-*trans*-decahydroquinoline

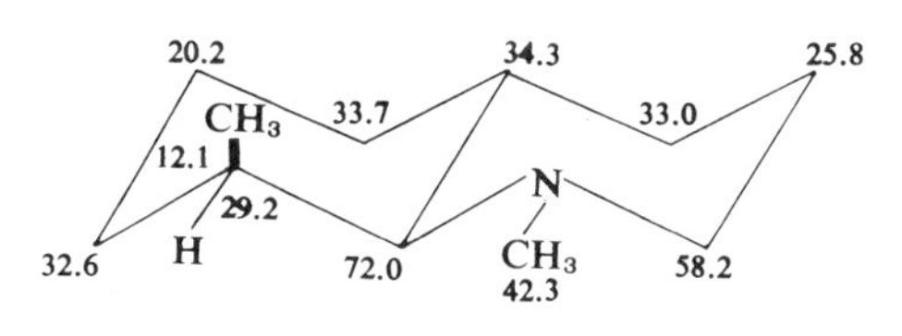

Ref. 57

291 **10-Methyl-*trans*-decahydroquinoline**

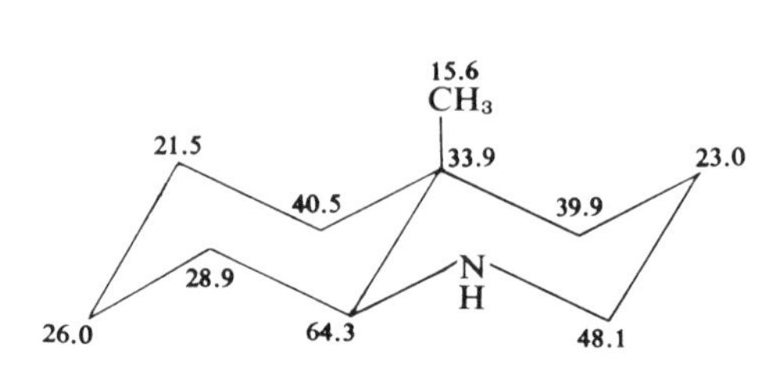

Ref. 57

292 **1,10-Dimethyl-*trans*-decahydroquinoline**

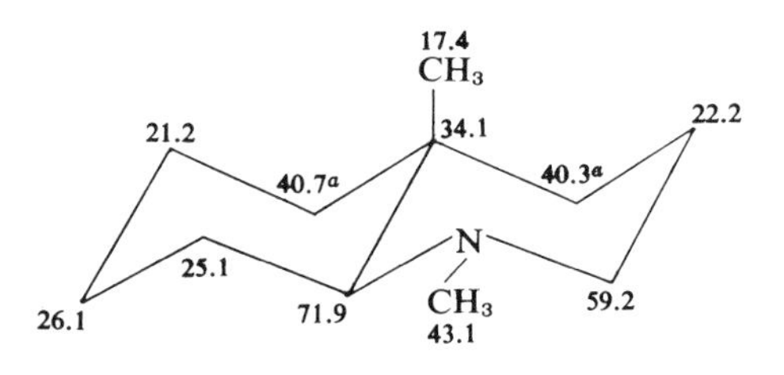

Ref. 57

293 **8α,10-Dimethyl-*trans*-decahydroquinoline**

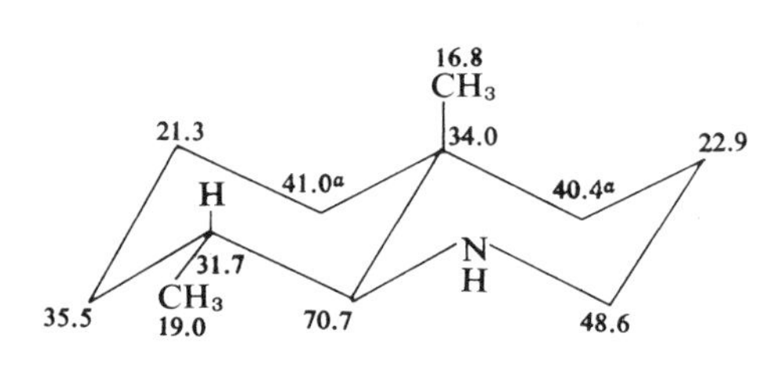

Ref. 57

294 1,8α,10-Trimethyl-*trans*-decahydroquinoline

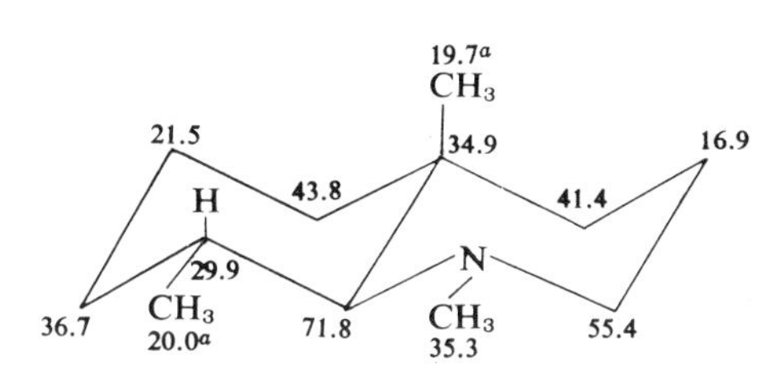

Ref. 57

Reduced Benzoquinolines[43,58]

295 *trans-syn-trans*-Perhydroacridine

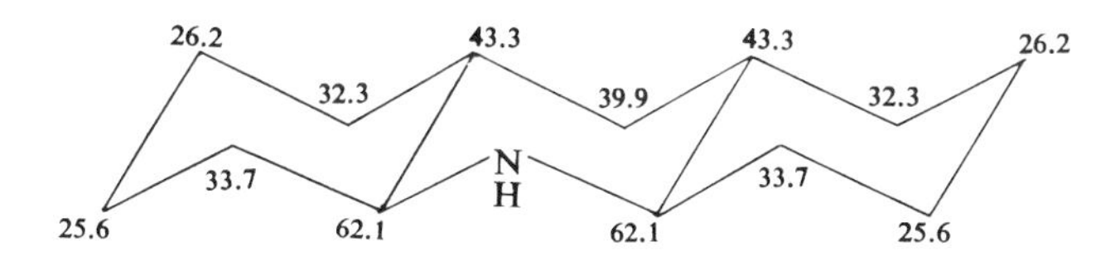

Ref. 57

296 *N*-Methyl-*trans-syn-trans*-perhydroacridine

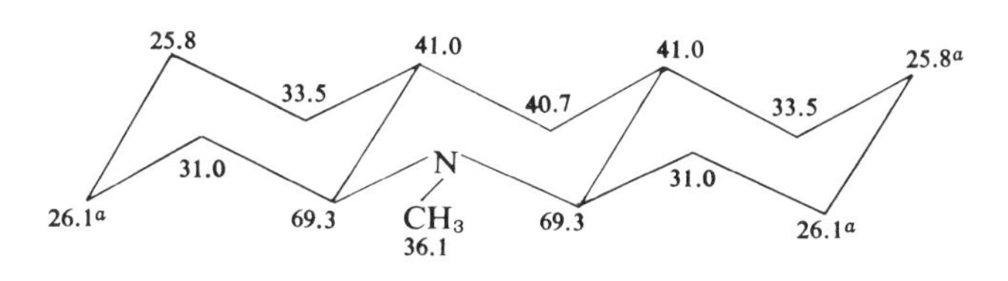

Ref. 57

297 *trans-anti-trans*-Perhydrobenzo-[h]-quinoline

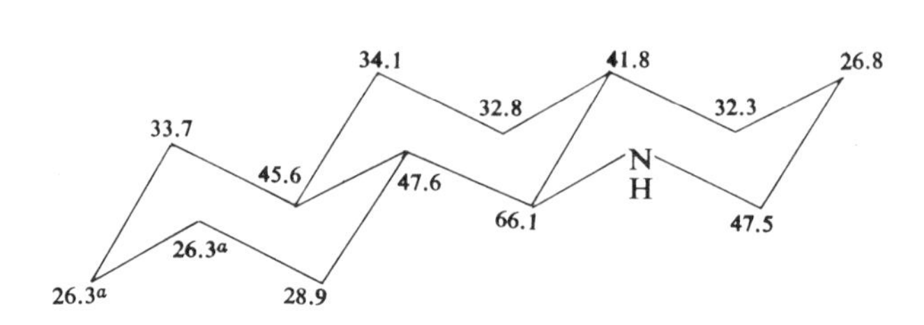

Ref. 57

298 *N*-Methyl-*trans-anti-trans*-perhydrobenzo-[h]-quinoline

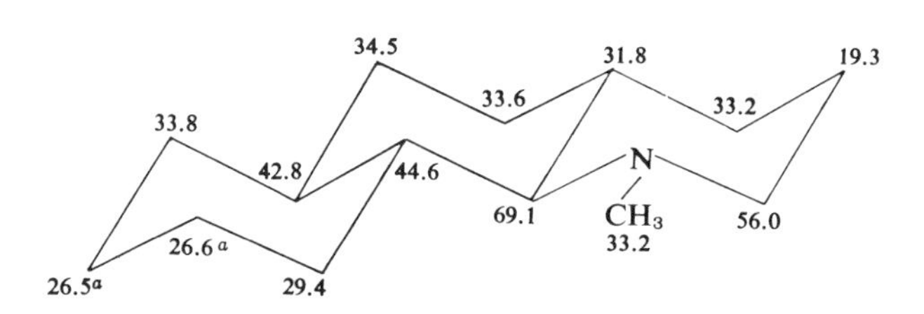

Ref. 57

299 *trans-anti-cis*-Perhydrobenzo-[h]-quinoline

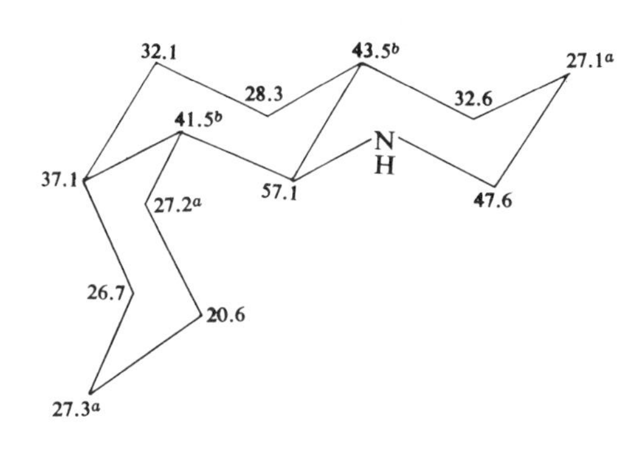

Ref. 57

300 ***N*-Methyl-*trans-anti-cis*-perhydrobenzo-[h]-quinoline**

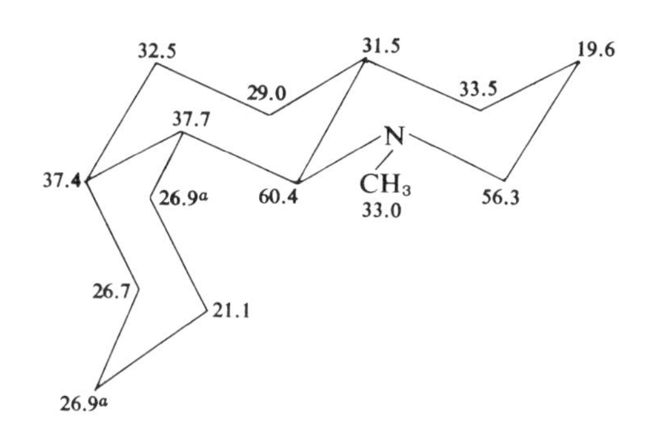

Ref. 57

301 ***cis-syn-cis*-Perhydrobenzo-[h]-quinoline**

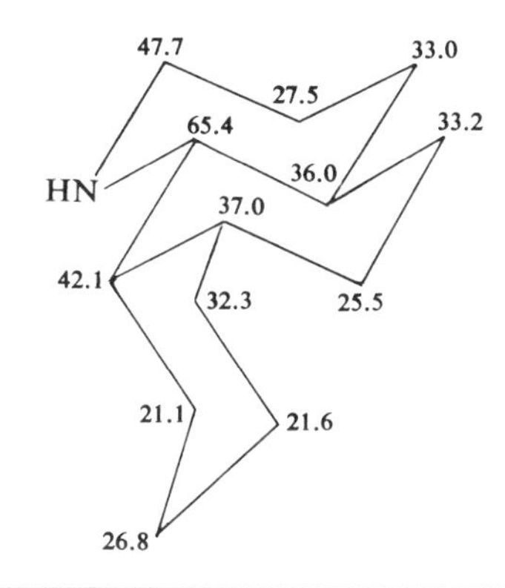

Ref. 57

302 ***N*-Methyl-*cis-syn-cis*-perhydrobenzo-[h]-quinoline**

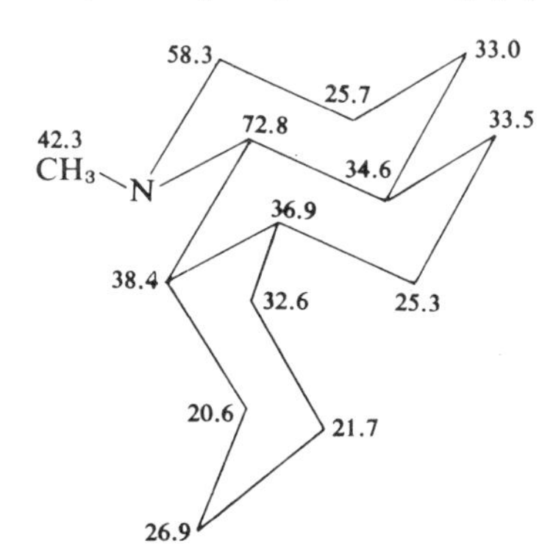

Ref. 57

10. PHENETHYLAMINES

303 Homoveratrylamine
(acetone-d_6)

55.9[a] CH_3O 149.8
113.4
133.7
36.8
44.0
NH_2
121.3
55.9[a] CH_3O 148.3
112.6

Ref. 59

304 Hordenine

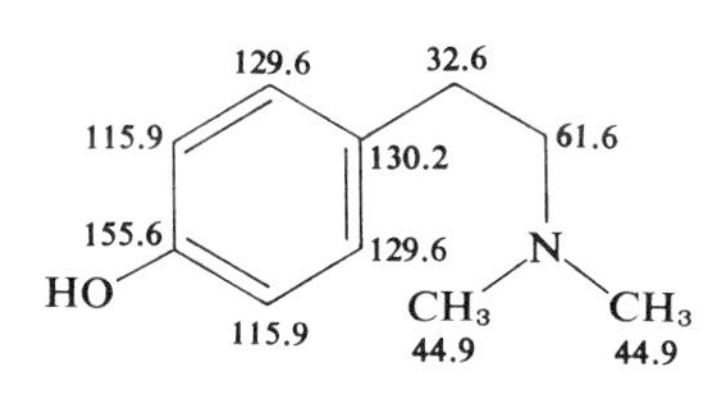

Ref. 30

11. ISOQUINOLINES

a. Simple Isoquinolines

305 Isoquinoline

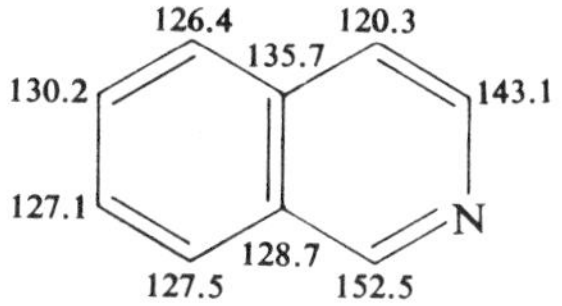

126.4 120.3 135.7 130.2 143.1 127.1 N 128.7 127.5 152.5

Ref. 50

306 5-Aminoisoquinoline

(acetone-d_6)

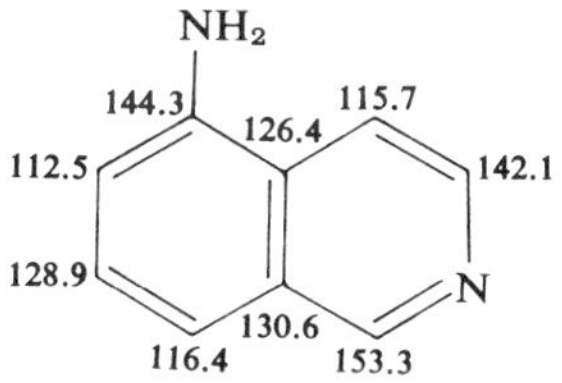

NH_2 144.3 115.7 126.4 112.5 142.1 128.9 N 130.6 116.4 153.3

Ref. 52

307 6,7-Dimethoxyisoquinoline

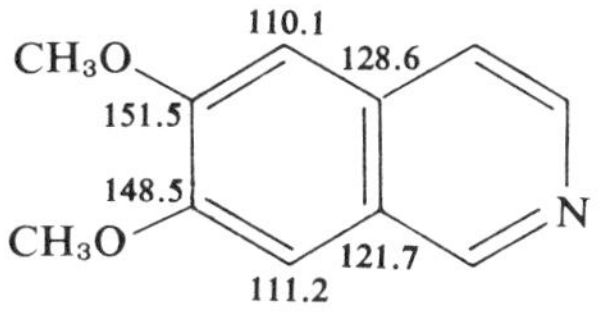

110.1 CH_3O 128.6 151.5 148.5 N CH_3O 121.7 111.2

Ref. 60

308 6,7-Dimethoxy-3,4-dihydroisoquinoline

56.0[a]
CH_3O
110.5
24.7
129.8
47.4
151.3
147.9
N
CH_3O
121.6
56.1[a]
110.5
159.5

Ref. 60

309 6,7-Dimethoxy-3,4-dihydroisoquinoline methiodide

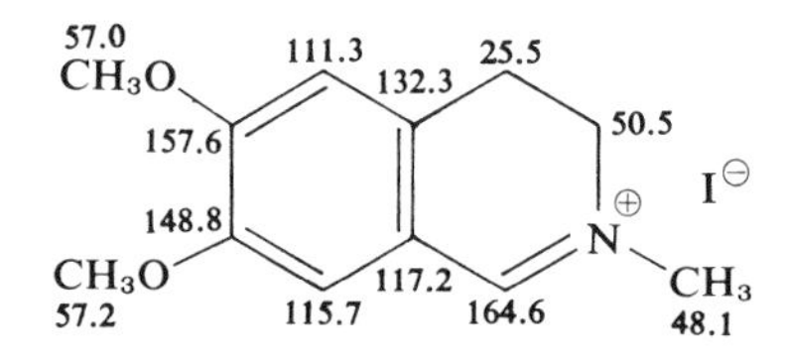

Ref. 60

310 1,2,3,4-Tetrahydroisoquinoline

129.2
29.1
136.1
125.6
43.8
125.9
N
H
134.8
126.1
48.2

Ref. 60

311 7,8-Dimethoxy-1,2,3,4-tetrahydroisoquinoline

Ref. 60

312 6,7-Dimethoxy-1,2,3,4-tetrahydroisoquinoline

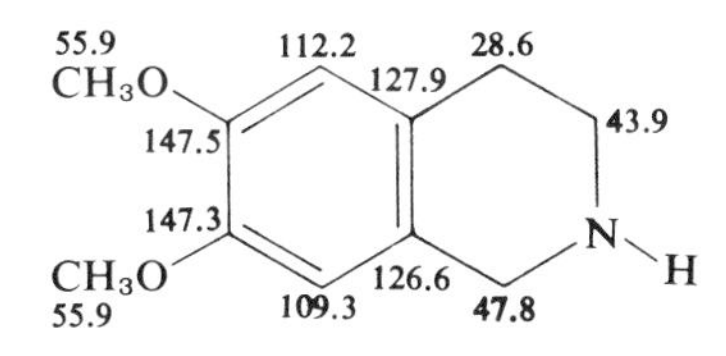

Ref. 60

313 *N*-Methyl-6,7-dimethoxy-1,2,3,4-tetrahydroisoquinoline

Ref. 60

314 Hydrastinine

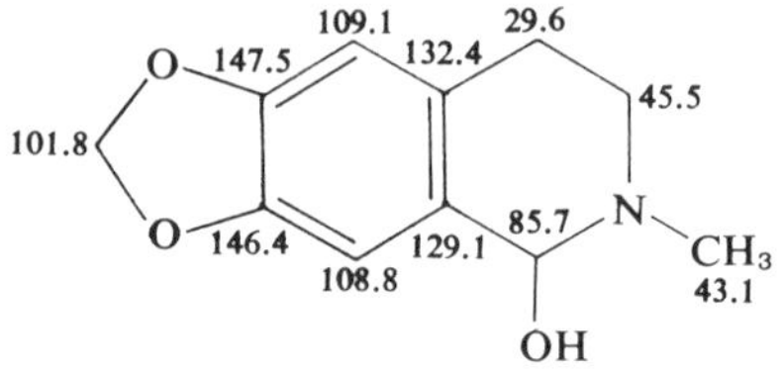

Ref. 68

315 Noroxyhydrastinine

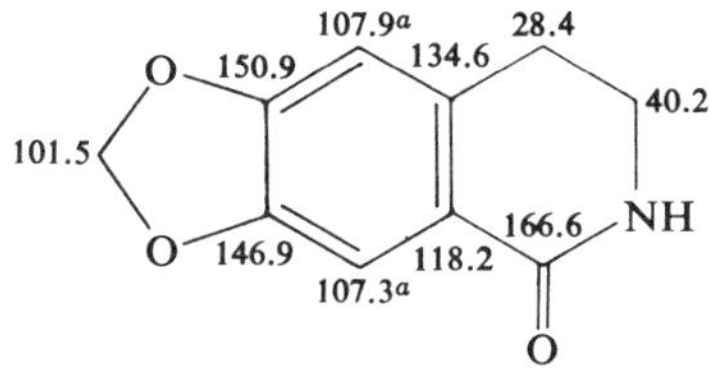

Ref. 59

316 Methyl anhydroberberilate

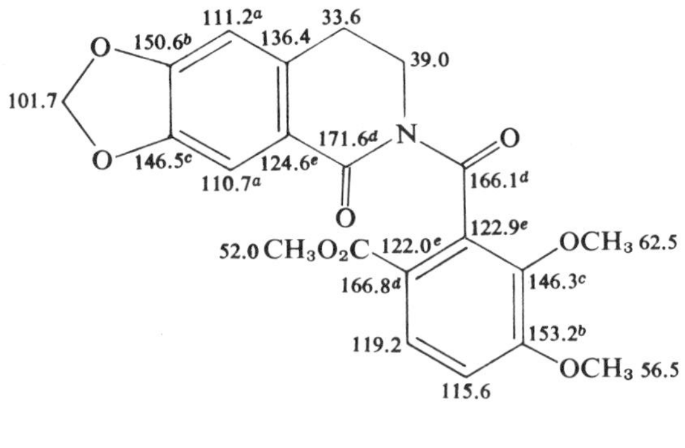

Ref. 59

317 Methyl isoanhydroberberilate

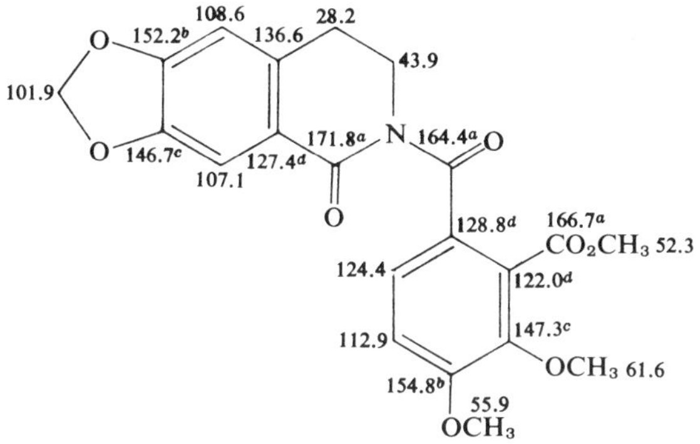

Ref. 59

Benzylisoquinolines

318 3′,4′-Dimethoxytetrahydrobenzylisoquinoline

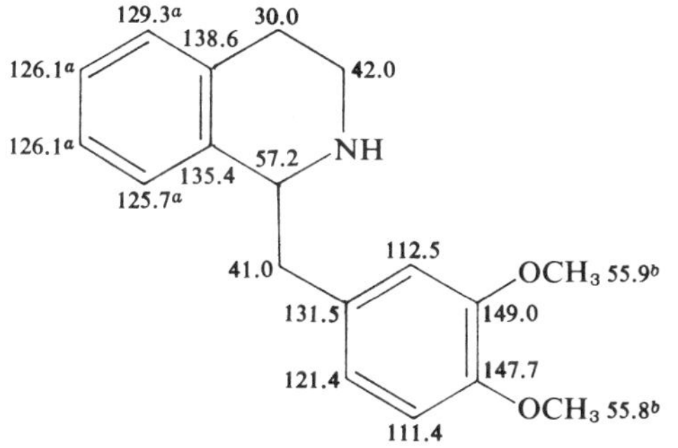

Ref. 59

319 Norlaudanosine

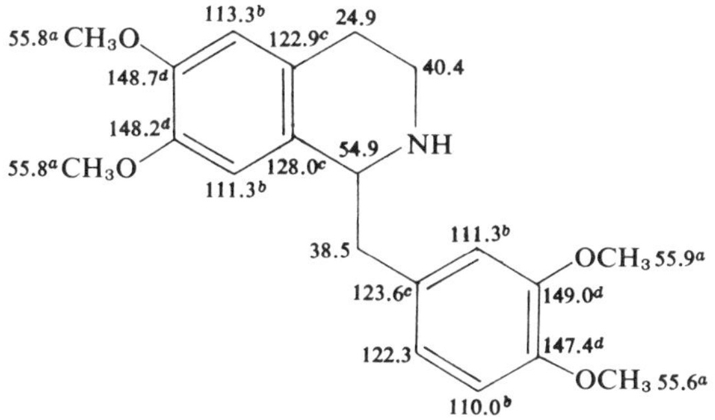

Ref. 59

320 Laudanosine

55.5[a] CH_3O
146.9
146.9
55.5[a] CH_3O
112.8
125.8
25.3
46.8
N
42.4
CH_3
65.5
132.2
110.7
40.4
110.7
OCH_3 55.3[a]
148.3
129.0
121.5
146.0
OCH_3 55.3[a]
110.7

Ref. 61;
see also Ref. 62

321 *cis*-Laudanosine *N*-oxide

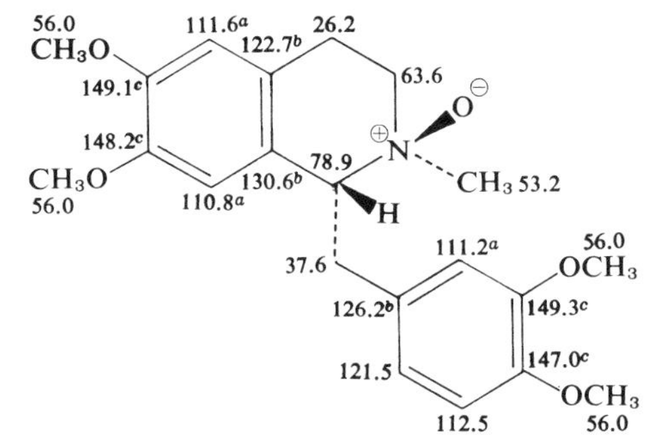

Ref. 62

322 *trans*-Laudanosine *N*-oxide

56.4
CH_3O
111.4[a]
123.2[b]
27.2
60.1
55.8
CH_3
148.1[c]
148.1[c]
$\oplus$N
79.4
$O^{\ominus}$
CH_3O
56.4
130.4[b]
108.6[a]
H
109.5[a]
56.4
OCH_3
38.8
126.3[b]
148.5[c]
147.5[c]
120.3
OCH_3
113.7
56.4

Ref. 62

323

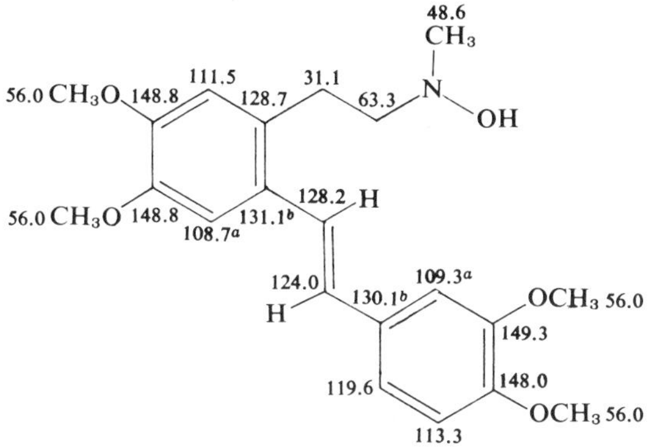

56.0 CH_3O 148.8; 111.5; 128.7; 31.1; 63.3; N; 48.6 CH_3; OH
56.0 CH_3O 148.8; 108.7[a]; 131.1[b]; 128.2 H; 124.0 H; 130.1[b]; 109.3[a]; OCH_3 56.0; 149.3; 148.0; OCH_3 56.0; 119.6; 113.3

Ref. 62

324 Reticuline

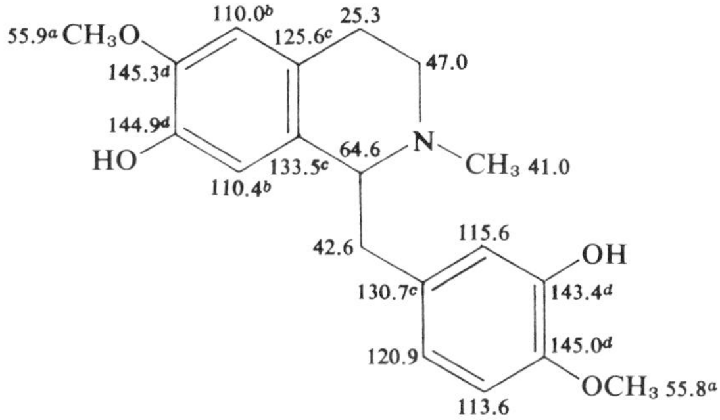

55.9[a] CH_3O; 145.3[d]; 144.9[d] HO; 110.0[b]; 110.4[b]; 125.6[c]; 133.5[c]; 25.3; 47.0; N; CH_3 41.0; 64.6; 42.6; 130.7[c]; 115.6; OH; 143.4[d]; 145.0[d]; OCH_3 55.8[a]; 120.9; 113.6

Ref. 59

325

100.4; O 147.1[a]; O 146.0[a]; 108.8[b]; 108.5[b]; 129.5[c]; 132.9[c]; 29.9; 40.8; NH; 59.7 H; 74.0 H; HO; 130.1[c]; 127.7[c]; CH_3 11.5; OCH_3 60.0; 145.0[a]; 151.9; OCH_3 55.5; 122.7; 109.6[b]

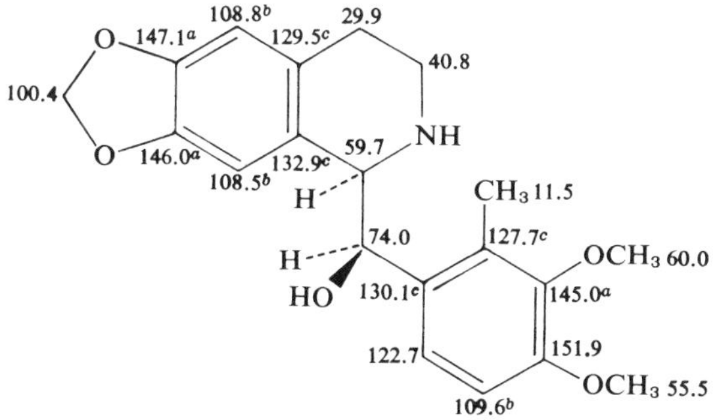

Ref. 59

326

108.0[a] 25.7
149.7 134.9
47.1
101.3
165.6 N
146.4[c] 120.5[b]
106.9
13.2 CH_3
194.9 133.0[b] OCH_3 60.0
O 129.1[b] 147.7[c]
129.7 156.3
OCH_3 55.6
108.4[a]

Ref. 59

c. Argemonine: A Pavine Alkaloid

327 Argemonine

CH_3O
66.2 109.9
129.7 OCH_3 55.4[b]
N 147.3[a]
CH_3
CH_3O 40.6 147.7[a]
123.7 OCH_3 55.8[b]
33.3 111.4

Ref. 61

d. Cularine

328 Cularine

124.3 26.0
126.3
110.4 47.5
148.9 56.7 N 42.4
CH_3O 144.8 132.5 CH_3
55.8[a]
O 35.5
148.4 118.3
105.1 113.6
147.3 144.8
CH_3O OCH_3
56.0[a] 56.0[a]

Ref. 61

Proaporphines

329 Glaziovine
(DMSO-d_6)

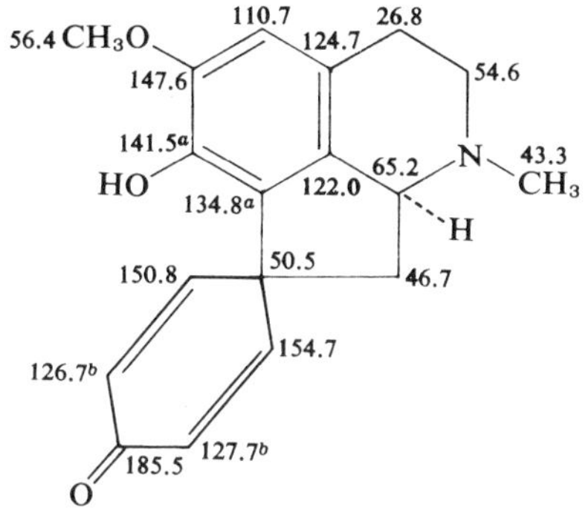

56.4 CH_3O; 110.7; 26.8; 124.7; 147.6; 54.6; 141.5[a]; HO; 65.2; N; 43.3 CH_3; 122.0; 134.8[a]; H; 50.5; 150.8; 46.7; 154.7; 126.7[b]; 185.5; 127.7[b]; O

Ref. 63

330 Pronuciferine
(DMSO-d_6)

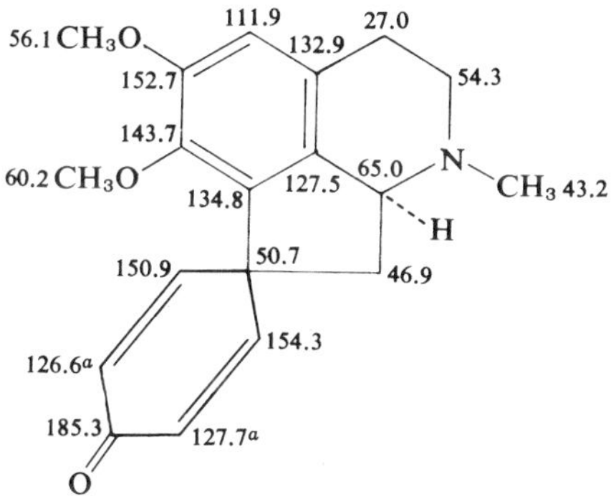

56.1 CH_3O; 111.9; 27.0; 132.9; 152.7; 54.3; 143.7; 60.2 CH_3O; 65.0; N; CH_3 43.2; 127.5; 134.8; H; 50.7; 150.9; 46.9; 154.3; 126.6[a]; 185.3; 127.7[a]; O

Ref. 63

331 *N*-Methylcrotsparinine
(DMSO-d_6)

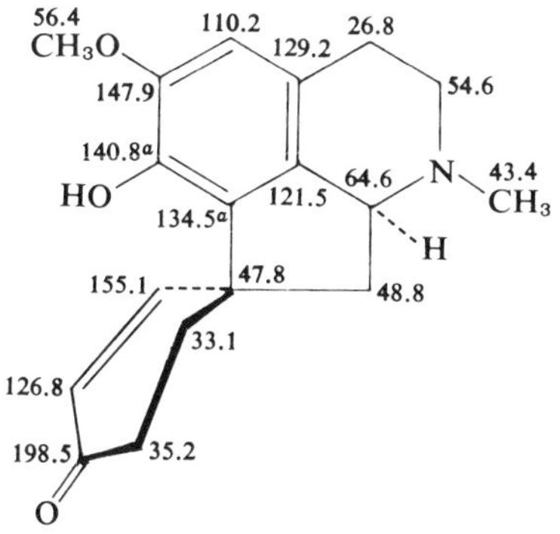

56.4 CH_3O; 110.2; 26.8; 129.2; 147.9; 54.6; 140.8[a]; HO; 64.6; N; 43.4 CH_3; 121.5; 134.5[a]; H; 47.8; 155.1; 48.8; 33.1; 126.8; 198.5; 35.2; O

Ref. 63

332 *O*-Acetyl-*N*-methylcrotsparinine
(DMSO-d_6)

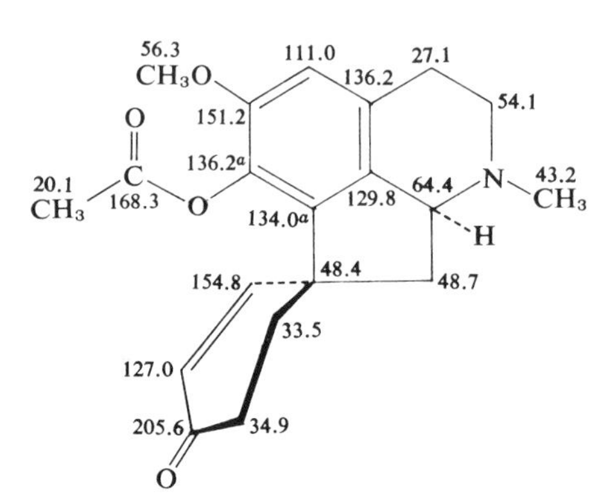

Ref. 63

333 *N*-Methylisocrotsparinine
(DMSO-d_6)

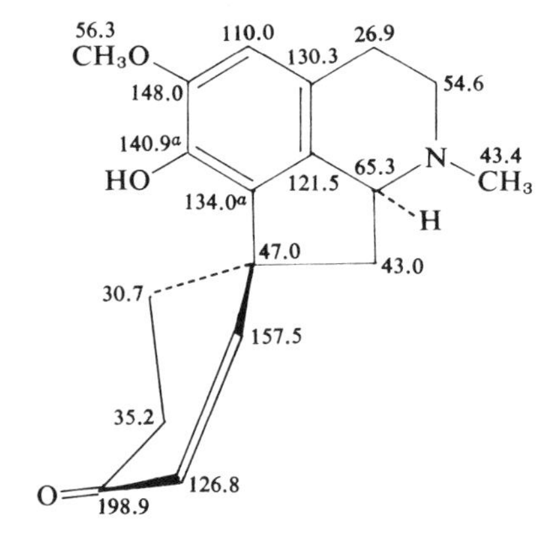

Ref. 63

334 ***O*-Acetyl-*N*-methylisocrotsparinine**
(DMSO-d_6)

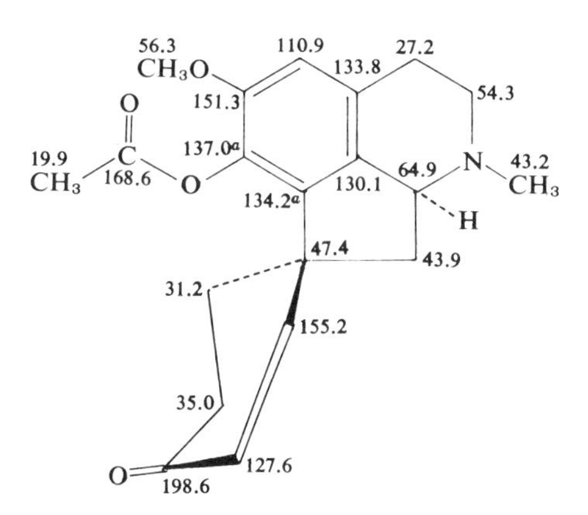

Ref. 63

335 **Amuronine**
(DMSO-d_6)

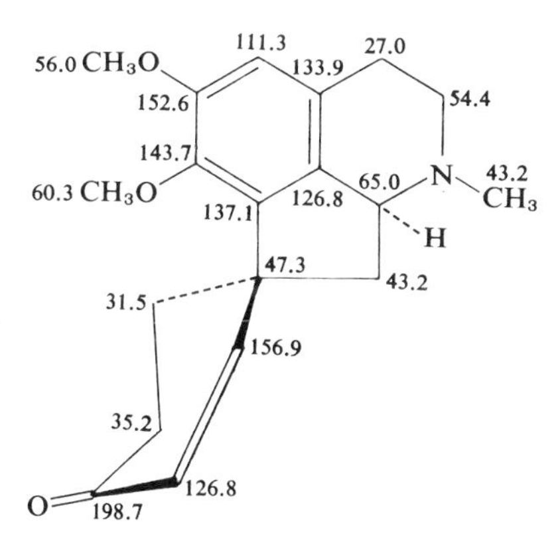

Ref. 63

336

(DMSO-d_6)

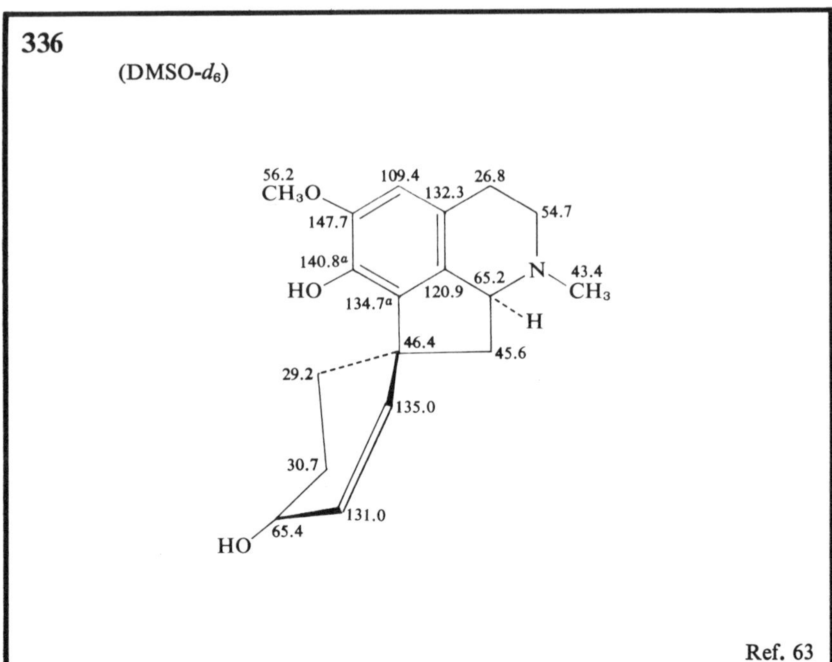

Ref. 63

337

(DMSO-d_6)

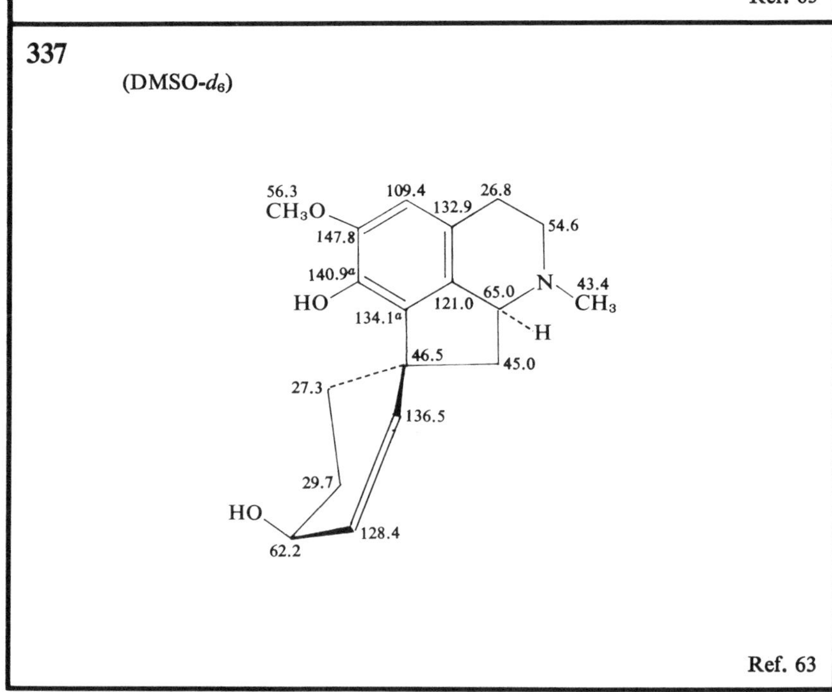

Ref. 63

338

(DMSO-d_6)

56.2 CH_3O
26.8
54.5
HO
64.6 N 43.3 CH_3
H
133.8 47.1 50.5
29.1
129.4
HO
29.4
62.5

Ref. 63

339

(DMSO-d_6)

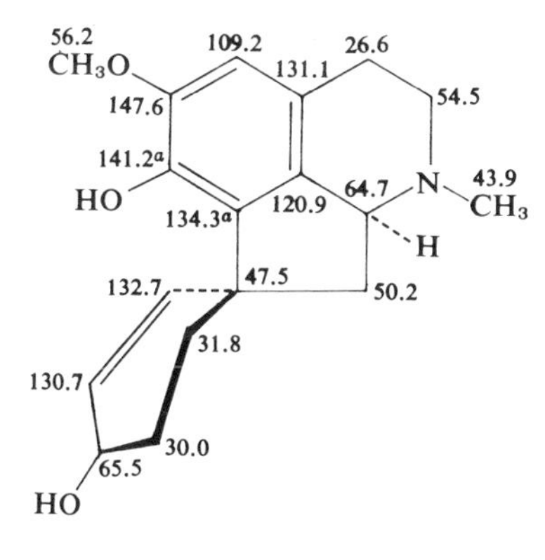

Ref. 63

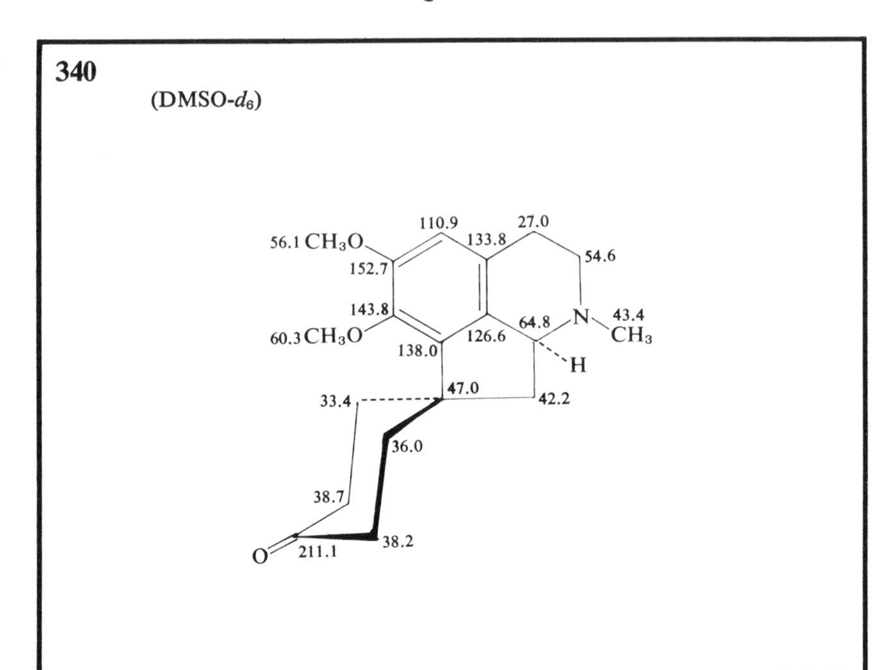

Ref. 63

341

(DMSO-d_6)

56.2 CH_3O
111.2
129.3
26.7
54.7
148.5
141.5[a]
HO
121.0
64.4
N
43.5
CH_3
134.1[a]
H
OH
51.4
40.4
26.7
75.5
H
38.4
O
209.2
46.0

Ref. 63

342

(DMSO-d_6)

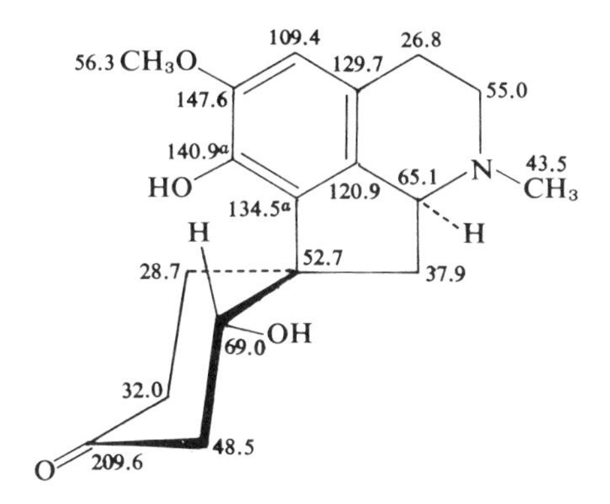

Ref. 63

343

(DMSO-d_6)

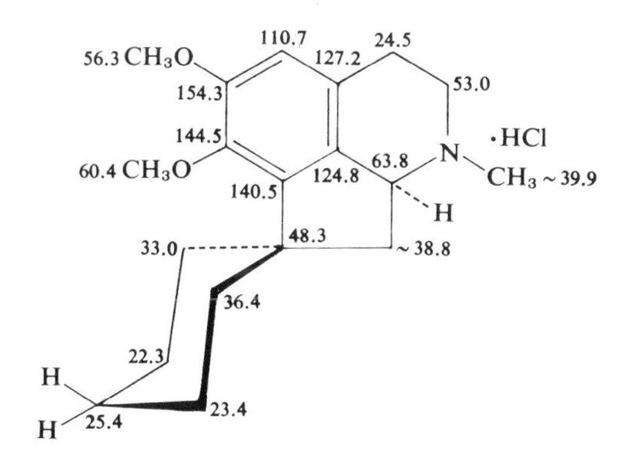

Ref. 63

344

(DMSO-d_6)

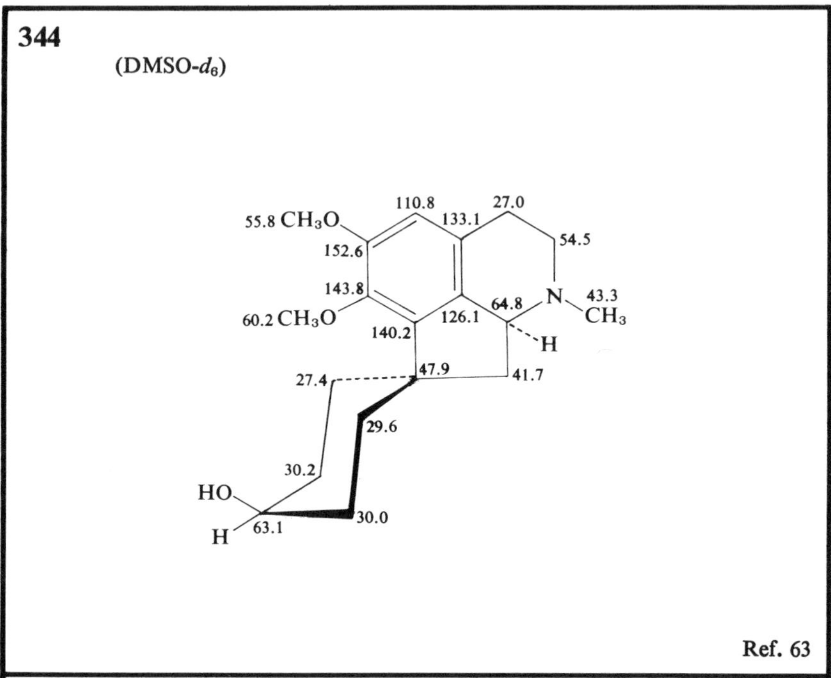

Ref. 63

345

(DMSO-d_6)

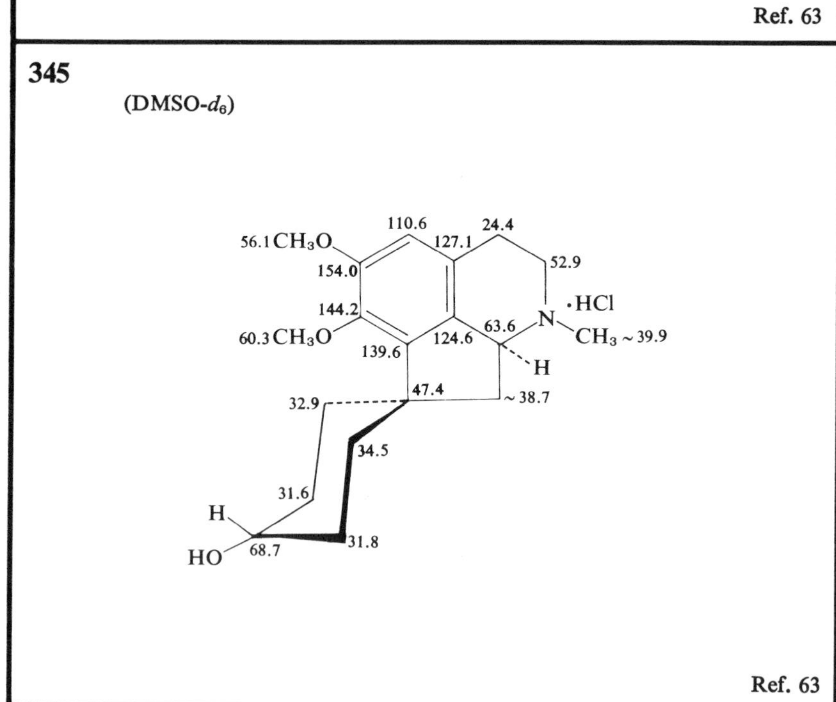

Ref. 63

C. Aporphines

346 Nuciferine

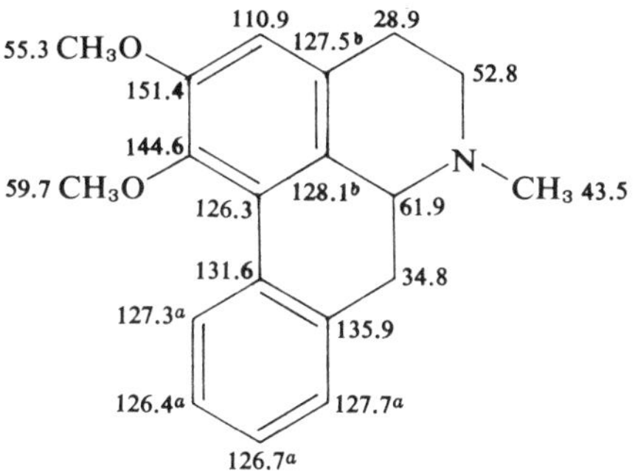

55.3 CH_3O, 151.4, 144.6, 59.7 CH_3O, 110.9, 127.5[b], 28.9, 52.8, N, CH_3 43.5, 128.1[b], 61.9, 126.3, 131.6, 34.8, 127.3[a], 135.9, 126.4[a], 127.7[a], 126.7[a]

Ref. 61

347 Glaucine

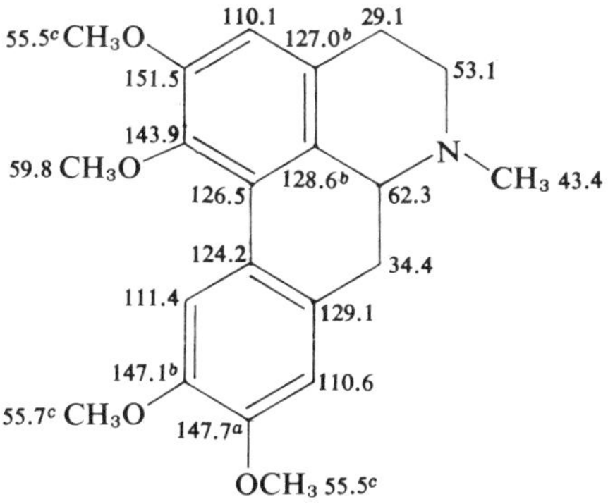

55.5[c] CH_3O, 151.5, 143.9, 59.8 CH_3O, 110.1, 127.0[b], 29.1, 53.1, N, CH_3 43.4, 128.6[b], 62.3, 126.5, 124.2, 34.4, 111.4, 129.1, 147.1[b], 110.6, 55.7[c] CH_3O, 147.7[a], OCH_3 55.5[c]

Ref. 61

348 Thaliporphine

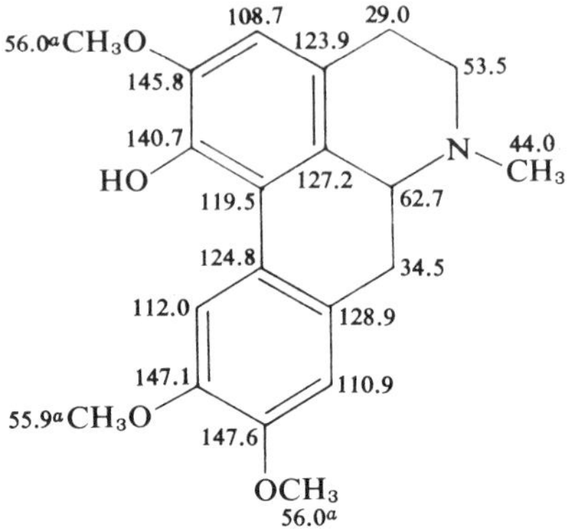

Ref 64;
see also Ref. 65

349 Isocorydine

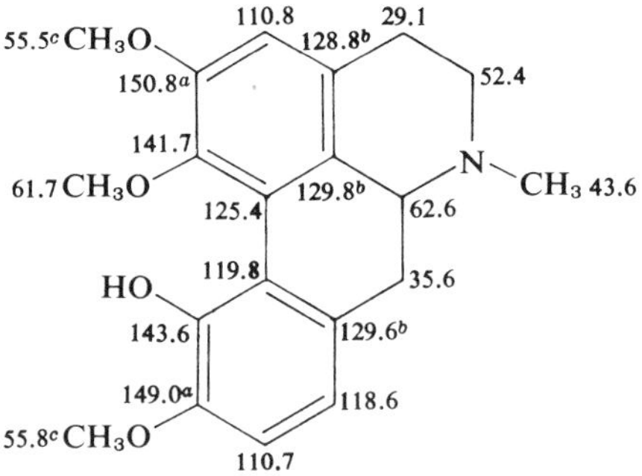

Ref. 61

350 Boldine

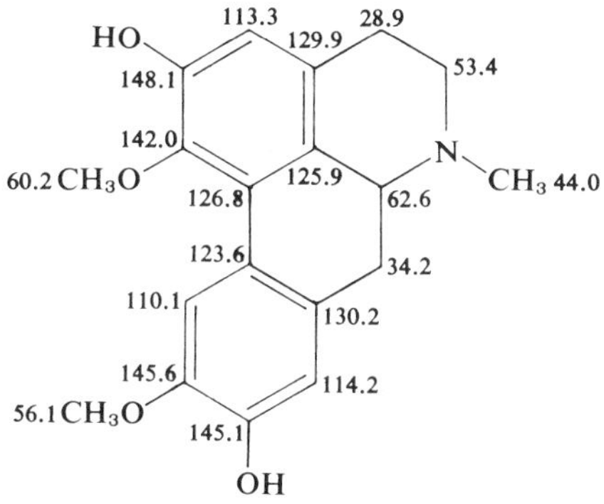

Ref. 64

351 Predicentrine

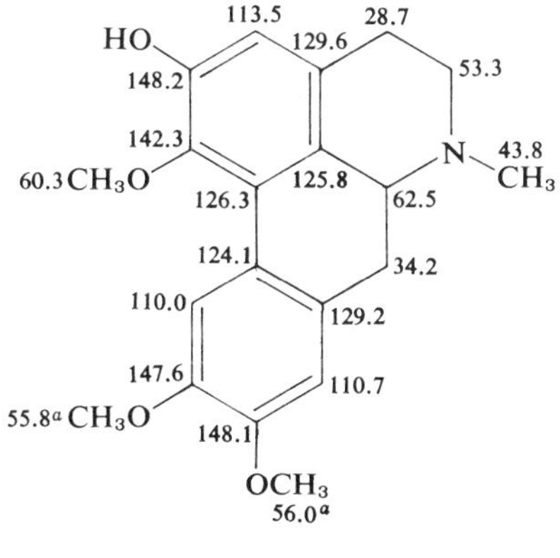

Ref. 64

352 Dicentrine

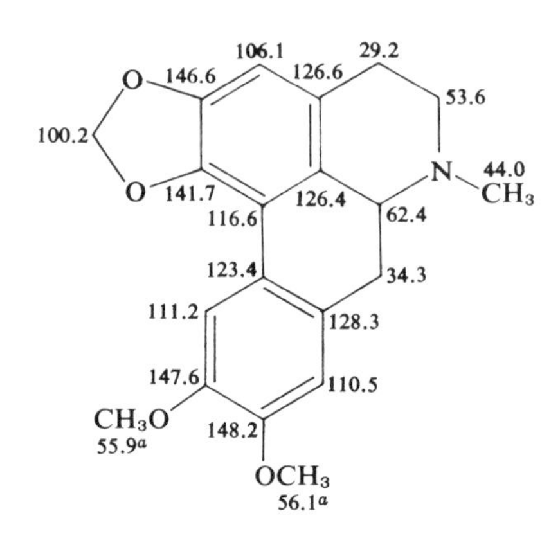

Ref. 64

353 Bulbocapnine

107.7
29.3
145.9
127.4[a]
53.0
100.2
N
44.0
CH_3
140.4
128.9[a]
62.8
125.8[b]
118.5[b]
35.4
HO
142.9
129.7[a]
148.3
119.2
56.2 CH_3O
110.8

Ref. 59

354 Nantenine

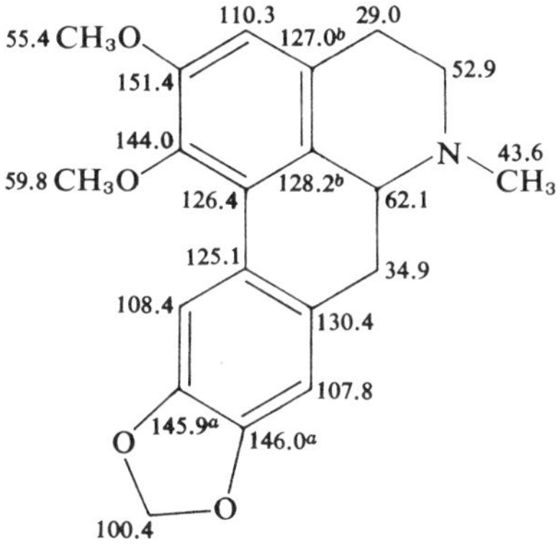

Ref. 61

355 Domesticine

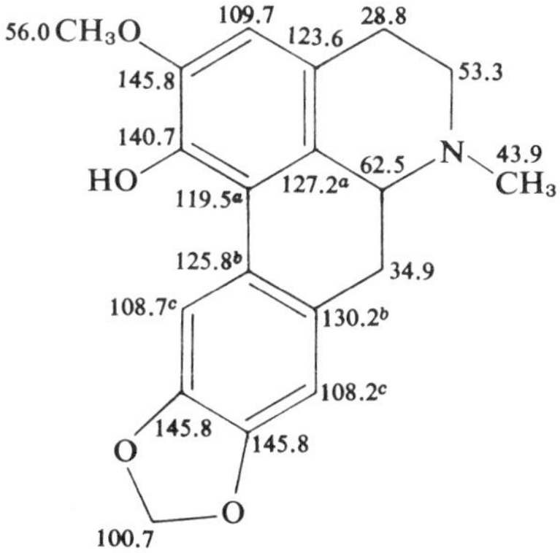

Ref. 65;
see also Ref. 64

g. Protoberberines

356 Berberine cation
(TFA-*d*)

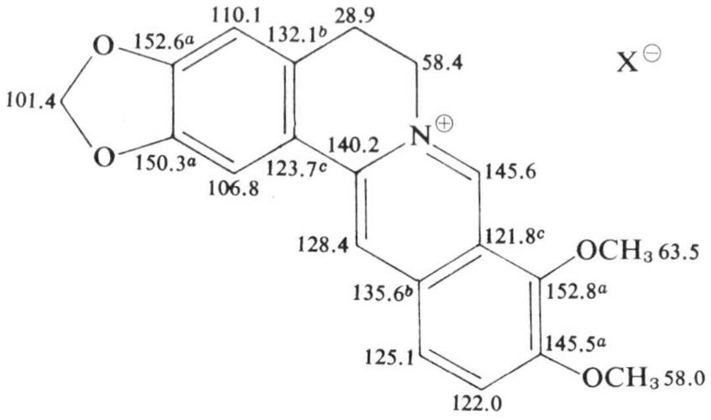

101.4, 152.6[a], 150.3[a], 110.1, 106.8, 132.1[b], 123.7[c], 28.9, 58.4, 140.2, 145.6, 128.4, 121.8[c], 135.6[b], 152.8[a], 145.5[a], 125.1, 122.0, OCH_3 63.5, OCH_3 58.0, $X^{\ominus}$

Ref. 59

357 *trans-N*-Methyltetrahydroprotoberberine chloride
(CD_3OD)

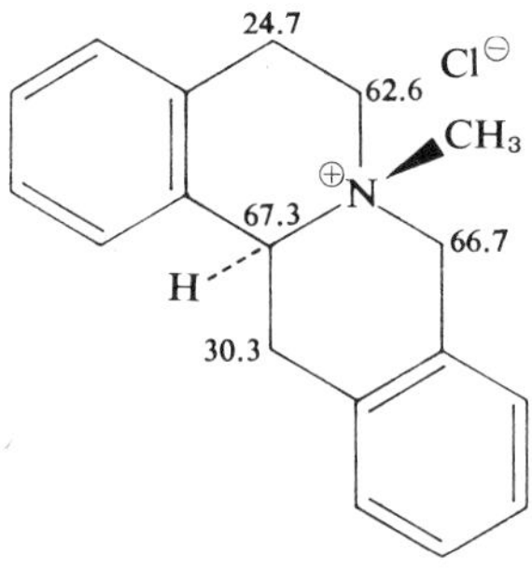

24.7, 62.6, $Cl^{\ominus}$, CH_3, 67.3, 66.7, 30.3

Ref. 66

358 ***cis-N*-Methyltetrahydroprotoberberine chloride**
(CD_3OD)

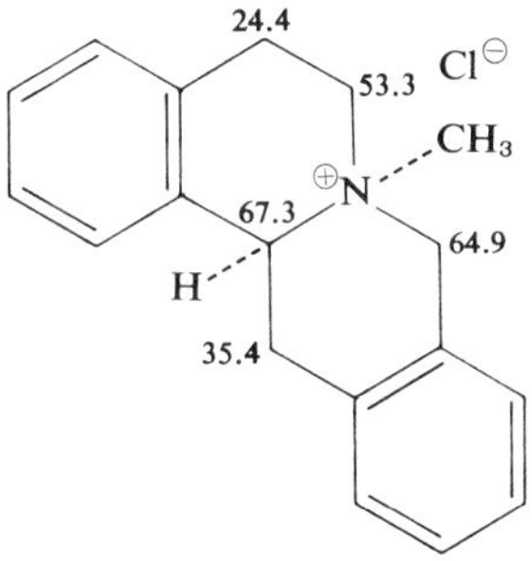

24.4
53.3 $Cl^{\ominus}$
CH_3
$^{\oplus}N$
67.3
64.9
H
35.4

Ref. 66

359 **Tetrahydroprotoberberine**

129.0 29.7
134.7
126.2^a 51.3
126.2^a N
60.1
138.2 58.7
125.6^a
134.7
36.8
134.7 129.0
129.0 126.2^a
126.0^a

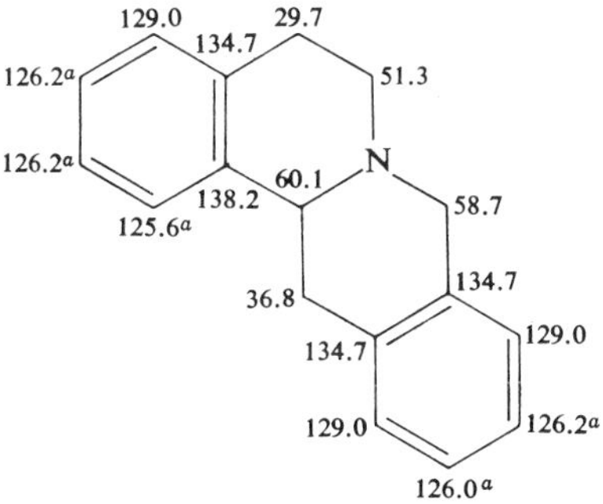

Ref. 67

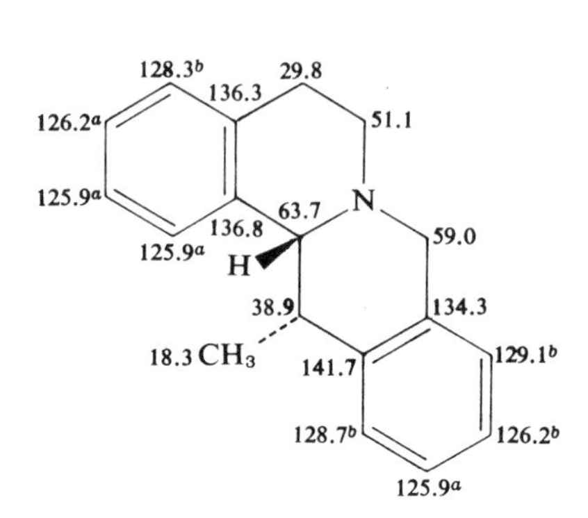

Ref. 67

361

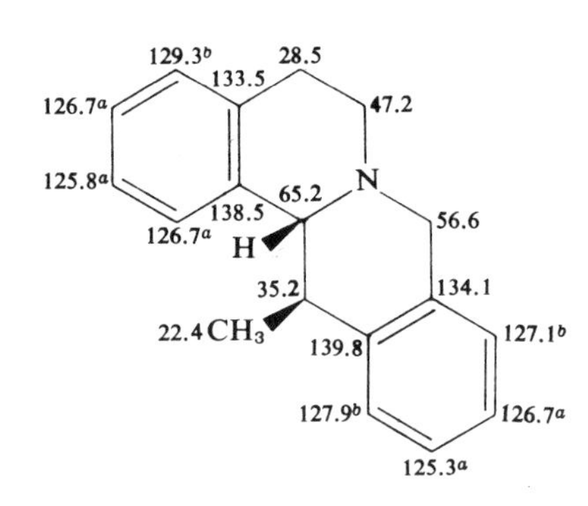

Ref. 67

362 Tetrahydropalmatine

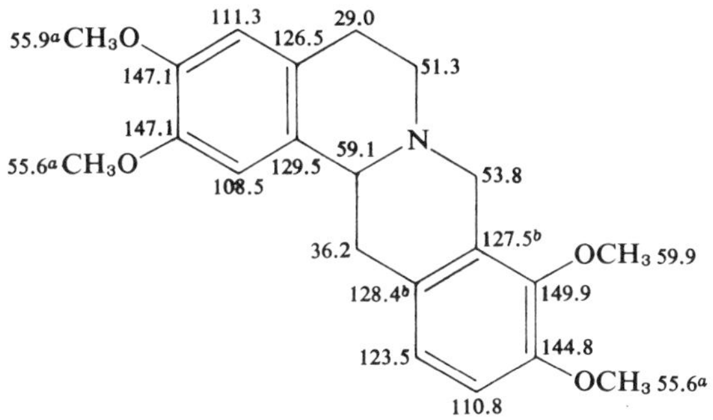

55.9[a] CH_3O 111.3 29.0 126.5 147.1 51.3 147.1 55.6[a] CH_3O 59.1 N 129.5 53.8 108.5 36.2 127.5[b] OCH_3 59.9 128.4[b] 149.9 123.5 144.8 OCH_3 55.6[a] 110.8

Ref. 61;
see also Ref. 60

363 Corynoxidine

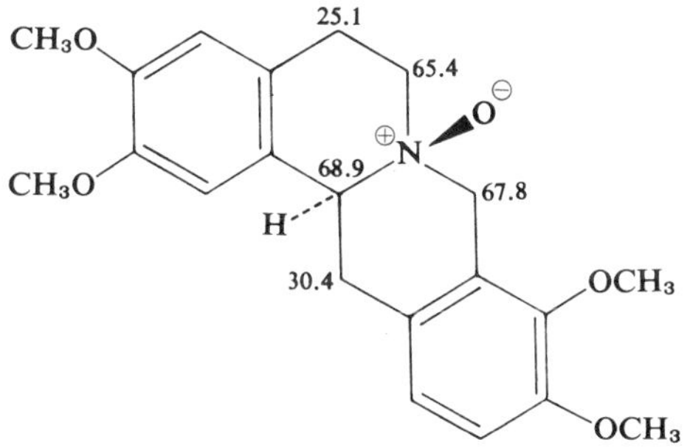

CH_3O 25.1 65.4 $O^{\ominus}$ CH_3O $^{\oplus}N$ 68.9 67.8 H 30.4 OCH_3 OCH_3

Ref. 66

364 Epicorynoxidine

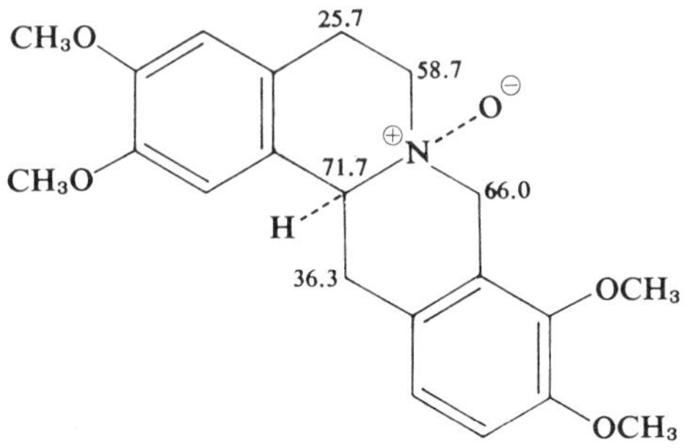

Ref. 66

365 Corydaline

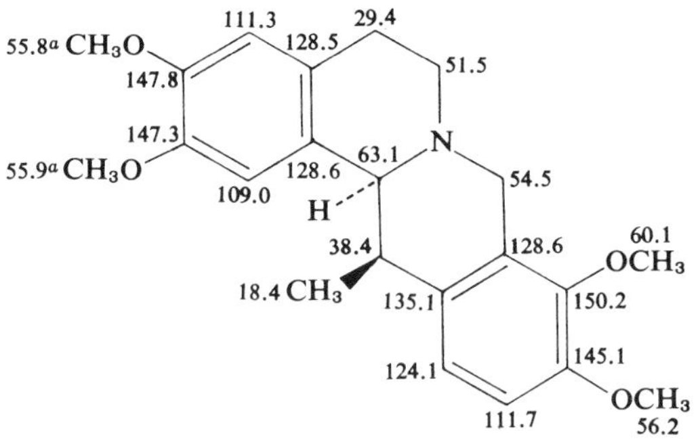

Ref. 60;
see also Ref. 68

366 Mesocorydaline

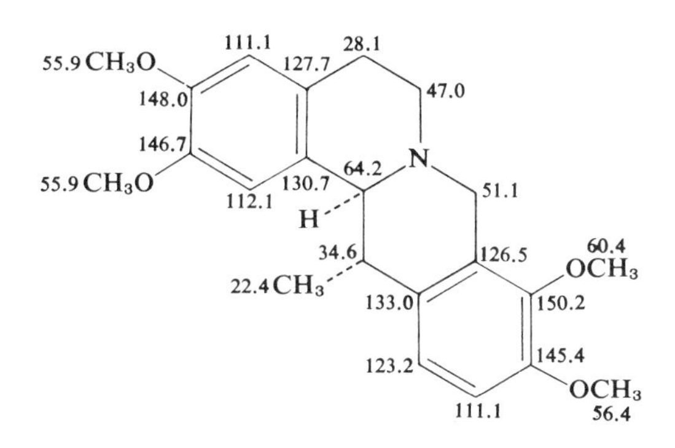

Ref. 60;
see also Ref. 68

367 Tetrahydropalmatrubine

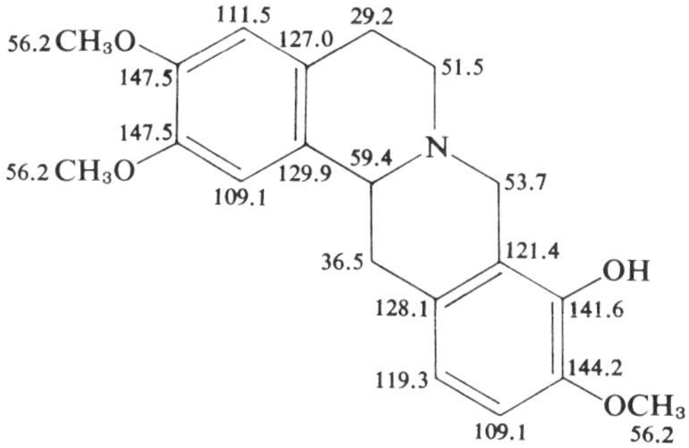

56.2 CH_3O
147.5
147.5
56.2 CH_3O
111.5
127.0
29.2
51.5
109.1
129.9
59.4
N
53.7
36.5
121.4
OH
128.1
141.6
144.2
119.3
OCH_3
109.1
56.2

Ref. 68a

368 Thalictrifoline

55.9[a] CH_3O
56.1[a] CH_3O
110.9
126.3
27.6
46.9
148.0
146.6
112.0
130.3
63.8
N
49.8
H
34.2
115.8
22.4 CH_3
133.5
O
144.8
101.1
143.7
120.3
O
106.8

Ref. 60

369 Cavidine

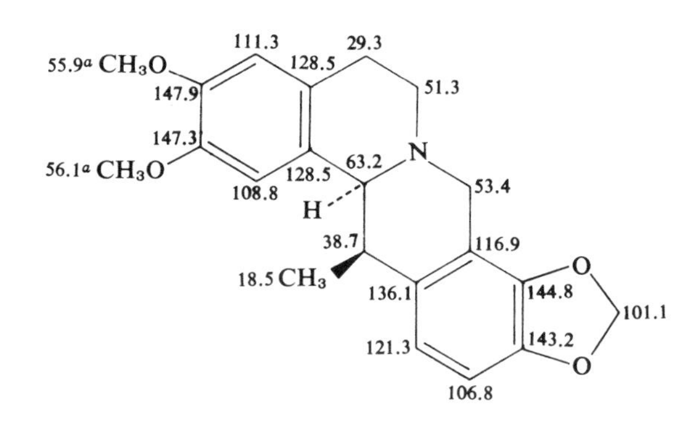

Ref. 60

370 Canadine

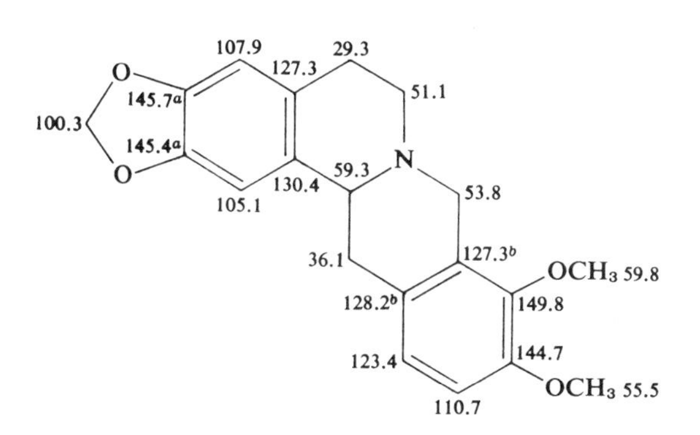

Ref. 61; see also Refs. 59, 60, 67, 68a

371 *trans*-Canadine *N*-oxide

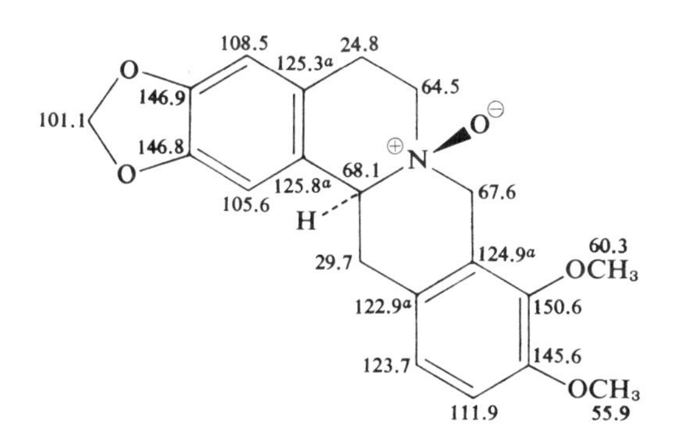

Ref. 59

372 Thalictricavine

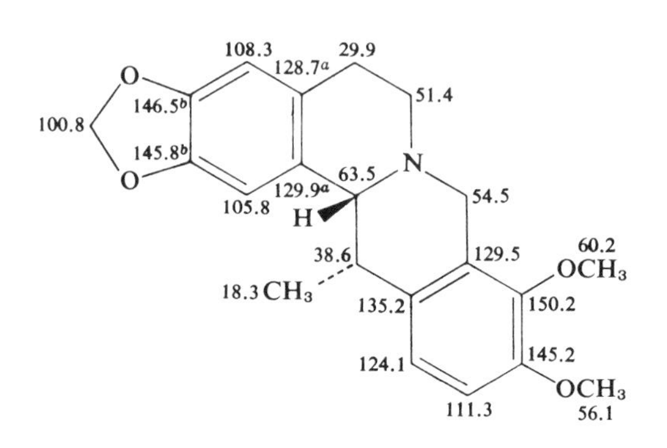

Ref. 67;
see also Ref. 59

373 Mesothalictricavine

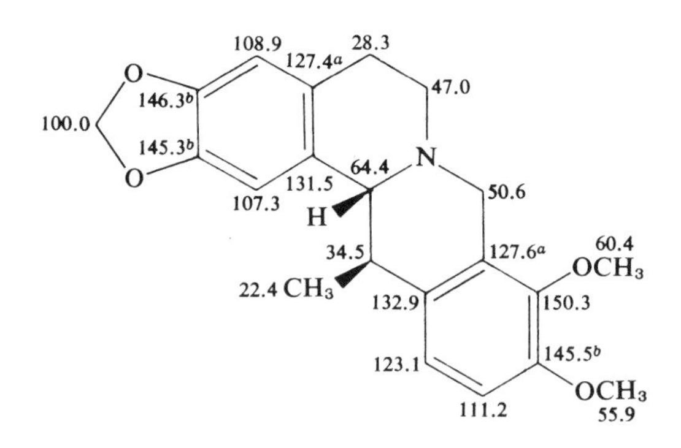

Ref. 67

374 Ophiocarpine

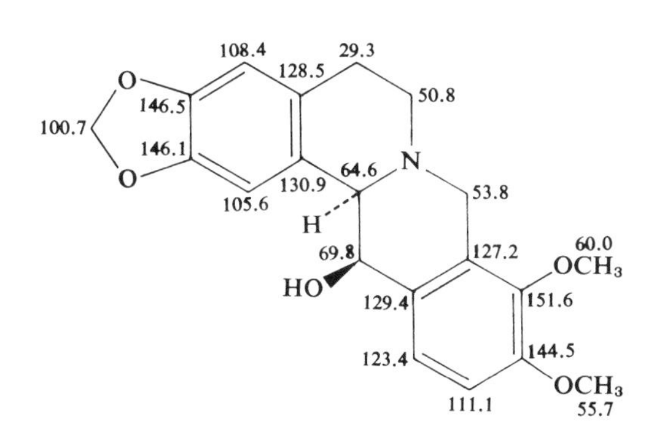

Ref. 68

375 8-(2′-Hydroxypropyl)canadine

108.3
29.9
127.8
O
48.7
146.1
100.8
38.8
23.6
145.9
N
CH$_2$
65.0
CH$_3$
O
58.9
CH
130.3[a]
61.5
OH
105.6
130.0[a]
60.2
37.2
OCH$_3$
130.2[a]
150.9
123.4
145.5
OCH$_3$
111.0
55.8

Ref. 59

376 Nandinine

108.5
128.0
29.7
O
146.2
51.4
100.8
146.1
N
O
59.7
131.1
53.5
105.7
36.5
121.4
OH
128.1
141.7
144.2
119.4
OCH_3
109.1
56.2

Ref. 60;
see also Ref. 59

377 Tetrahydrocoptisine

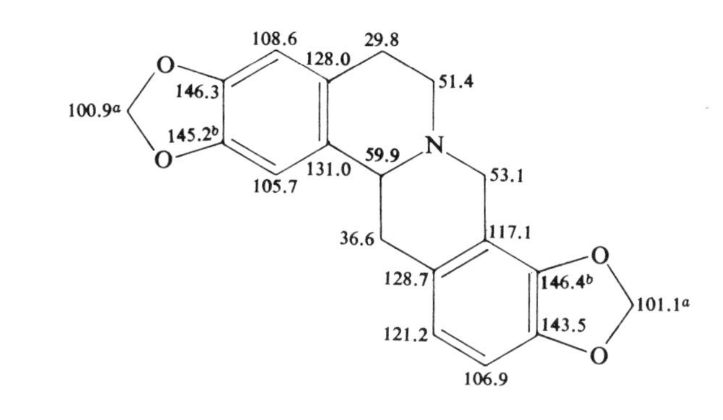

Ref. 67

378 Tetrahydrocorysamine

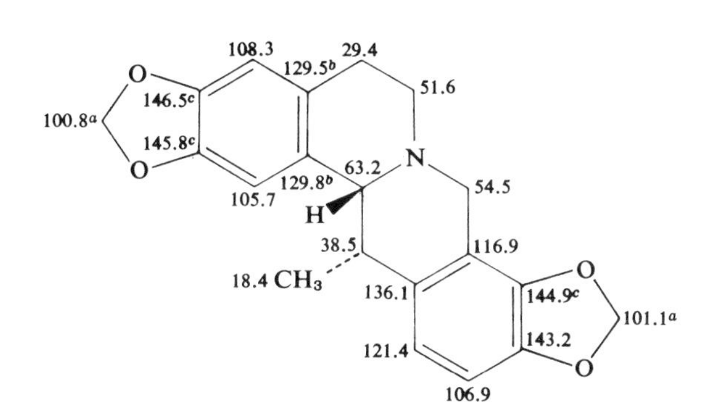

Ref. 67

379 Xylopinine

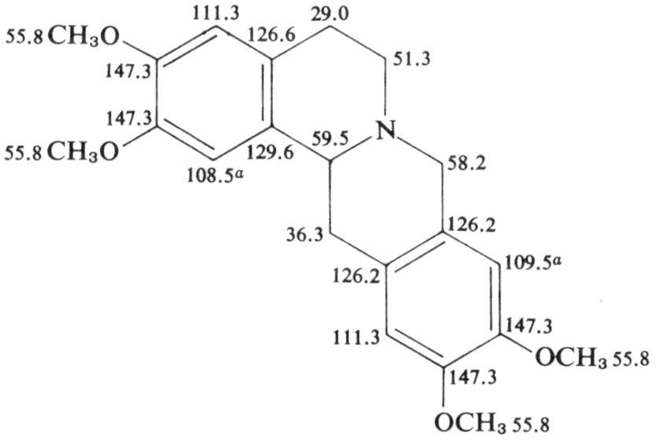

111.3
29.0
55.8 CH3O
126.6
51.3
147.3
147.3
N
59.5
55.8 CH3O
129.6
58.2
108.5a
36.3
126.2
109.5a
126.2
147.3
111.3
OCH3 55.8
147.3
OCH3 55.8

Ref. 68a

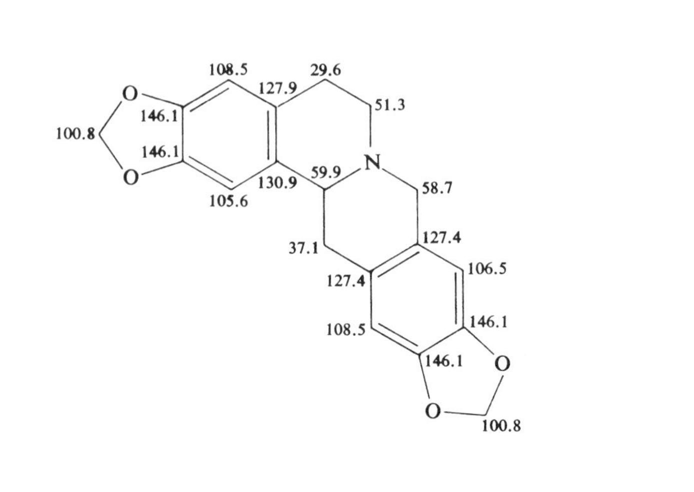

Ref. 68a

381

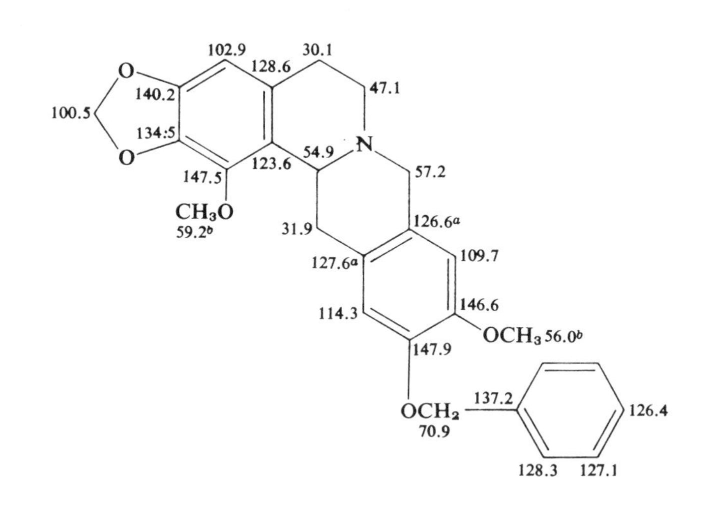

Ref. 68a

382

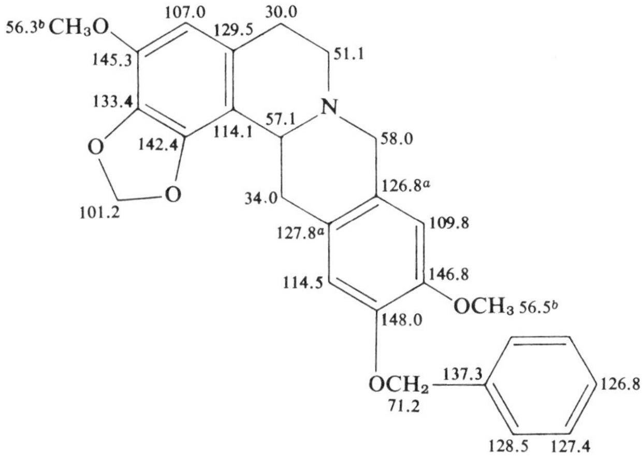

56.3[b] CH_3O, 145.3, 133.4, 107.0, 129.5, 30.0, 51.1, N, 57.1, 114.1, 142.4, 101.2, 58.0, 34.0, 126.8[a], 127.8[a], 109.8, 114.5, 146.8, OCH_3 56.5[b], 148.0, OCH_2 71.2, 137.3, 126.8, 128.5, 127.4

Ref. 68a

383 Capaurine

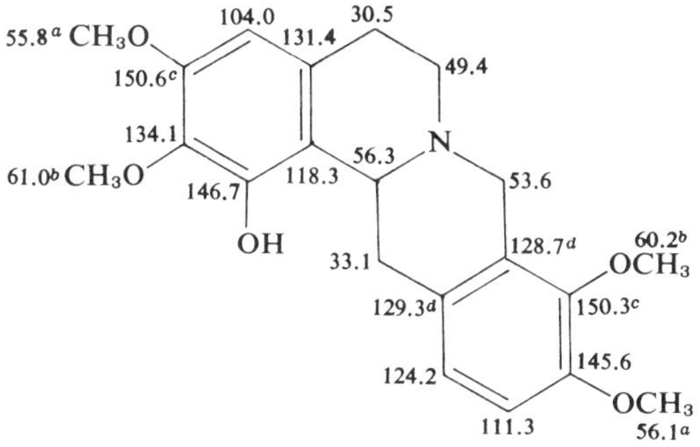

55.8[a] CH_3O, 150.6[c], 134.1, 61.0[b] CH_3O, 146.7, OH, 104.0, 131.4, 118.3, 30.5, 49.4, N, 56.3, 53.6, 33.1, 128.7[d], 60.2[b] OCH_3, 150.3[c], 129.3[d], 124.2, 111.3, 145.6, OCH_3 56.1[a]

Ref. 67

384 *O*-Methylcapaurine

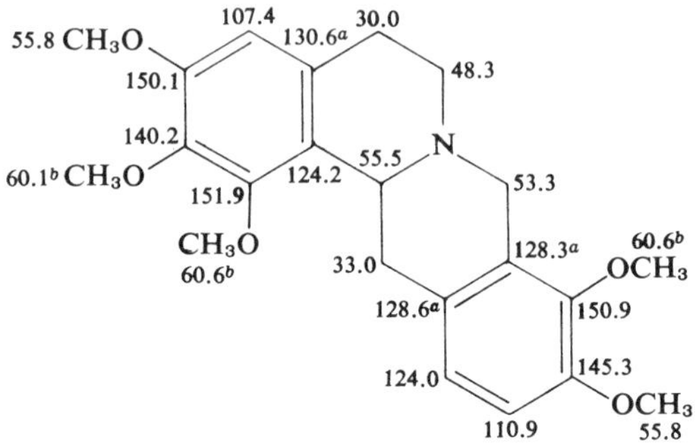

55.8 CH_3O; 150.1; 107.4; 130.6[a]; 30.0; 48.3; 140.2; 60.1[b] CH_3O; 151.9; 124.2; 55.5; N; 53.3; CH_3O 60.6[b]; 33.0; 128.3[a]; 60.6[b] OCH_3; 128.6[a]; 150.9; 124.0; 145.3; 110.9; OCH_3 55.8

Ref. 68a

385 Capaurimine

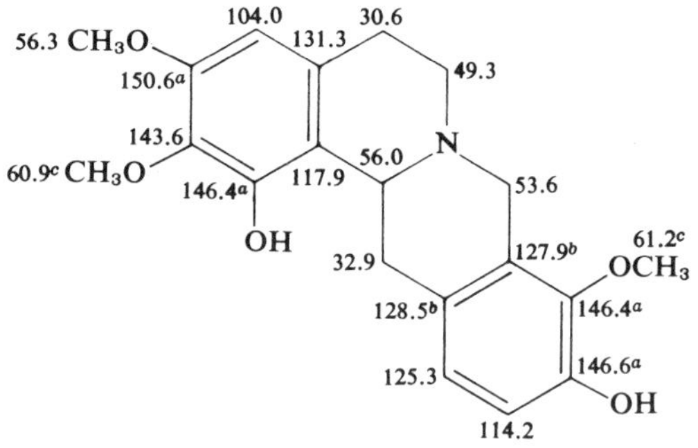

56.3 CH_3O; 150.6[a]; 104.0; 131.3; 30.6; 49.3; 143.6; 60.9[c] CH_3O; 146.4[a]; 117.9; 56.0; N; 53.6; OH; 32.9; 127.9[b]; 61.2[c] OCH_3; 128.5[b]; 146.4[a]; 125.3; 146.6[a]; OH; 114.2

Ref. 68a

386

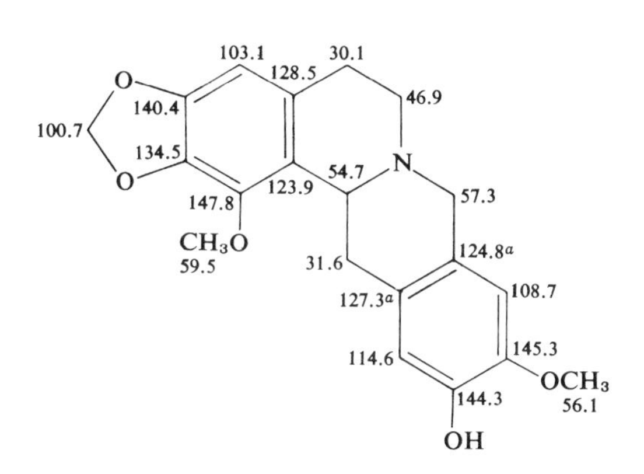

Ref. 68a

387 ***O*-Acetylcapaurine**

110.8
30.3
60.3[a] CH_3O
131.1
48.5
151.8
139.6
N
55.7
60.8[a] CH_3O
123.6
141.8
53.4
20.8 CH_3
168.8
O
C
O
33.3
128.4[b]
60.3[a]
OCH_3
128.7[b]
150.5
124.0
145.8
OCH_3
111.4
56.1

Ref. 67

388 *O,O*-Diacetylcapaurimine

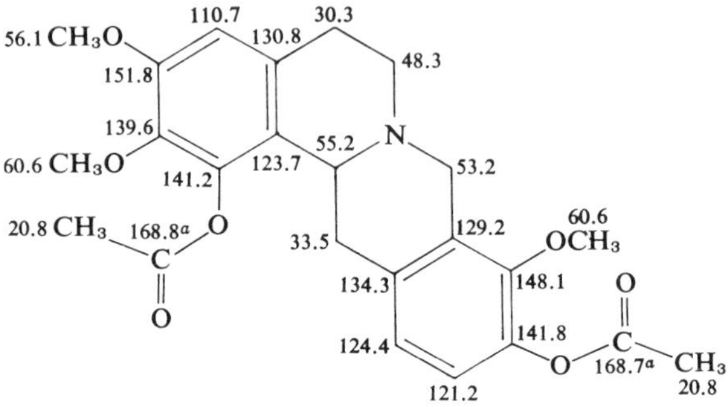

Ref. 67

389 8,13-Dioxo-14-methoxycanadine

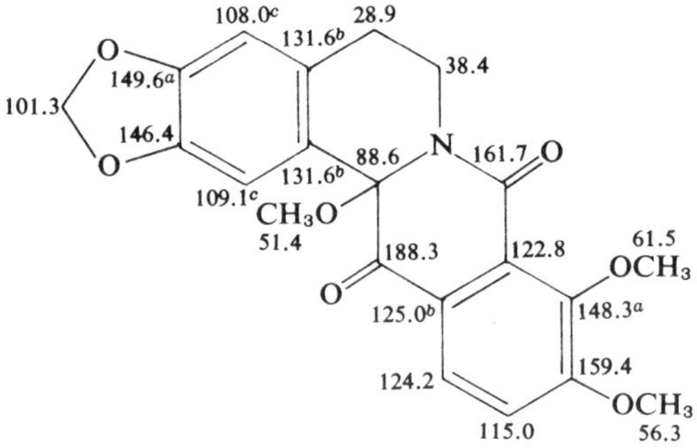

Ref. 59

390 **13-Methyldihydroberberine methiodide**

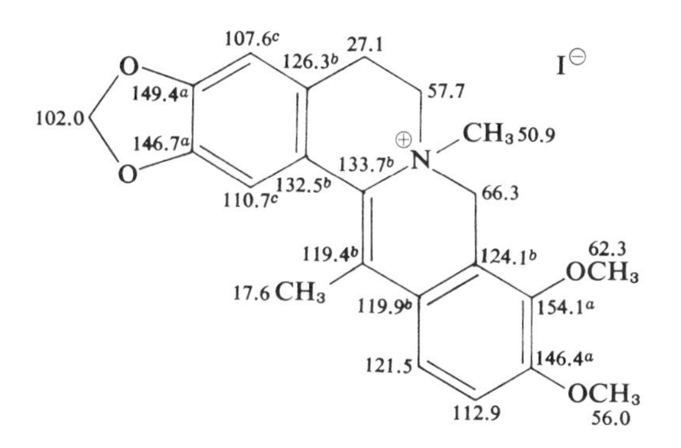

Ref. 59

391 **8-Acetyldihydroberberine**

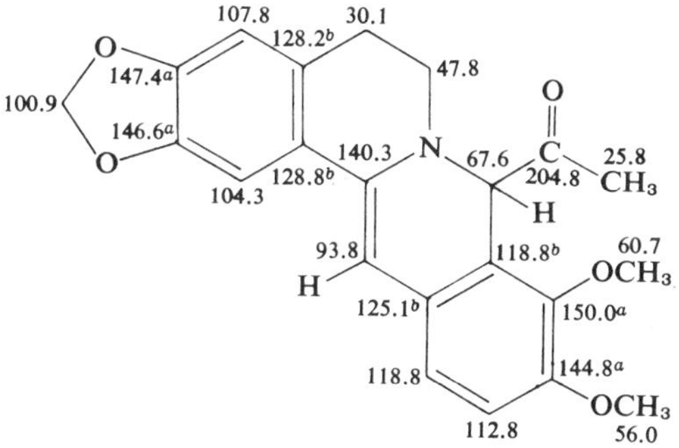

107.8
30.1
128.2b
O
147.4a
47.8
O
100.9
146.6a
N
140.3
67.6
C
25.8
O
128.8b
204.8
CH3
104.3
H
93.8
118.8b
60.7
H
OCH3
125.1b
150.0a
118.8
144.8a
OCH3
112.8
56.0

Ref. 59

392 8,13-Diacetyldihydroberberine

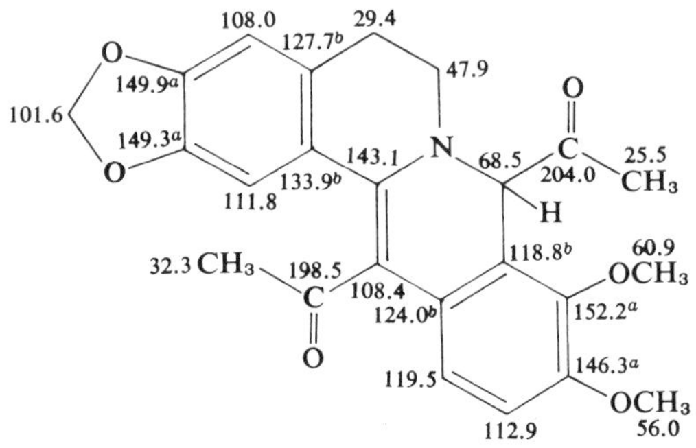

Ref. 59

393 Berberine–acetone

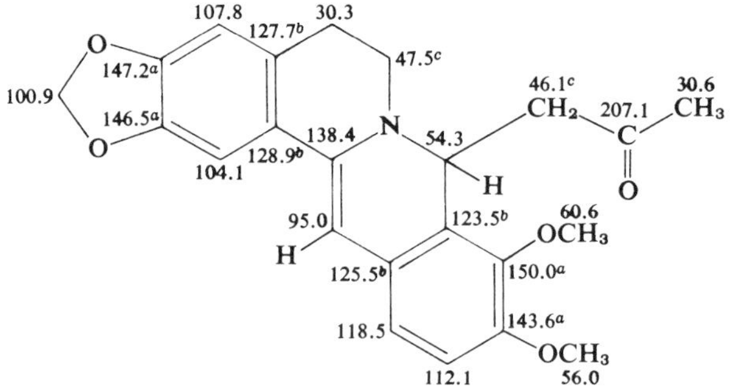

Ref. 59

394 Berberine–chloroform

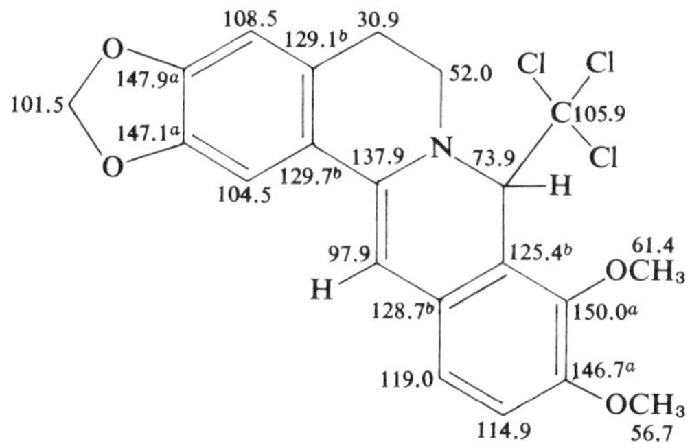

Ref. 59

395 Oxyberberine

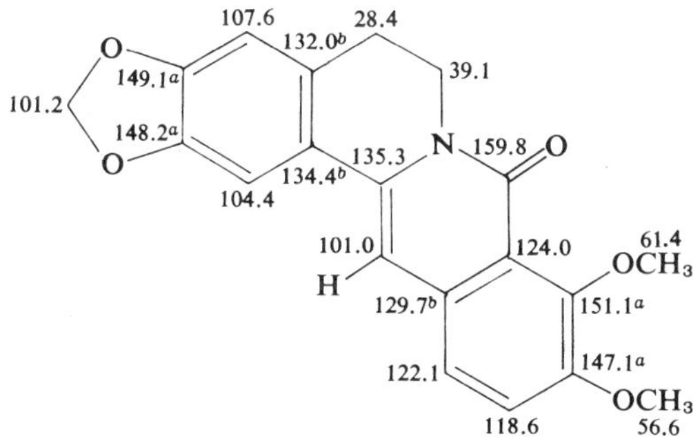

Ref. 59

396 13-Hydroxyoxyberberine

(TFA-*d*)

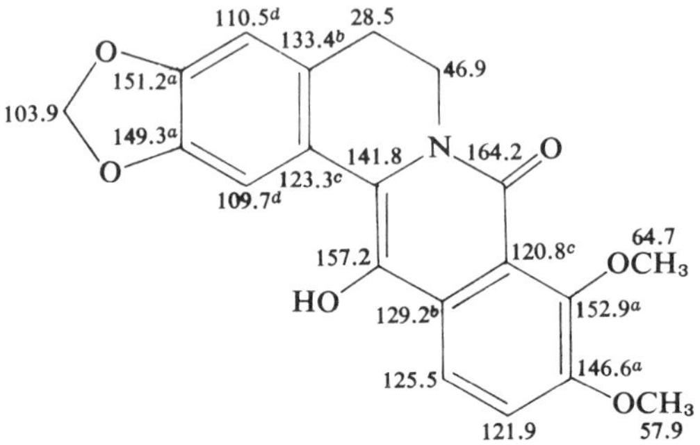

103.9, 151.2[a], 149.3[a], 110.5[d], 109.7[d], 133.4[b], 123.3[c], 28.5, 46.9, 141.8, N, 164.2, O, 157.2, HO, 129.2[b], 120.8[c], 64.7 OCH_3, 152.9[a], 146.6[a], 125.5, 121.9, OCH_3 57.9

Ref. 59

397 13-Methoxyoxyberberine

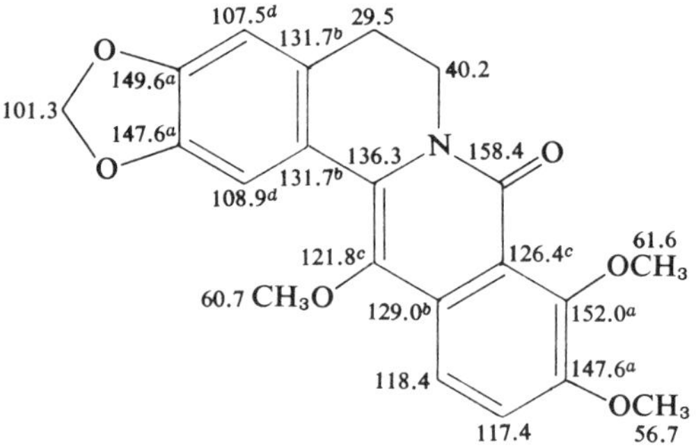

101.3, 149.6[a], 147.6[a], 107.5[d], 108.9[d], 131.7[b], 131.7[b], 29.5, 40.2, 136.3, N, 158.4, O, 121.8[c], 60.7 CH_3O, 129.0[b], 126.4[c], 61.6 OCH_3, 152.0[a], 147.6[a], 118.4, 117.4, OCH_3 56.7

Ref. 59

398 13-Acetoxyoxyberberine

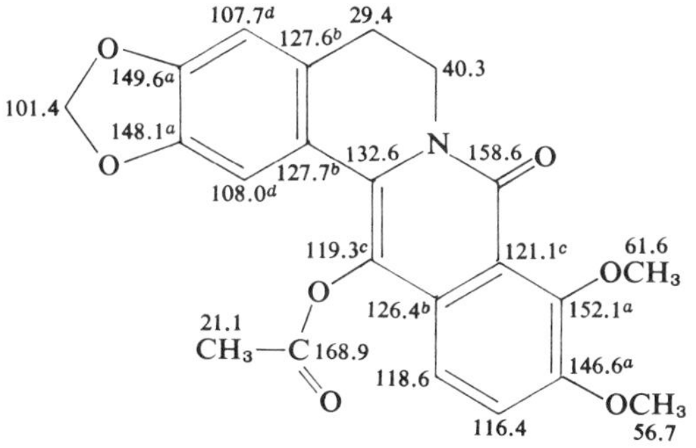

Ref. 59

Aporhoeadanes and Related Structures

399

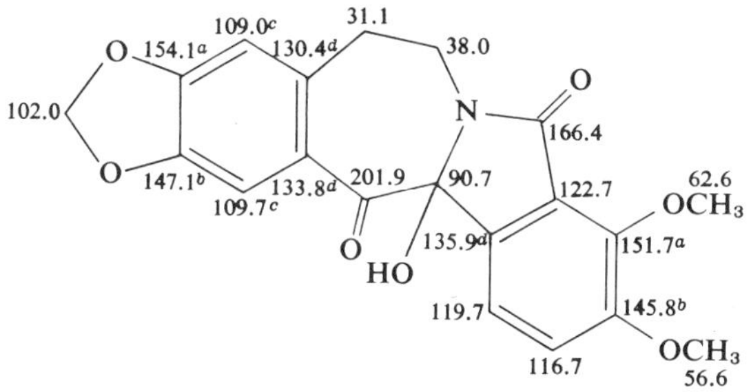

Ref. 59

400

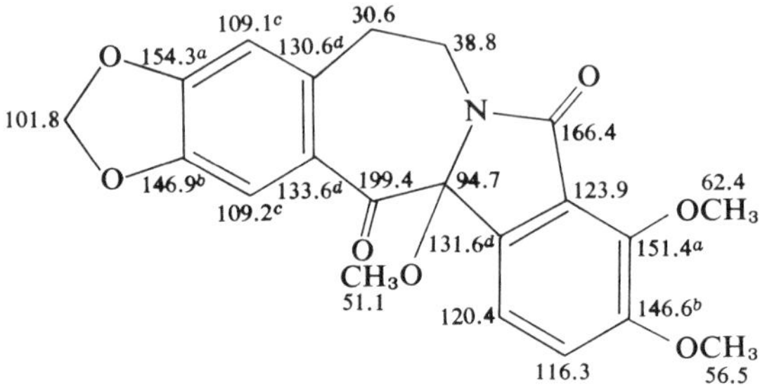

109.1[c] 30.6
154.3[a] 130.6[d] 38.8
O O
101.8 N
166.4
O 146.9[b] 133.6[d] 199.4 94.7 123.9 62.4
109.2[c] OCH$_3$
O 131.6[d] 151.4[a]
CH$_3$O
51.1 146.6[b]
120.4
OCH$_3$
116.3 56.5

Ref. 59

401

111.3[c] 34.1
156.0[a] 137.3[d] 41.4
O O
102.1 N
166.8
125.3[e]
O 146.5[b] CO$_2$H 80.6 126.5[e] 60.5
111.5[c] 169.5 OCH$_3$
HO 137.9[d] 150.7[a]
145.6[b]
118.8
OCH$_3$
113.9 56.2

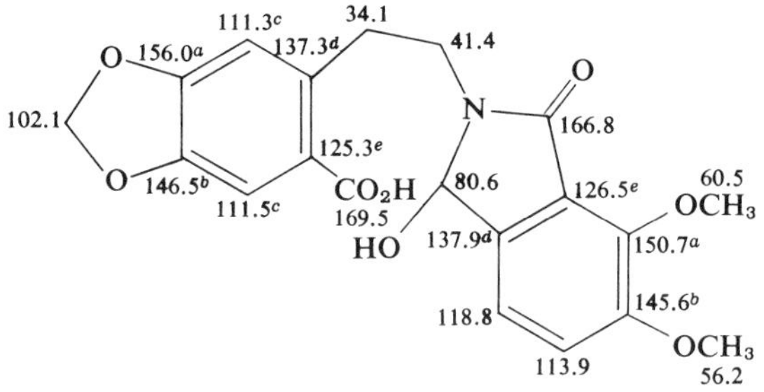

Ref. 59

402

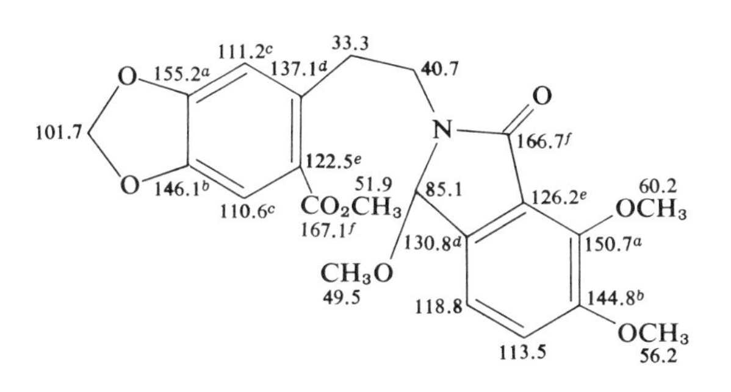

Ref. 59

403

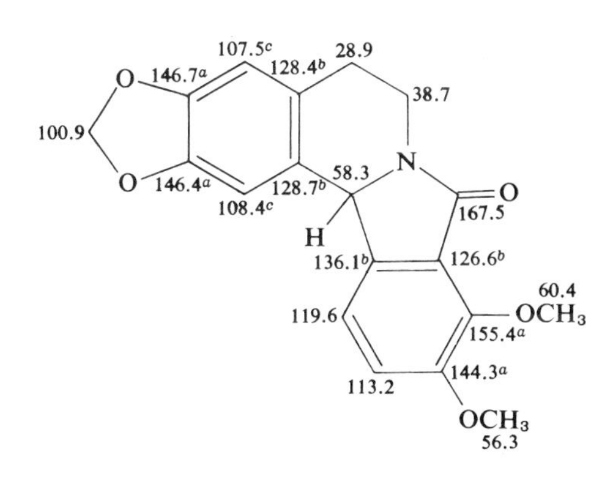

Ref. 59

i. Protopines

404 Protopine

(values adjusted from C_6H_{12} reference)

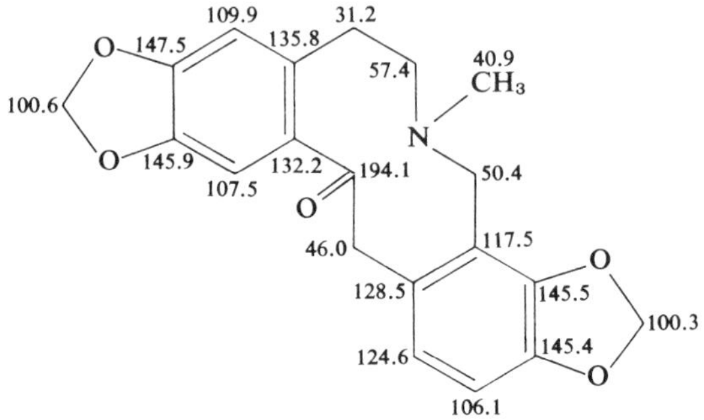

Ref. 69

405 Cryptopine

(values adjusted from C_6H_{12} reference)

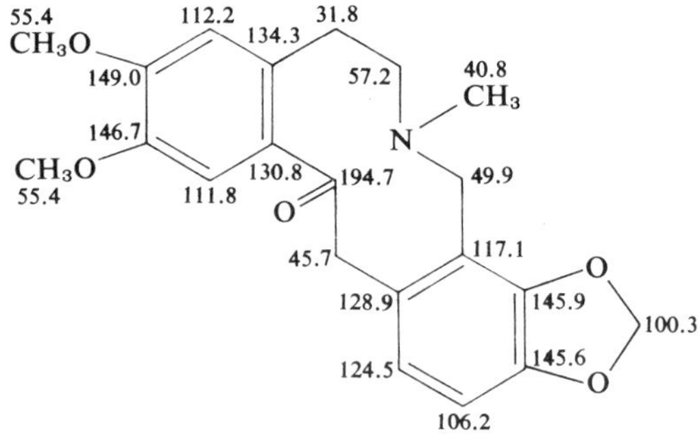

Ref. 69

406 Allocryptopine

(values adjusted from C_6H_{12} reference)

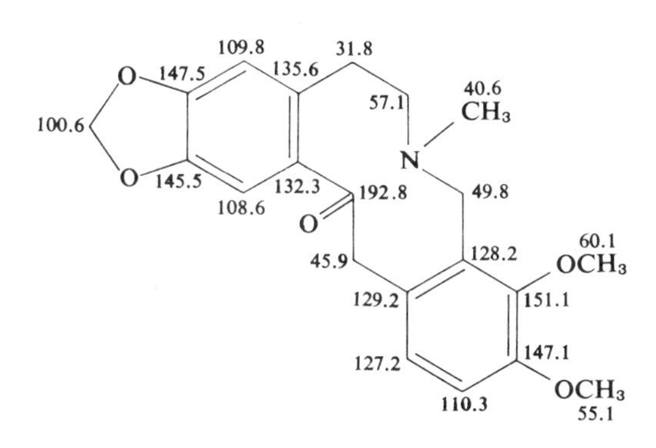

Ref. 69

407 Muramine

(values adjusted from C_6H_{12} reference)

55.4
112.3
31.9
CH3O
134.2
149.0
57.1
40.7
CH3
146.9
N
CH3O
130.9
193.0
49.5
55.4
112.6
O
60.1
45.7
128.4
OCH3
128.9
151.1
147.1
127.1
OCH3
110.3
55.1

Ref. 69

408 Hunnemanine

(values adjusted from C_6H_{12} reference)

109.7
31.0
O 147.5
135.0
56.7
40.5
CH_3
100.6
N
O 145.5
131.8
173.8
49.8
108.1
O
45.4
120.7
OH
128.0
145.0
122.3
143.5
OCH_3
108.5
55.4

Ref. 69

j. Phthalideisoquinolines

409 α-Hydrastine

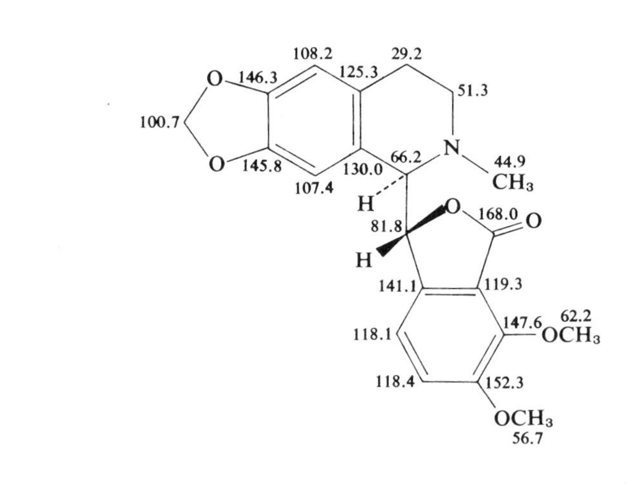

Ref. 60

410 β-Hydrastine

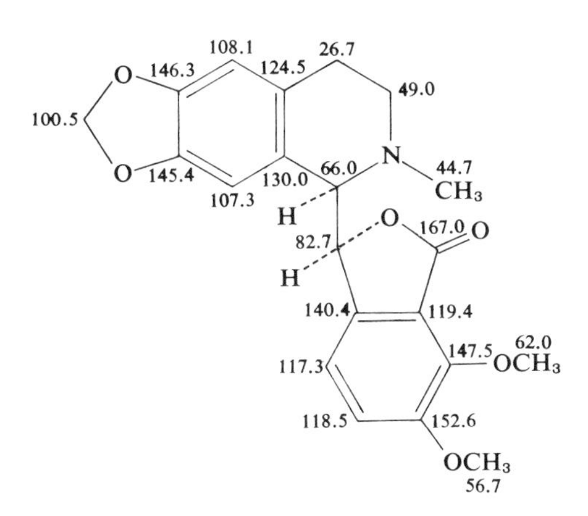

Ref. 60

411 Corlumine

55.9
CH_3O
111.3
123.4
26.5
148.2
49.5
147.2
N
CH_3O
129.5
65.7
45.1
CH_3
55.9
110.7
H
O
167.2
O
84.9
H
140.8
110.3
115.5
144.5
O
113.1
149.1
103.3
O

Ref. 60

412 Adlumine

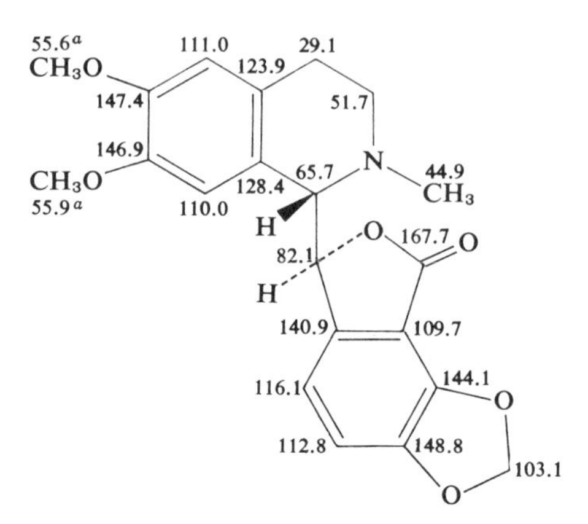

Ref. 60

413 Cordrastine

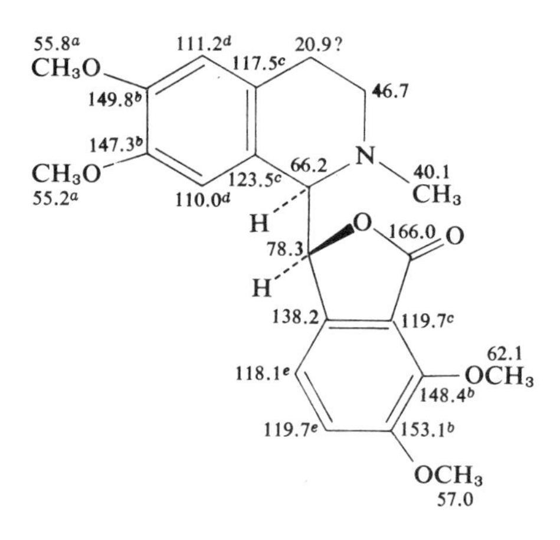

Ref. 59

414 Capnoidine

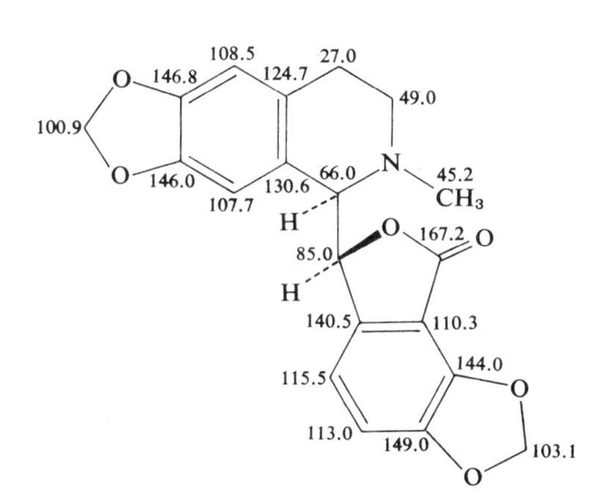

Ref. 59

415 9-Hydroxy-Δ^1-norhydrasteine

108.1
26.7
O
149.8[b]
134.4[a]
46.2
101.4
N
161.7[c]
O
146.3[b]
120.2
106.4
O
167.5[c]
O
119.3
HO
138.3[a]
120.2
62.4
117.7
OCH_3
147.8[b]
108.1
154.5
OCH_3
56.6

Ref. 59

416 Dehydronorhydrastine methyl ester

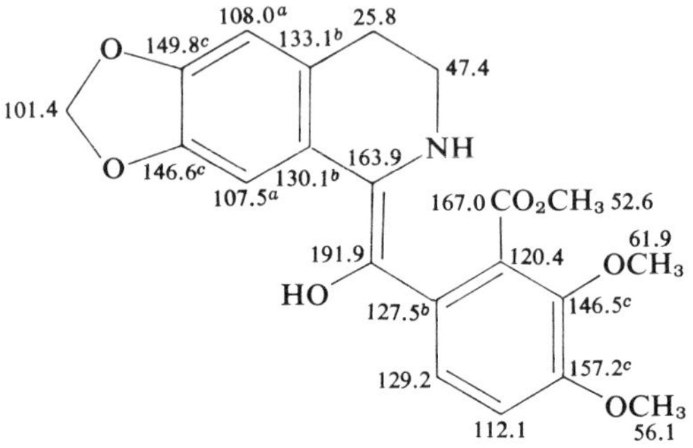

Ref. 59

417

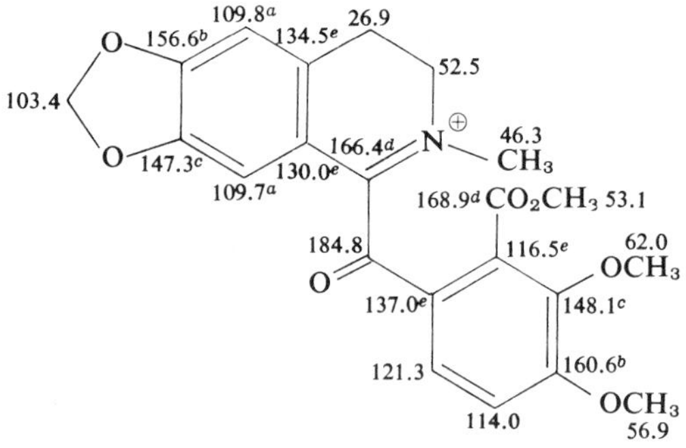

Ref. 59

Spirobenzylisoquinolines

418

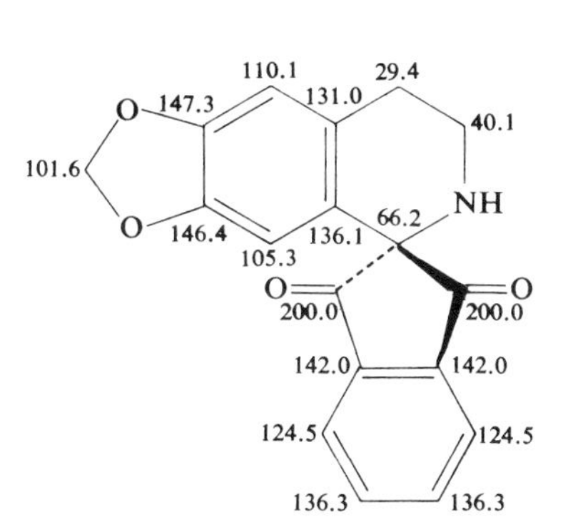

Ref. 70

419

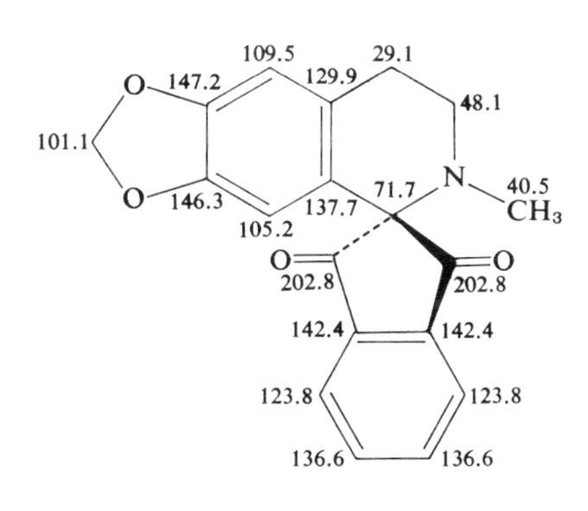

Ref. 70

420 Ochotensimine

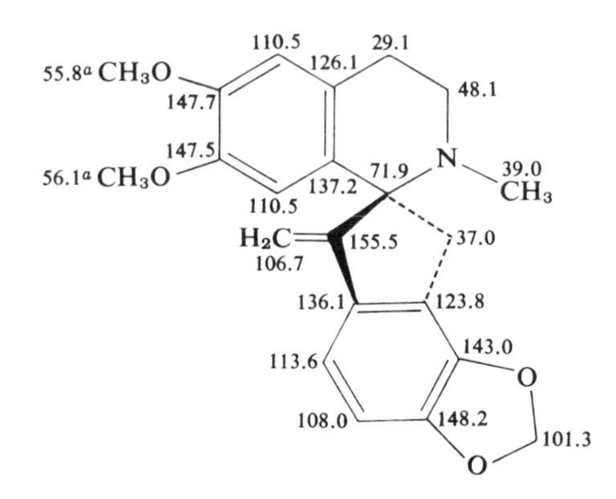

Ref. 70

421

Ref. 70

422

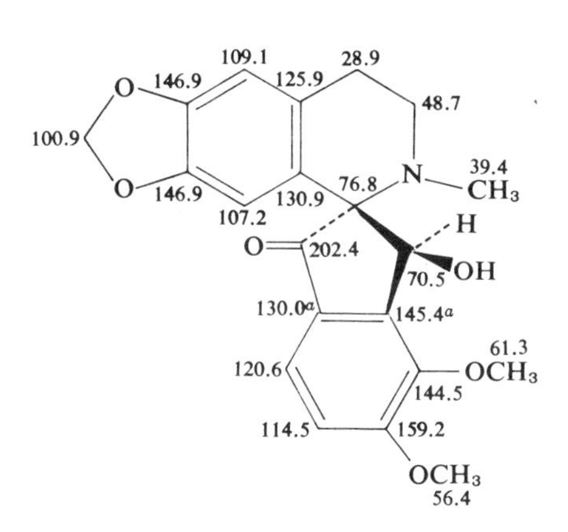

Ref. 70

423

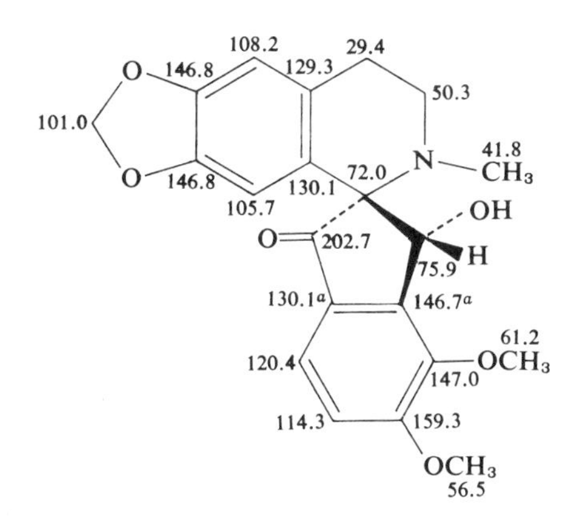

Ref. 70

424 Raddeanone

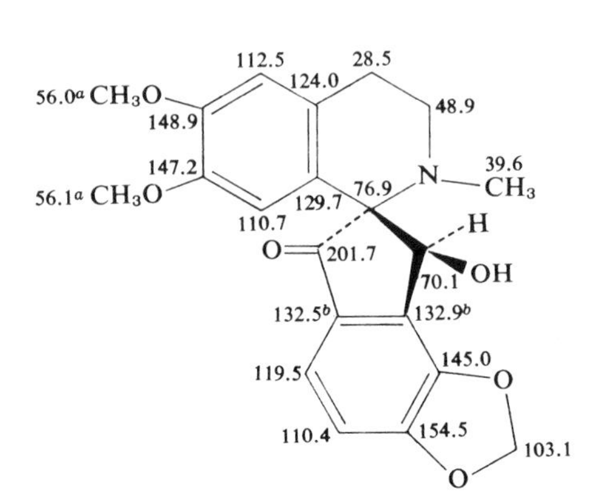

Ref. 70

425 Yenhusomidine

111.4
29.3
56.5[a] CH_3O
128.7
50.3
148.6
148.5
41.9
N
56.1[a] CH_3O
CH_3
72.0
128.7
110.7
OH
O
202.7
75.1
H
131.3[b]
134.6[b]
144.4
119.6
O
109.5
154.6
103.2
O

Ref. 70

426 Sibiricine

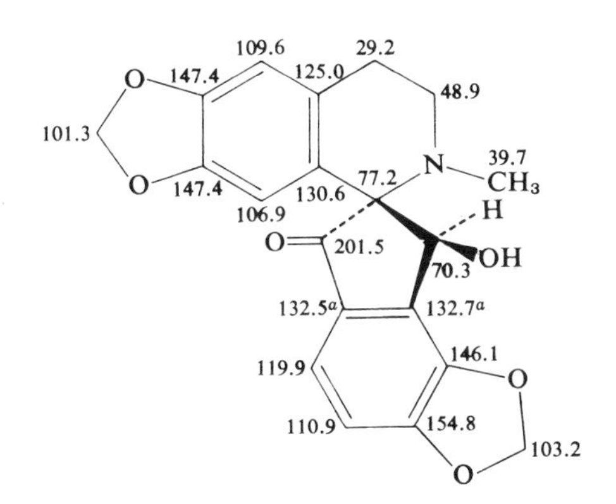

Ref. 70

427 Corydaine

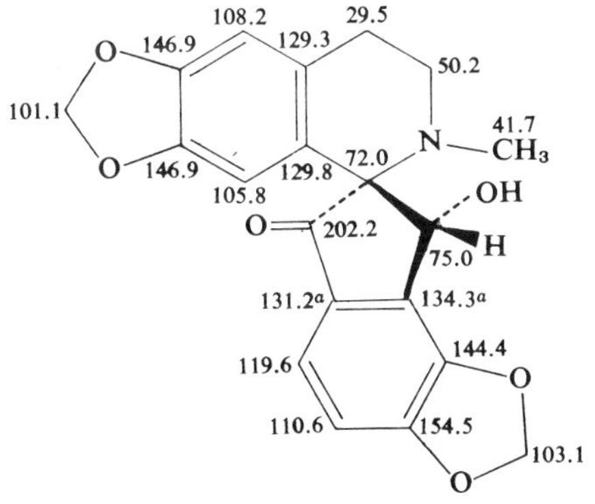

Ref. 70

428 Ochrobirine

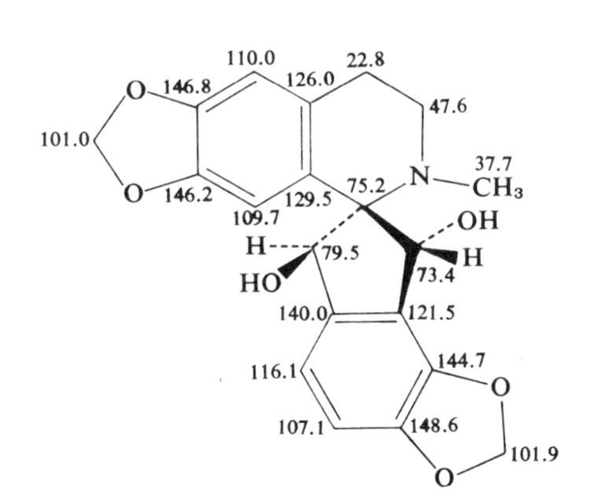

Ref. 70

429 Yenhusomine

113.0
22.0
124.9
55.5[a] CH_3O
47.8
148.3
147.3
N
37.9
75.2
56.0[a] CH_3O
CH_3
128.3
110.1
OH
H
79.0
HO
73.4
H
140.9
121.5
144.9
115.7
O
109.7
148.4
101.8
O

Ref. 70

l. Rhoeadine

430 Rhoeadine

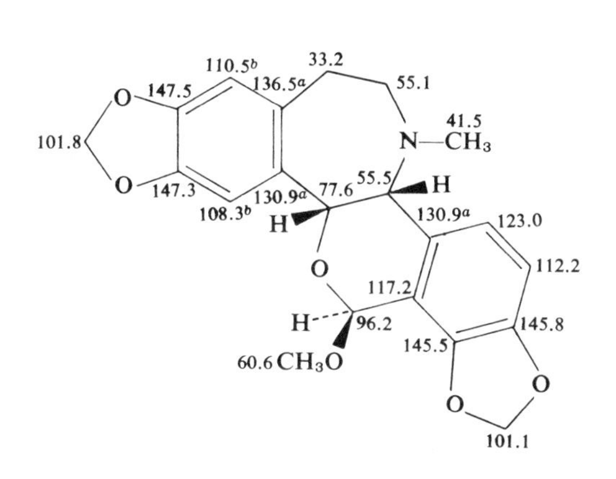

Ref. 71

Emetines and Related Structures

431

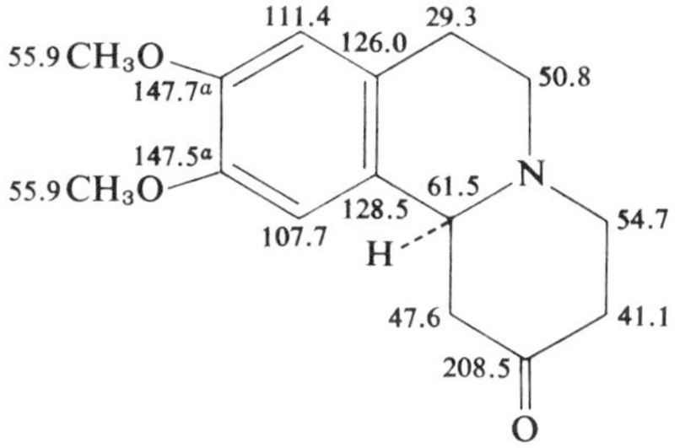

Ref. 72

432

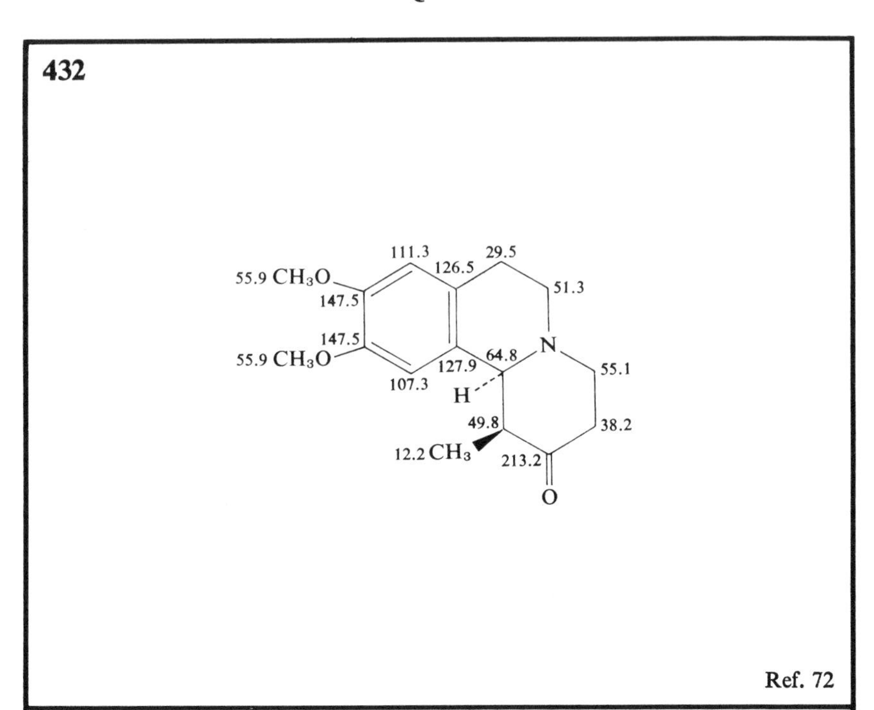

Ref. 72

433

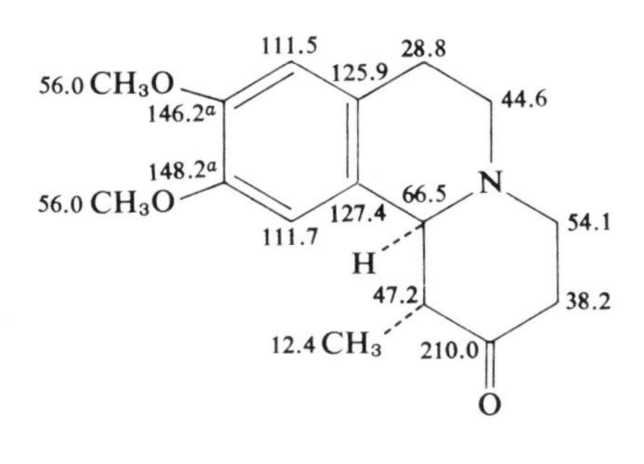

Ref. 72

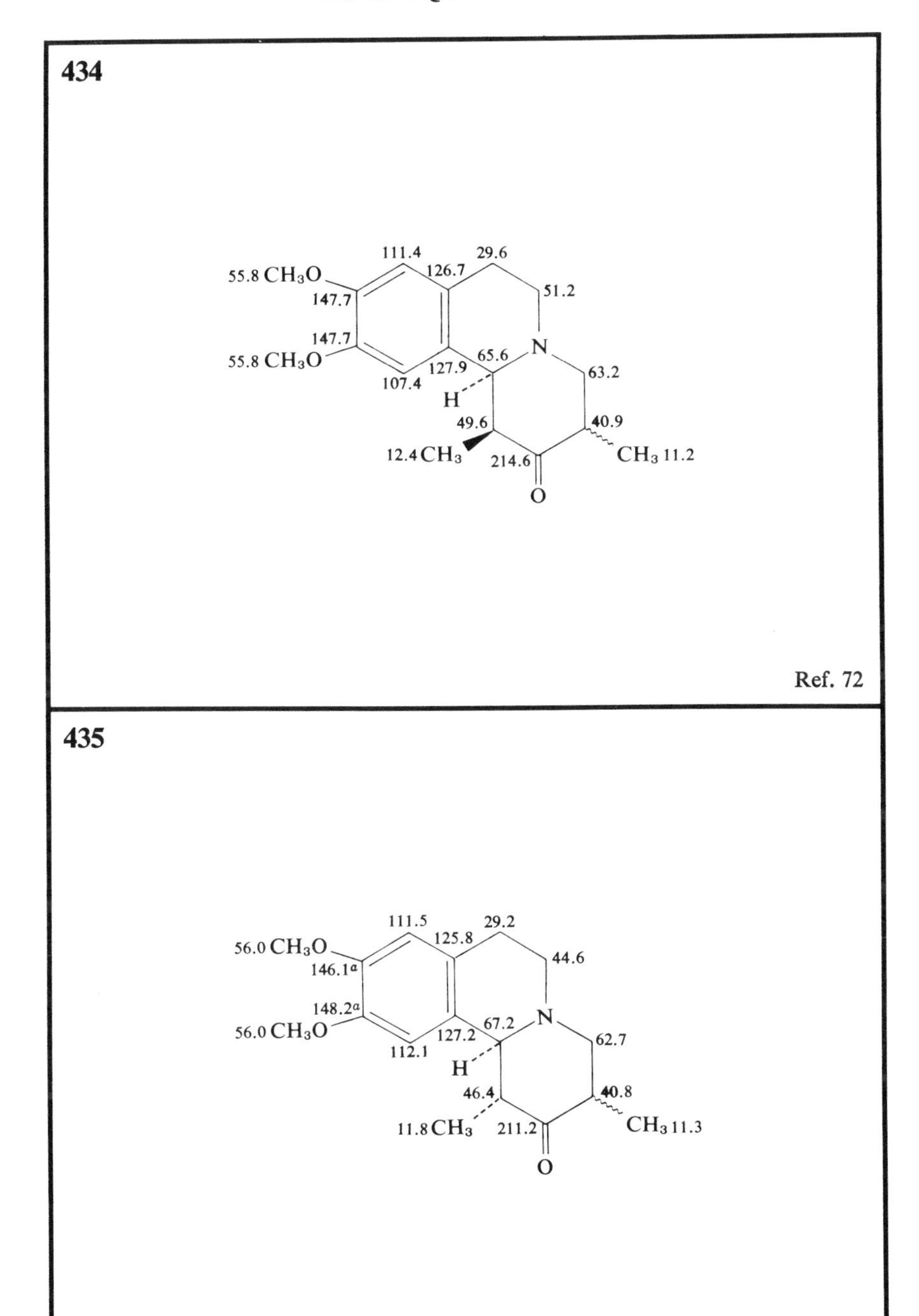
434
111.4
29.6
55.8 CH3O
126.7
51.2
147.7
147.7
N
55.8 CH3O
127.9
65.6
63.2
107.4
H
49.6
40.9
12.4 CH3
214.6
CH3 11.2
O
Ref. 72
435
111.5
29.2
56.0 CH3O
125.8
44.6
146.1a
148.2a
N
56.0 CH3O
127.2
67.2
62.7
112.1
H
46.4
40.8
11.8 CH3
211.2
CH3 11.3
O
Ref. 72

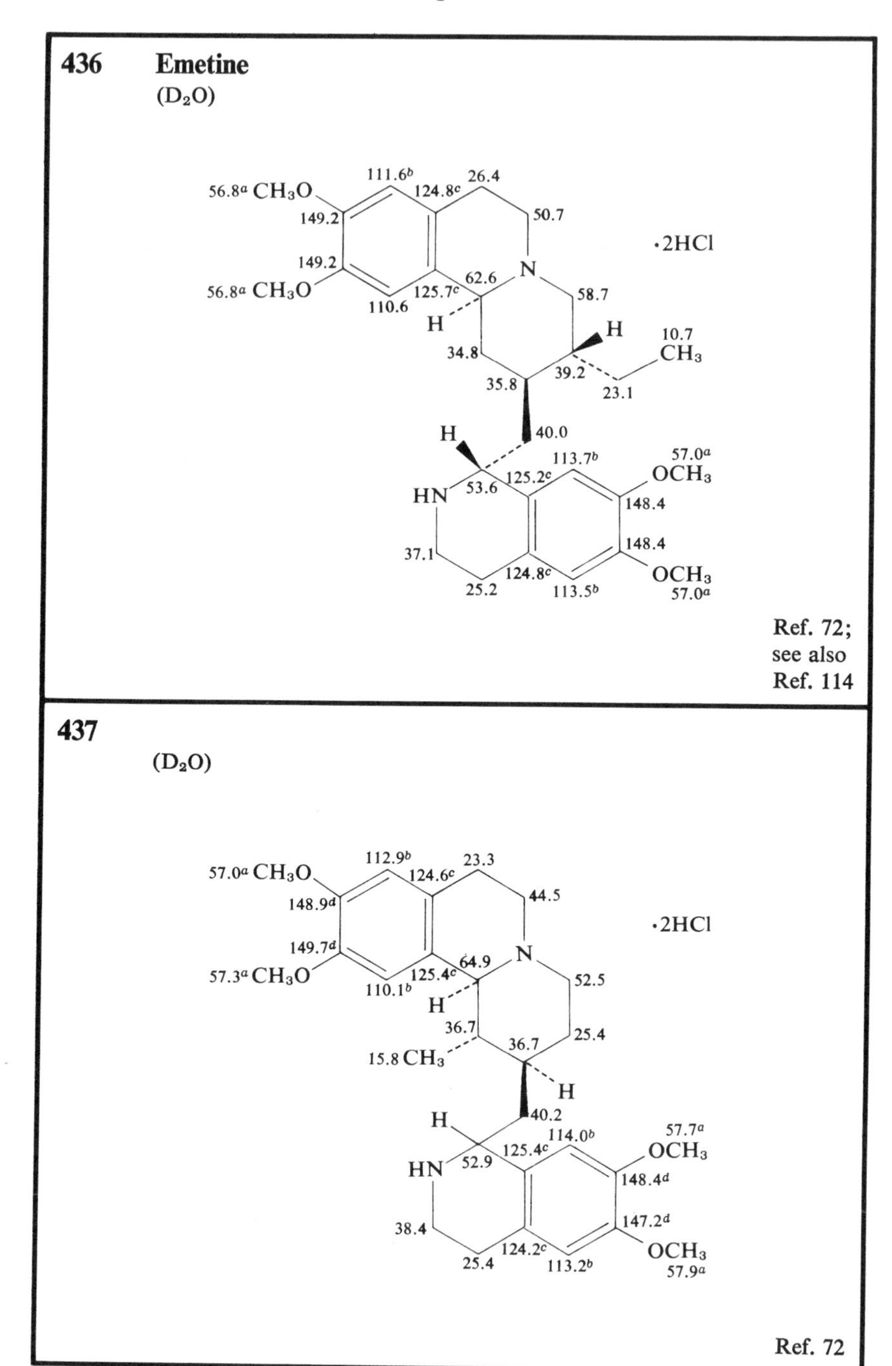

436 Emetine
(D2O)
56.8a CH3O
149.2
149.2
56.8a CH3O
111.6b
124.8c
26.4
50.7
N
62.6
125.7c
110.6
H
58.7
H
10.7
CH3
34.8
39.2
35.8
23.1
•2HCl
H
40.0
113.7b
57.0a
OCH3
HN
53.6
125.2c
148.4
148.4
37.1
124.8c
OCH3
25.2
113.5b
57.0a
Ref. 72;
see also
Ref. 114
437
(D2O)
57.0a CH3O
148.9d
149.7d
57.3a CH3O
112.9b
124.6c
23.3
44.5
N
64.9
125.4c
110.1b
H
52.5
36.7
25.4
36.7
15.8 CH3
H
•2HCl
H
40.2
114.0b
57.7a
OCH3
HN
52.9
125.4c
148.4d
147.2d
38.4
124.2c
OCH3
25.4
113.2b
57.9a
Ref. 72

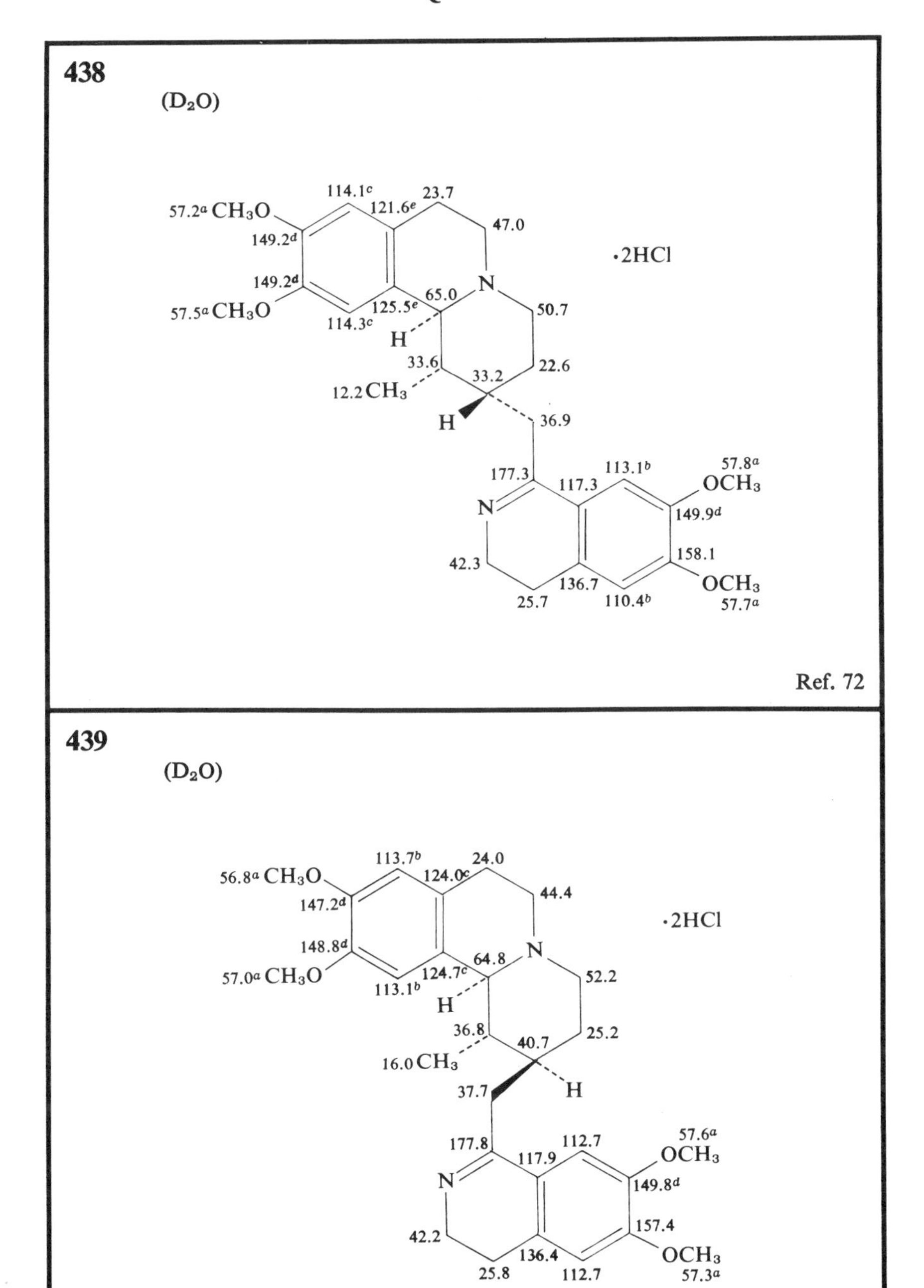
438
(D2O)
57.2a CH3O
114.1c
121.6e
23.7
47.0
149.2d
149.2d
•2HCl
57.5a CH3O
114.3c
125.5e
65.0
N
50.7
H
33.6
33.2
22.6
12.2 CH3
H
36.9
177.3
117.3
113.1b
57.8a
OCH3
N
149.9d
42.3
158.1
136.7
OCH3
25.7
110.4b
57.7a
Ref. 72
439
(D2O)
56.8a CH3O
113.7b
124.0c
24.0
44.4
147.2d
•2HCl
148.8d
57.0a CH3O
113.1b
124.7c
64.8
N
52.2
H
36.8
40.7
25.2
16.0 CH3
37.7
H
177.8
117.9
112.7
57.6a
OCH3
N
149.8d
42.2
157.4
136.4
OCH3
25.8
112.7
57.3a
Ref. 72

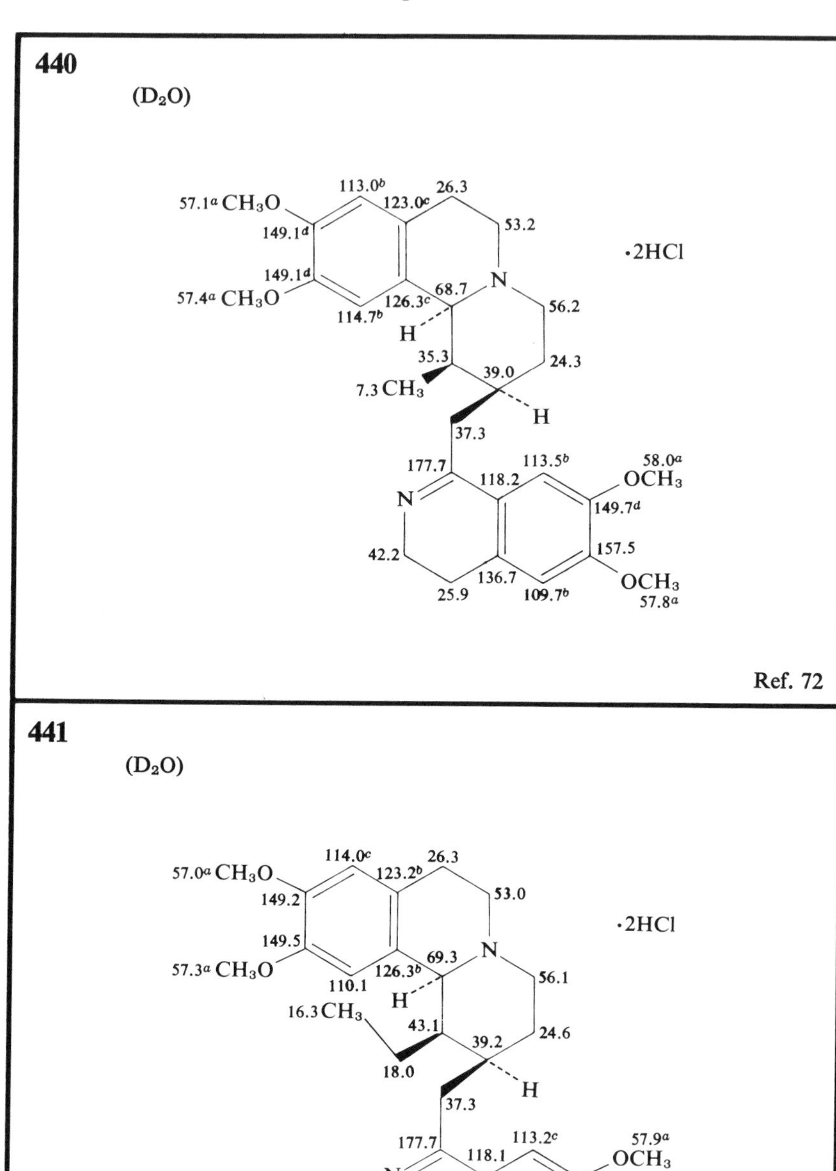
440
(D_2O)
57.1[a] CH_3O
113.0[b]
123.0[c]
26.3
53.2
149.1[d]
149.1[d]
57.4[a] CH_3O
114.7[b]
126.3[c]
68.7
N
56.2
H
35.3
24.3
39.0
7.3 CH_3
H
37.3
·2HCl
177.7
118.2
113.5[b]
58.0[a]
OCH_3
N
149.7[d]
157.5
42.2
136.7
OCH_3
25.9
109.7[b]
57.8[a]
Ref. 72
441
(D_2O)
57.0[a] CH_3O
114.0[c]
123.2[b]
26.3
53.0
149.2
149.5
57.3[a] CH_3O
110.1
126.3[b]
69.3
N
56.1
H
16.3 CH_3
43.1
39.2
24.6
18.0
H
37.3
·2HCl
177.7
118.1
113.2[c]
57.9[a]
OCH_3
N
148.7
157.9
38.4
136.6
OCH_3
25.9
112.9[c]
57.7[a]
Ref. 72

n. Yolantinine: A Bisphenethylisoquinoline Alkaloid

442 Yolantinine

(structural assignment will have to be modified for biogenetic reasons)

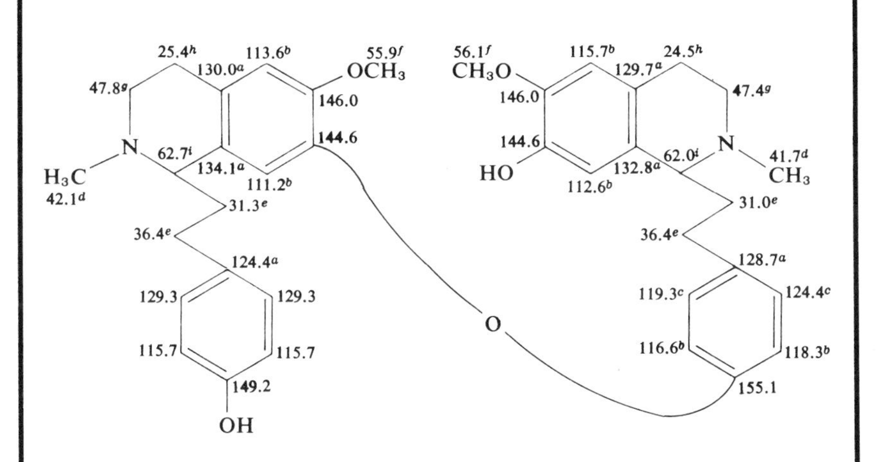

Ref. 73

o. 1-Phenyl- and 1-Naphthylisoquinolines

443 Cryptostyline II

55.8 CH3O
110.4[a]
28.9
126.5
52.5
149.1[b]
147.4[b]
N
55.8 CH3O
71.0
130.6[c]
CH3
111.5[a]
44.3
136.4[c]
112.0[a]
122.1
148.3[b]
110.8[a]
CH3O
55.8
147.0[b]
OCH3
55.8

Ref. 59

444 Norcryptostyline II

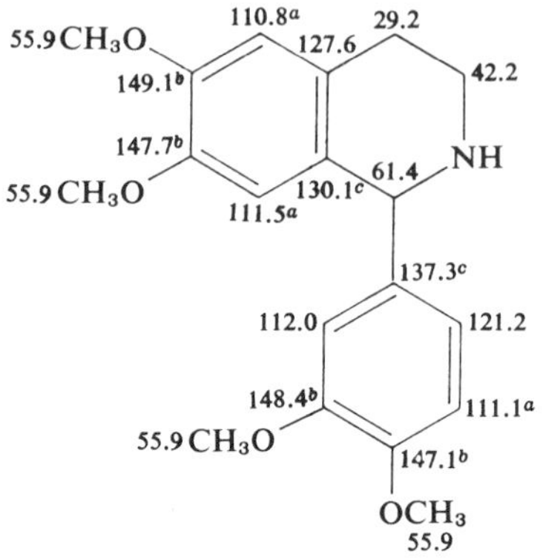

55.9 CH_3O
110.8[a]
29.2
127.6
149.1[b]
42.2
147.7[b]
NH
61.4
55.9 CH_3O
130.1[c]
111.5[a]
137.3[c]
112.0
121.2
148.4[b]
111.1[a]
55.9 CH_3O
147.1[b]
OCH_3
55.9

Ref. 59

445

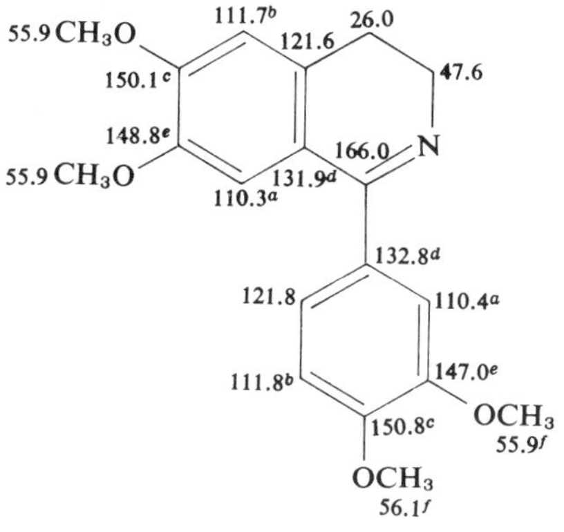

55.9 CH_3O
111.7[b]
26.0
121.6
150.1[c]
47.6
148.8[e]
166.0
N
55.9 CH_3O
131.9[d]
110.3[a]
132.8[d]
121.8
110.4[a]
111.8[b]
147.0[e]
OCH_3
150.8[c]
55.9[f]
OCH_3
56.1[f]

Ref. 59

446

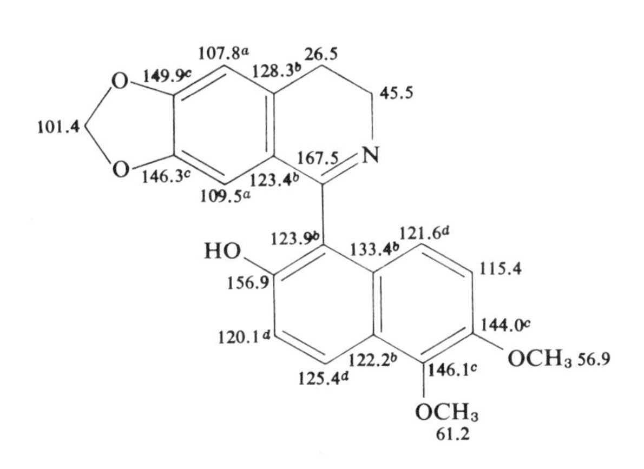

Ref. 59

447

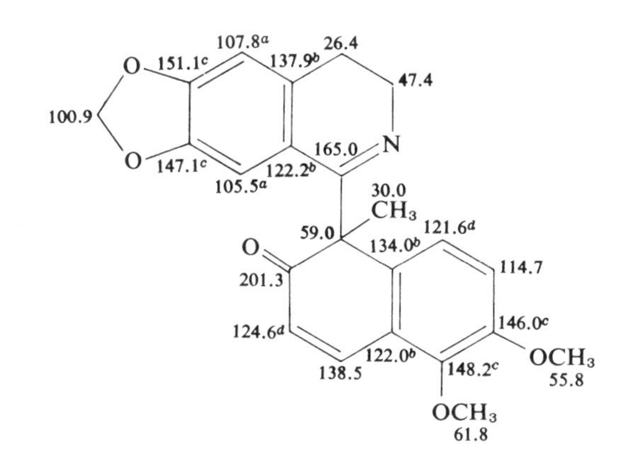

Ref. 59

448

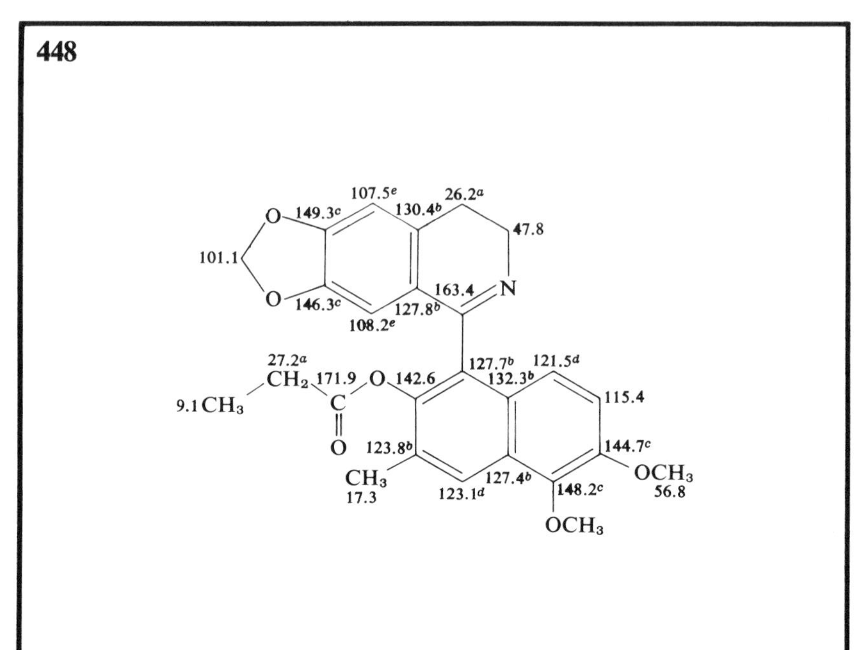
107.5[e]
26.2[a]
149.3[c]
130.4[b]
47.8
101.1
163.4
N
146.3[c]
127.8[b]
108.2[e]
27.2[a]
127.7[b]
121.5[d]
CH_2
171.9
O
142.6
132.3[b]
115.4
9.1 CH_3
C
O
123.8[b]
144.7[c]
CH_3
127.4[b]
OCH_3
17.3
123.1[d]
148.2[c]
56.8
OCH_3

Ref. 59

12. MORPHINES[140]

449 3-Methoxymorphinane

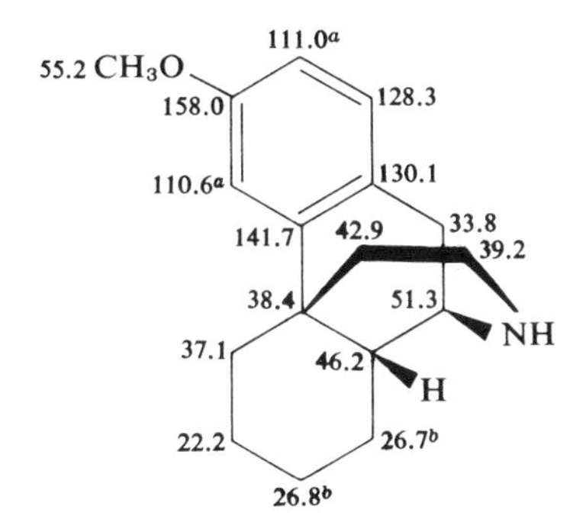

Ref. 74

450 Morphine

($DMSO$-d_6)

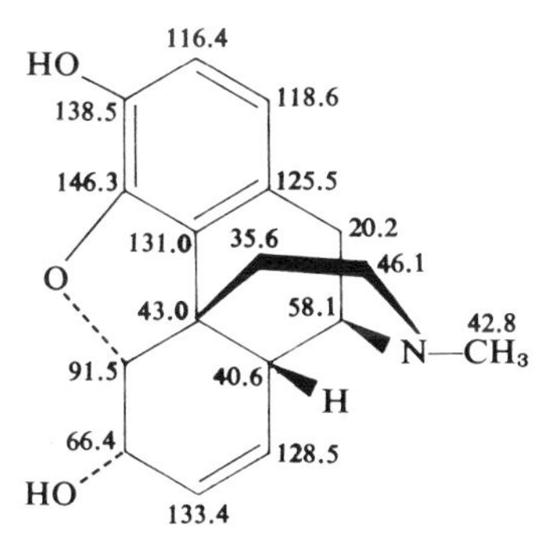

Ref. 75

451 Codeine

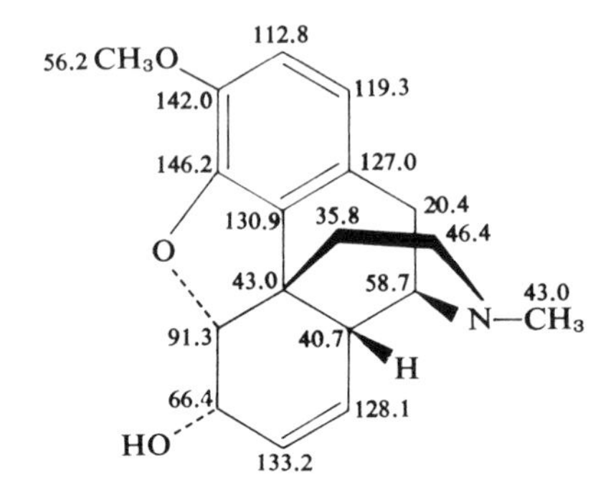

Ref. 74

452 Dihydrocodeinone

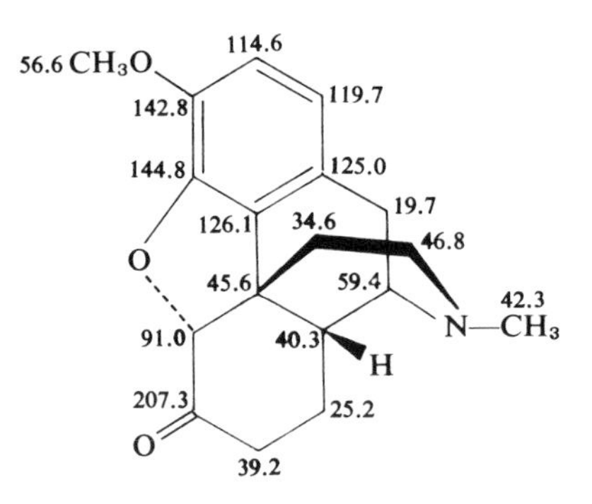

Ref. 75

453 Thebaine

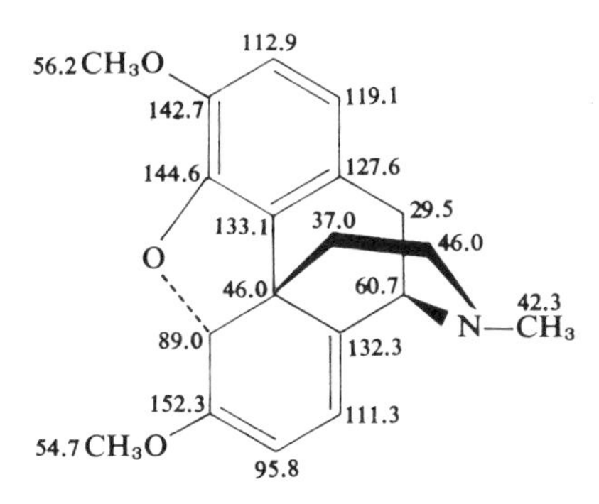

Ref. 74

454 Sinomenine

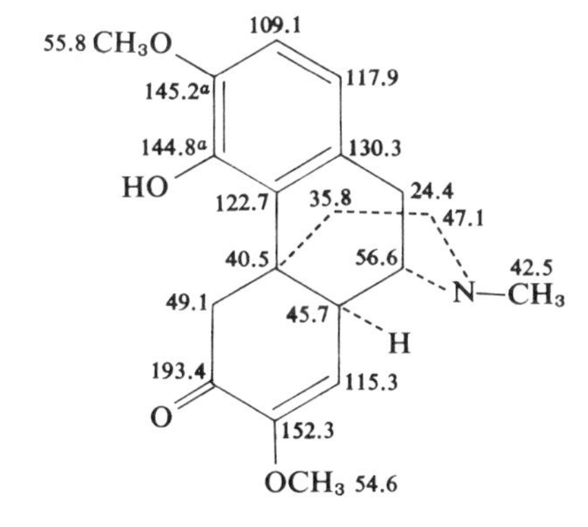

Ref. 74

455 6α-Naltrexol

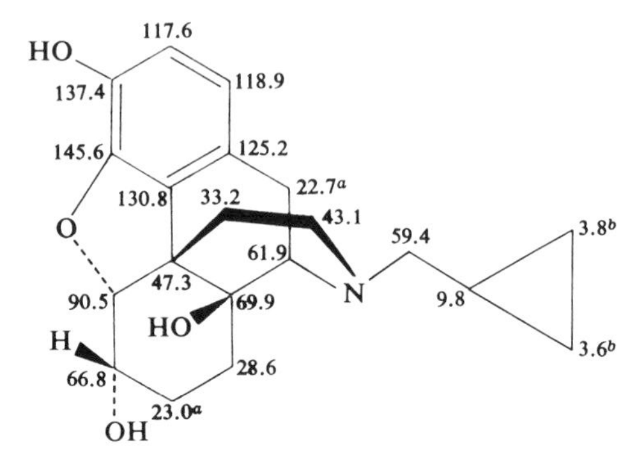

Ref. 75

456 6β-Naltrexol

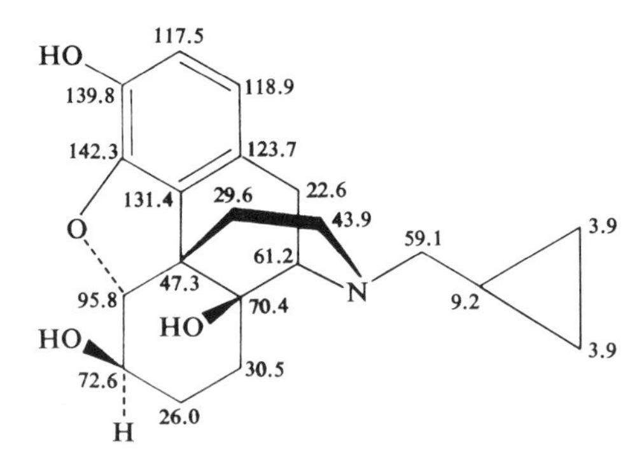

Ref. 75

457 Diprenorphine

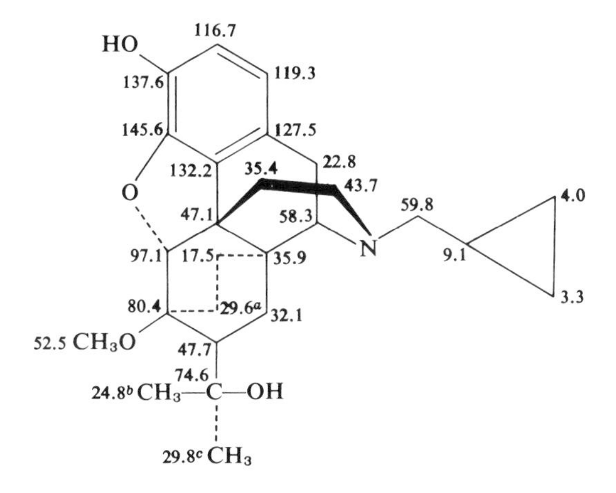

Ref. 75

458 Etorphine

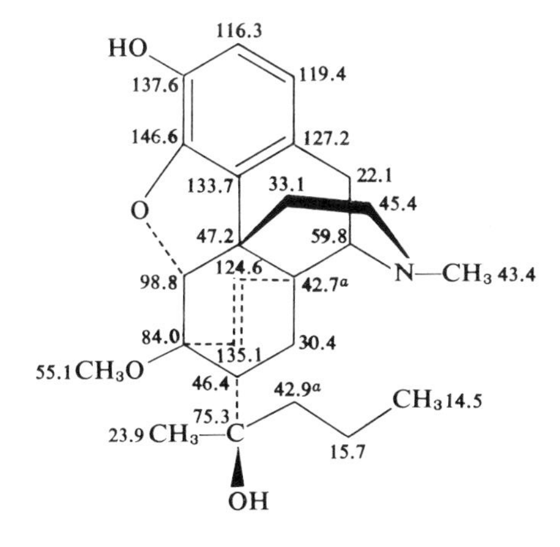

Ref. 75

13. COLCHICINE

459 **Colchicine**

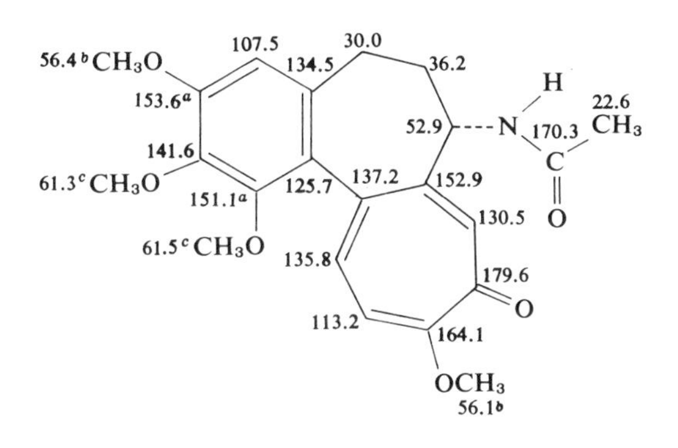

Ref. 76

14. AMARYLLIDACEAE ALKALOIDS

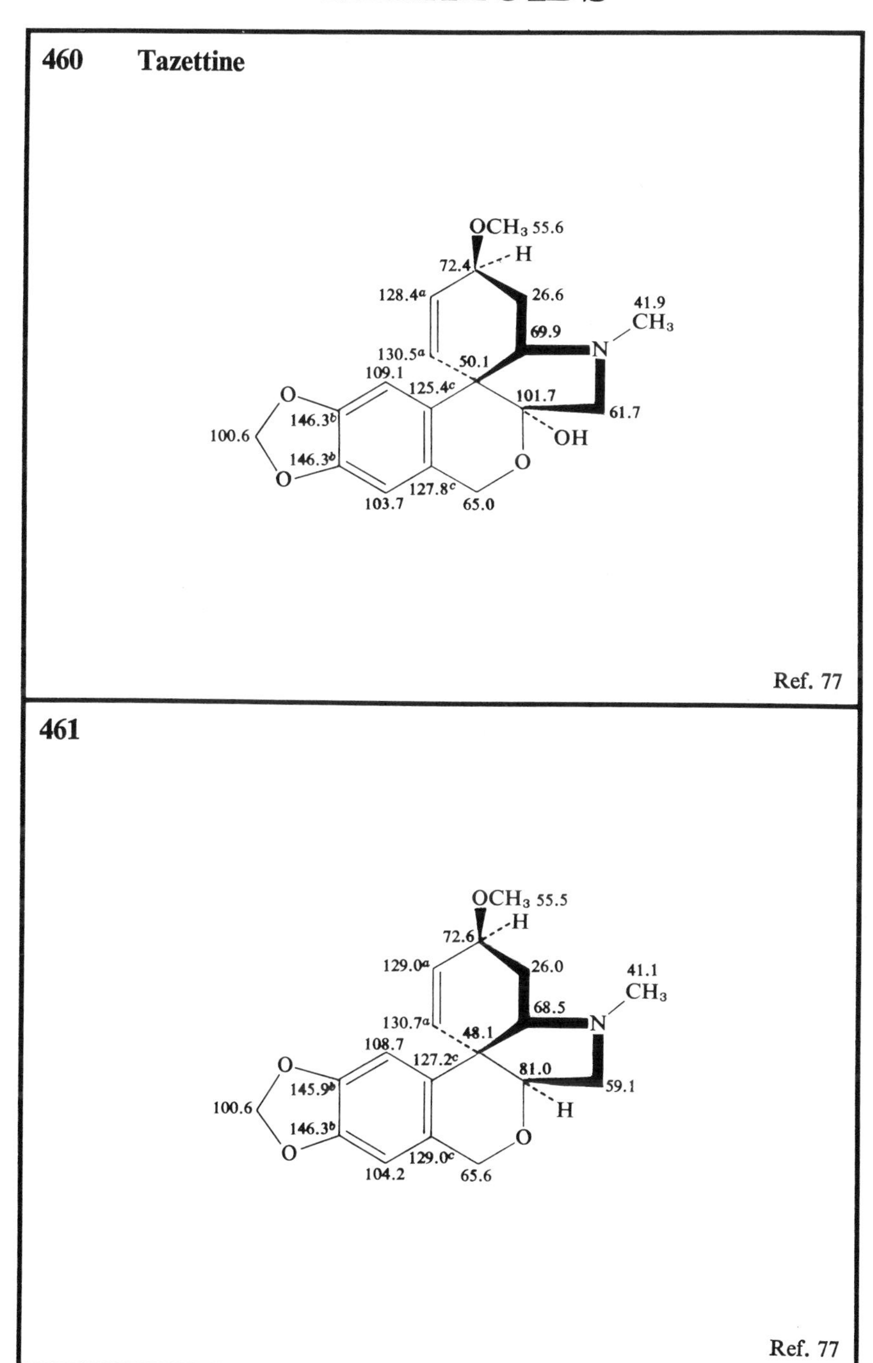

462 Galanthine

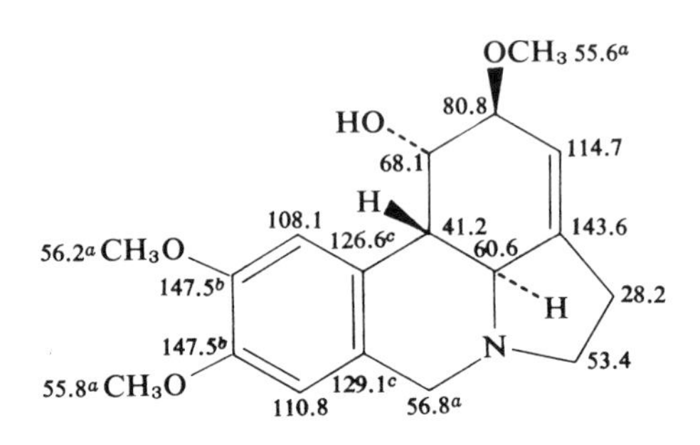

Ref. 77

463 Lycorenine

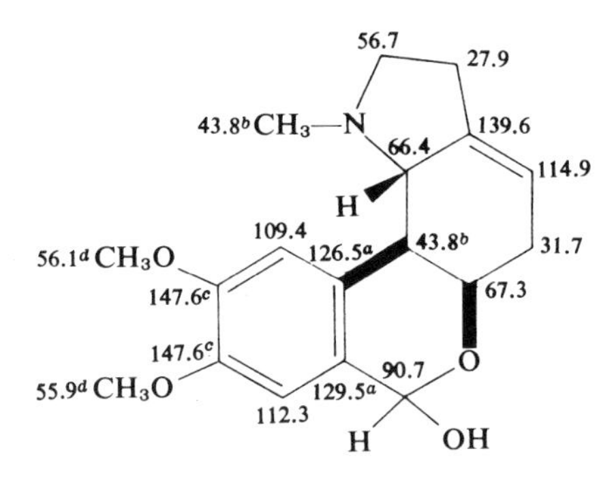

Ref. 77

464 Buphanamine

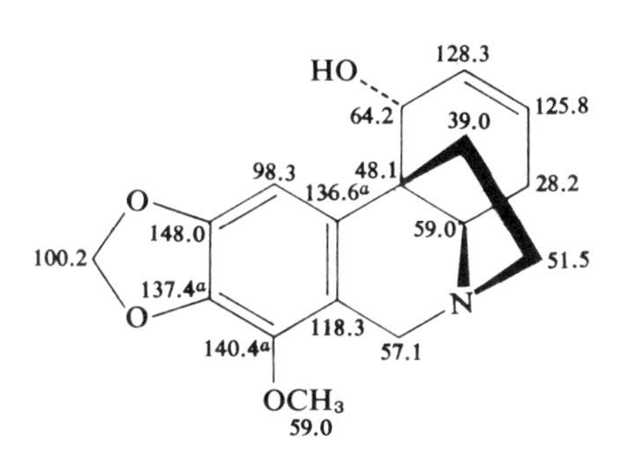

Ref. 77

465 Montanine

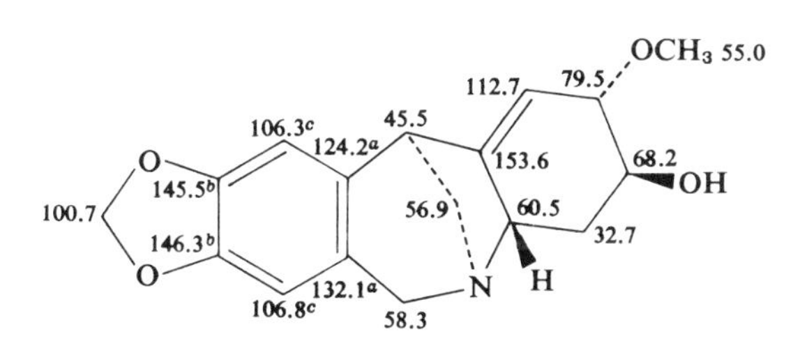

Ref. 77

466 Narciclasine tetraacetate

(values adjusted from CS_2 reference)

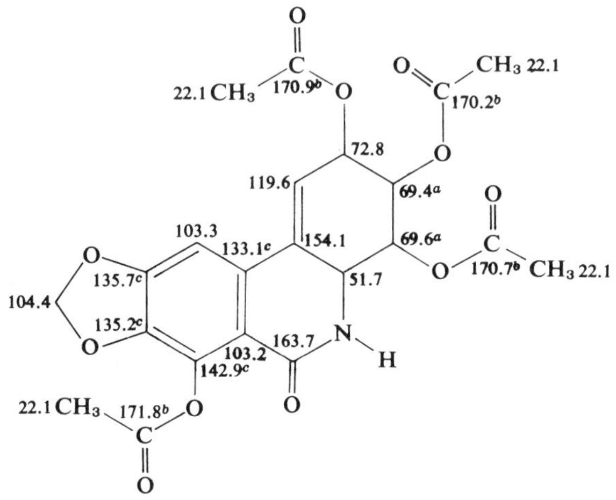

22.1 CH_3 170.9[b] O C O
O C CH_3 22.1 170.2[b] O
72.8
119.6
69.4[a]
O C
103.3
133.1[c]
154.1
69.6[a]
O 170.7[b] CH_3 22.1
O
135.7[c]
51.7
104.4
135.2[c]
O
N H
163.7
103.2
142.9[c]
22.1 CH_3 171.8[b] O C O
O

Ref. 78

15. FUSED AROMATIC AMINES

467 1-Aminonaphthalene
(acetone-d_6)

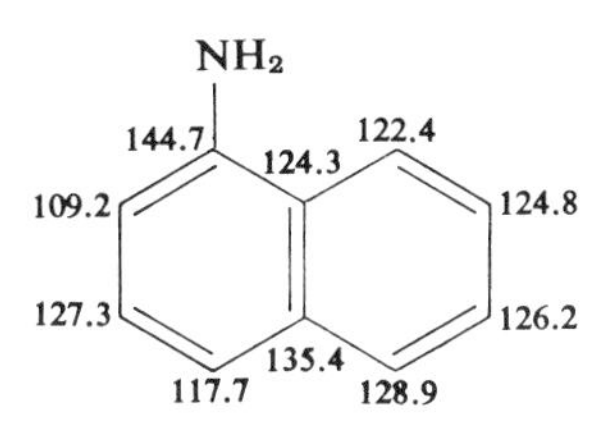

Ref. 52

468 5-Aminoacenaphthene
(acetone-d_6)

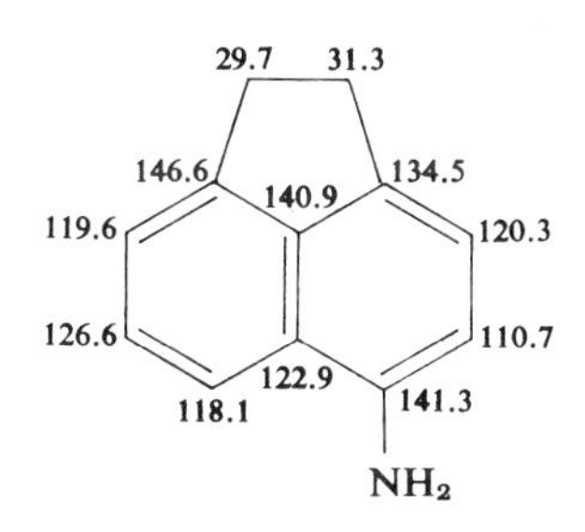

Ref. 52

469 3-Aminofluoranthene
(acetone-d_6)

125.6 128.0
122.0 123.6
138.7 140.8
137.2 125.7
134.1
120.8 120.4
126.3 109.9
121.4
122.2 147.4
NH_2

Ref. 52

16. INDOLES

a. Simple Indoles

470 Indole

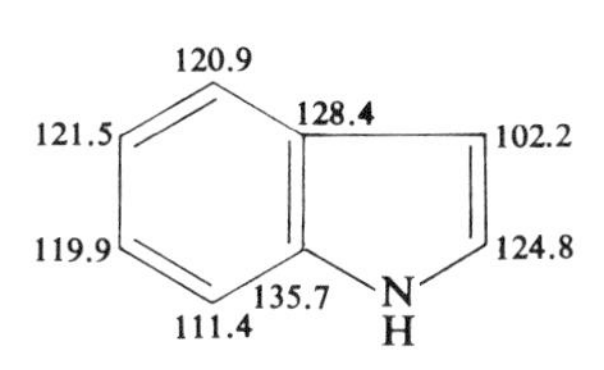

Ref. 25;
see also Ref. 79

471 2-Methylindole

(dioxane, values adjusted from CS_2 reference)

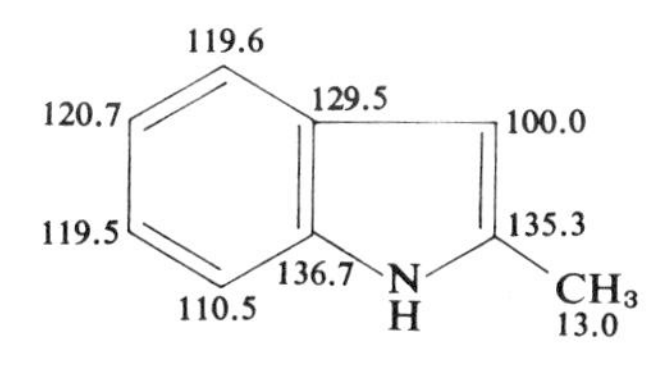

Ref. 80

472 Skatole (3-methylindole)

(dioxane, values adjusted from CS_2 reference)

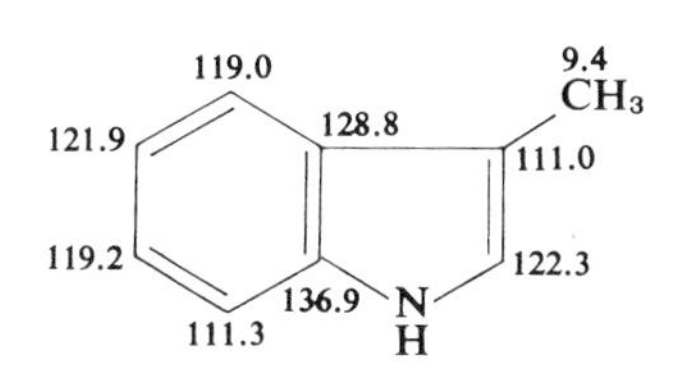

Ref. 80;
see also Ref. 25

473 4-Methylindole

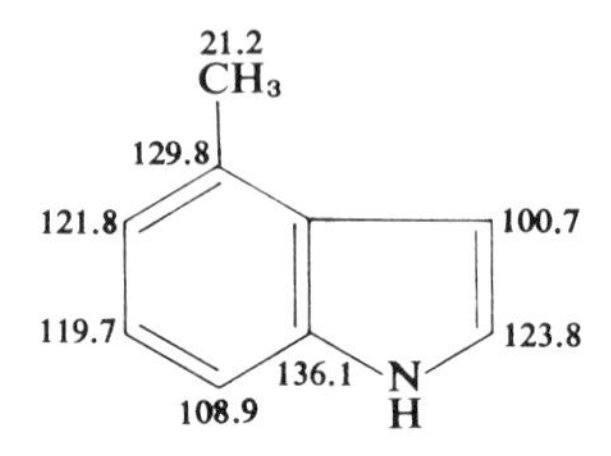

Ref. 25

474 5-Methylindole

(dioxane, values adjusted from CS_2 reference)

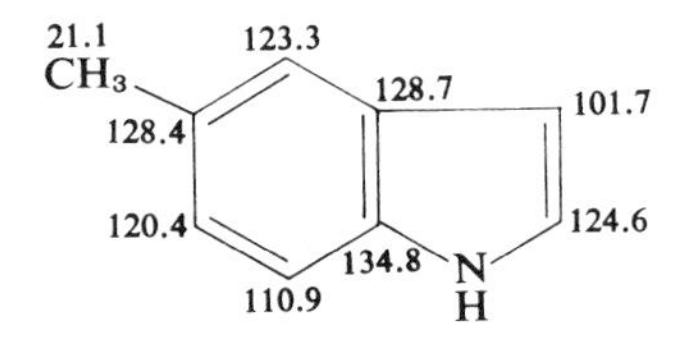

Ref. 80

475 6-Methylindole

(dioxane, values adjusted from CS_2 reference)

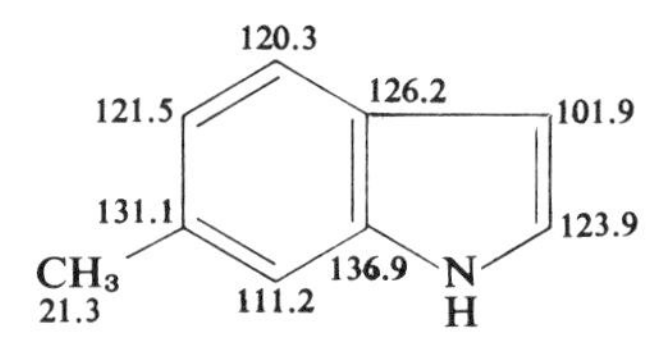

Ref. 80

476 **2-Carboxyindole**
(DMSO-d_6)

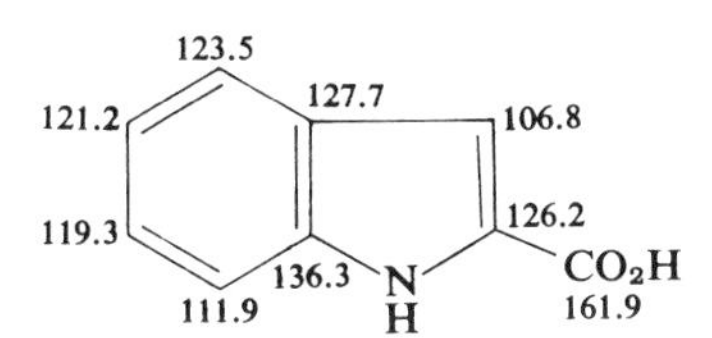

Ref. 79

477 **3-Formylindole**
(DMSO-d_6)

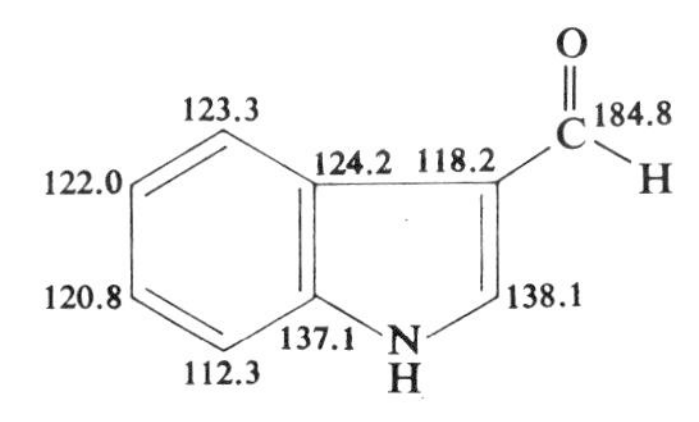

Ref. 79

478 **3-Acetylindole**
(DMSO-d_6)

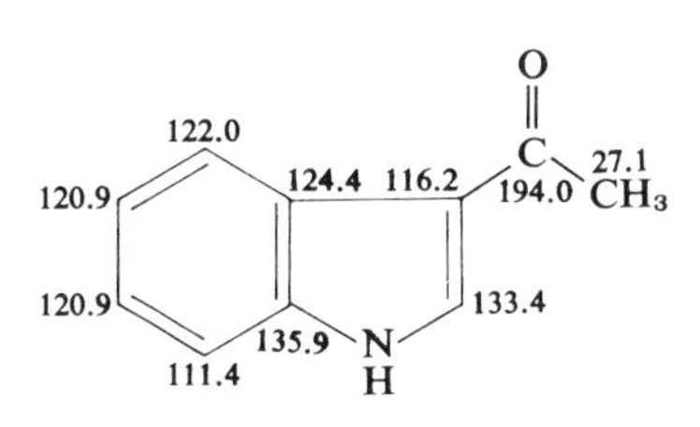

Ref. 79

479 3-Acetoxyindole
(DMSO-d_6)

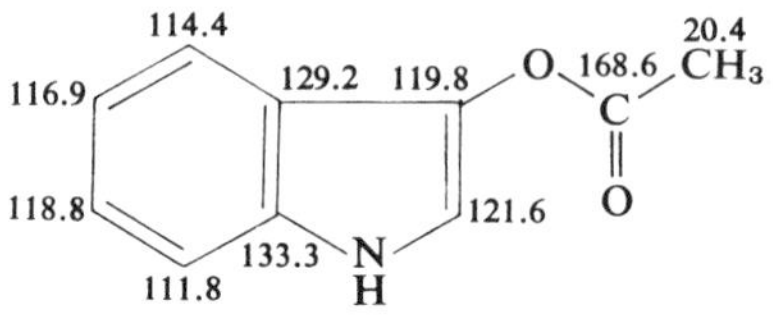

114.4, 116.9, 118.8, 111.8, 133.3, 129.2, 119.8, 121.6, N H, O, 168.6, C, O, 20.4 CH_3

Ref. 79

480 5-Methoxyindole

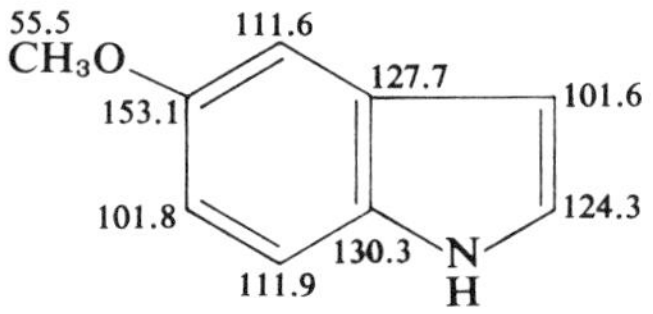

55.5 CH_3O, 153.1, 111.6, 127.7, 101.6, 124.3, 130.3, N H, 111.9, 101.8

Ref. 79

481 6-Methoxyindole

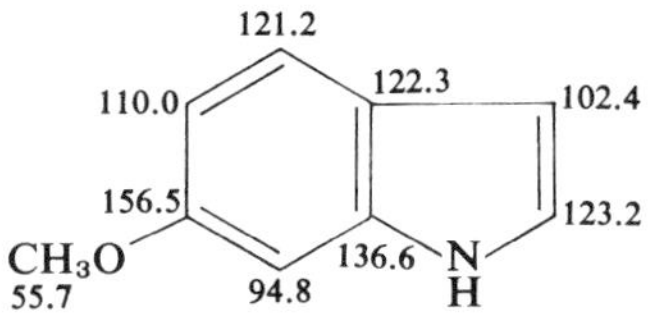

121.2, 110.0, 122.3, 102.4, 123.2, 136.6, N H, 94.8, 156.5, CH_3O 55.7

Ref. 79

482 7-Methoxyindole

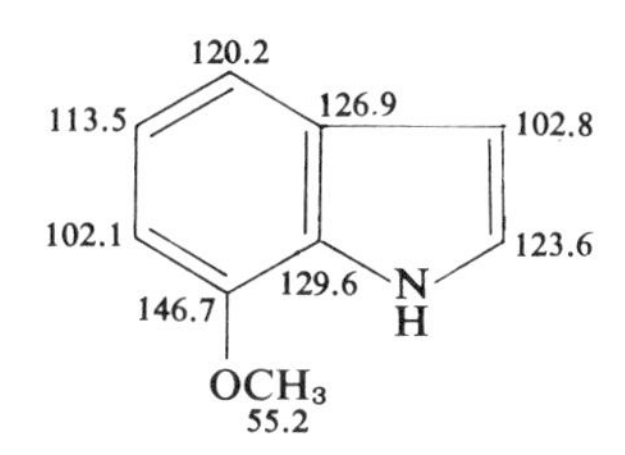

Ref. 79

Oxindoles

483 Oxindole

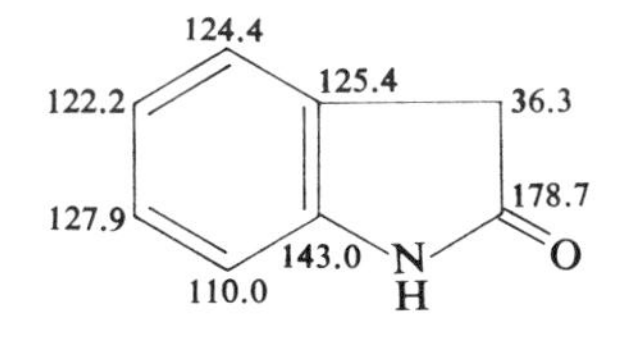

Ref. 84;
see also Ref. 56

484 Isatin
(DMSO)

Ref. 81

485 Phthalimide

O
123.5
132.5
167.6
134.1
NH
O

Ref. 82

486

(DMSO-d_6)

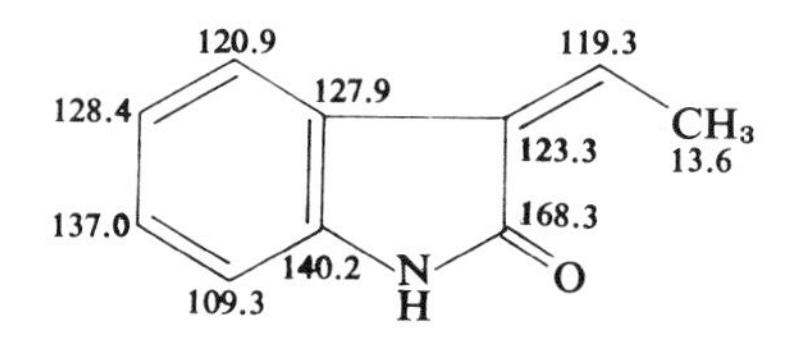

Ref. 83

487

(DMSO-d_6)

CH3 14.8
121.2
123.6
128.8
128.8
122.3
167.9
135.7
142.0
N
O
109.5
H

Ref. 83

c. Indolines

488 Indoline

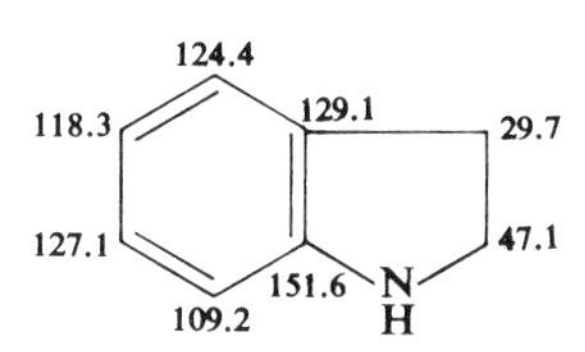

Ref. 84

489 *N*-Formylindoline

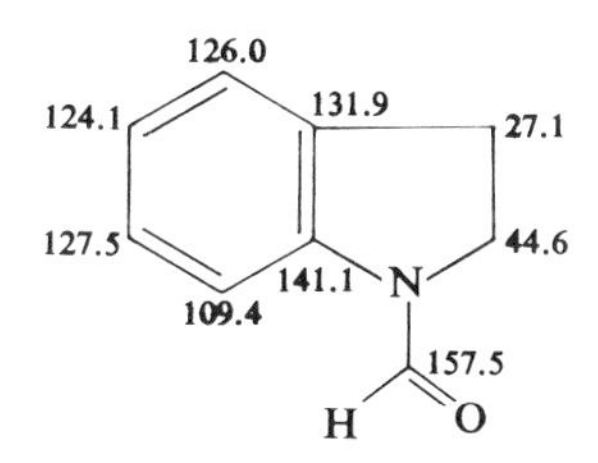

Ref. 84

490 *N*-Formyl-2-hydroxyindoline

126.1
124.6 129.3 36.3
127.7 81.2
139.1 N OH
109.2
158.9
H O

Ref. 84

491 ***N*-Acetylindoline**

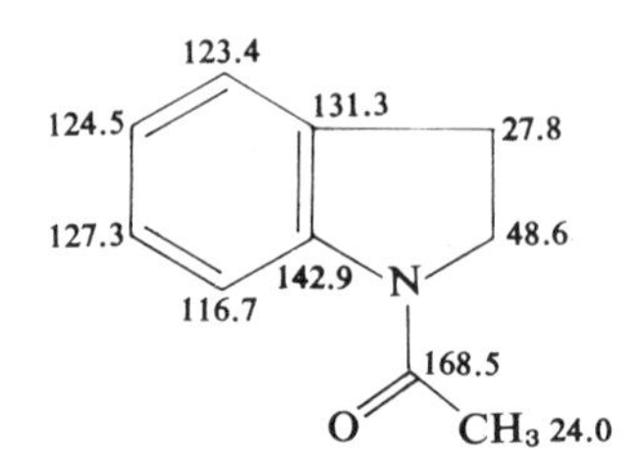

Ref. 84

492

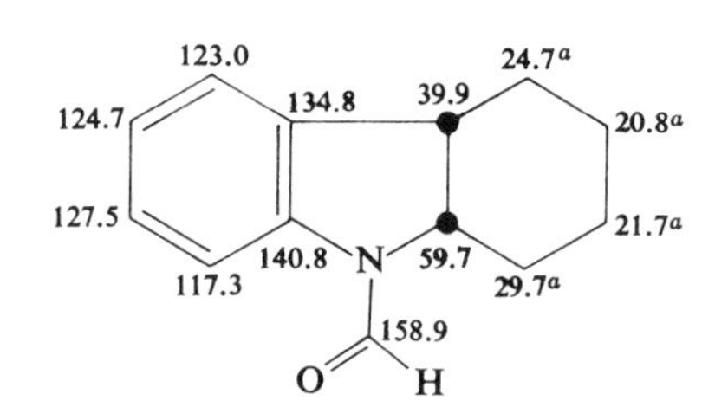

Ref. 84

493

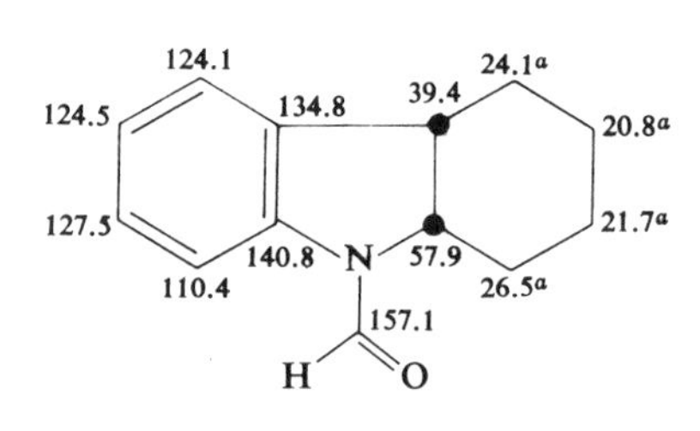

Ref. 84

494

($CHCl_3$)

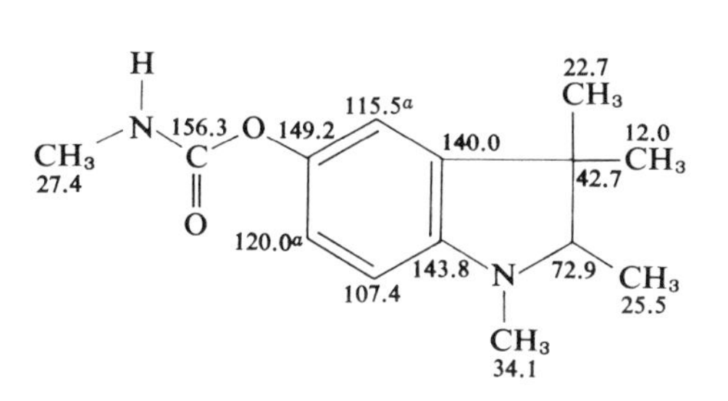

Ref. 85

495 ***N*$_8$-Norphysostigmine**

($CHCl_3$)

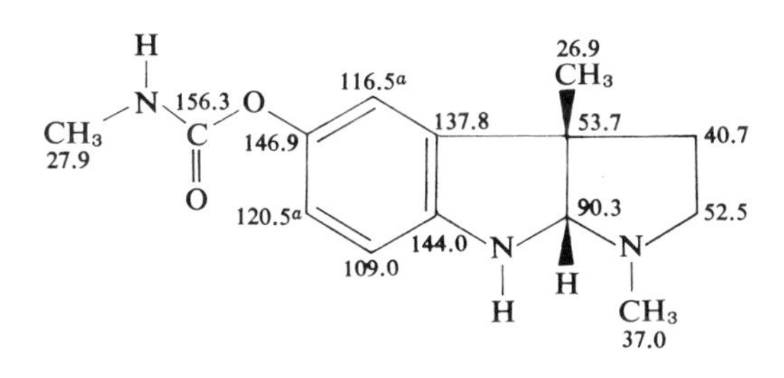

Ref. 85

496 **Physostigmine**

($CHCl_3$)

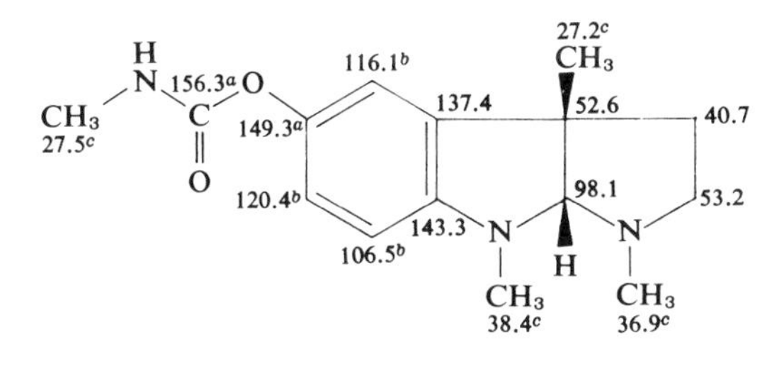

Ref. 85

497 Eseramine
(DMSO)

H
N
155.6
O
CH_3
26.9
C
O
147.4
120.7[a]
116.0[a]
135.1
142.8
105.8
26.9
CH_3
50.4
38.5
89.2
45.7
N
N
H
CH_3
33.6
157.7
C
O
N
H
CH_3
23.3

Ref. 85

498 Physovenine
($CHCl_3$)

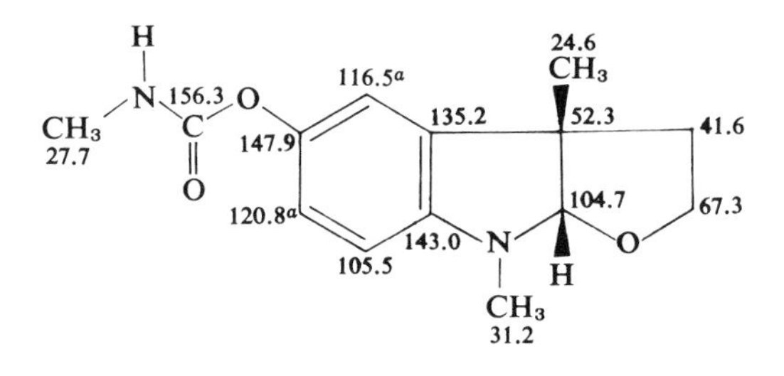

Ref. 85

17. INDOLE ALKALOIDS

a. Simple Indoles

499 Gramine

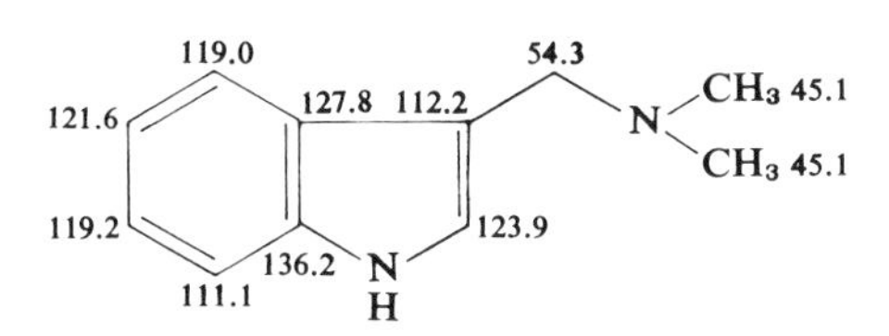

Ref. 21

500 N_b-Methyltryptamine

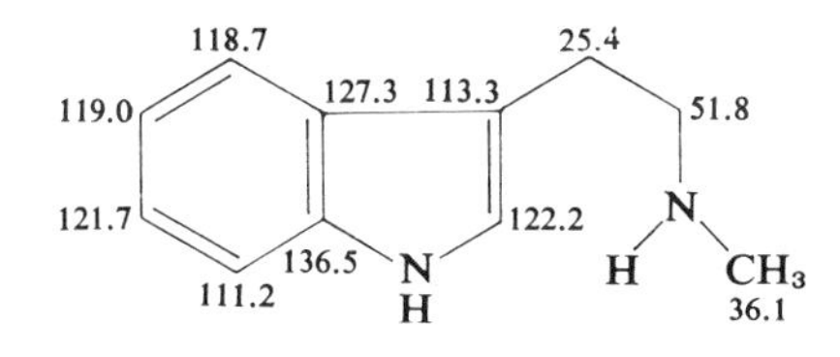

Ref. 86

501 *N*,*N*-Dimethyltryptamine

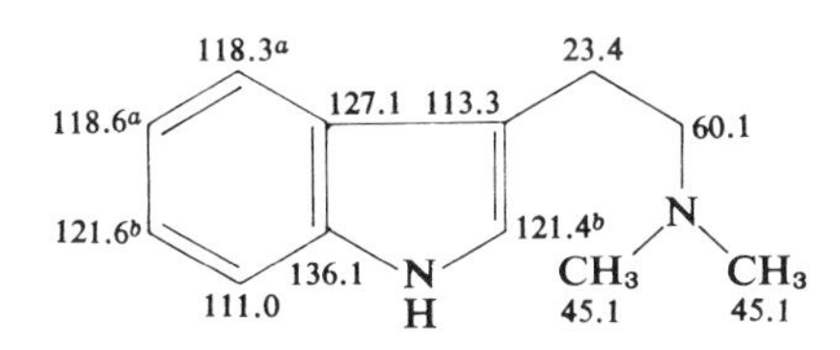

Ref. 88; see also Refs. 21, 86

502 2-Methyl-1,2,3,4-tetrahydro-β-carboline

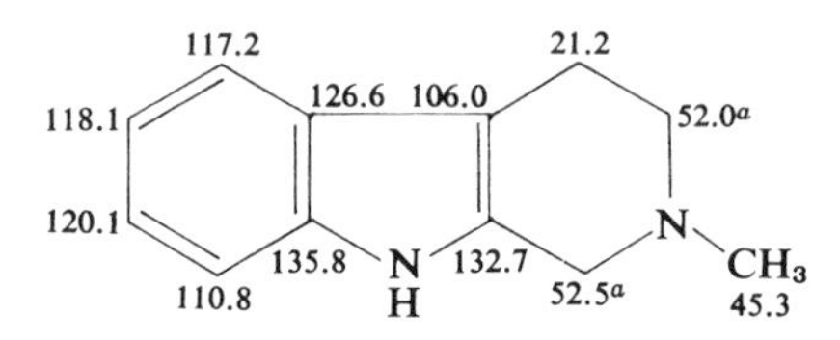

Ref. 86

503 N_b-Methyltetrahydroharmine

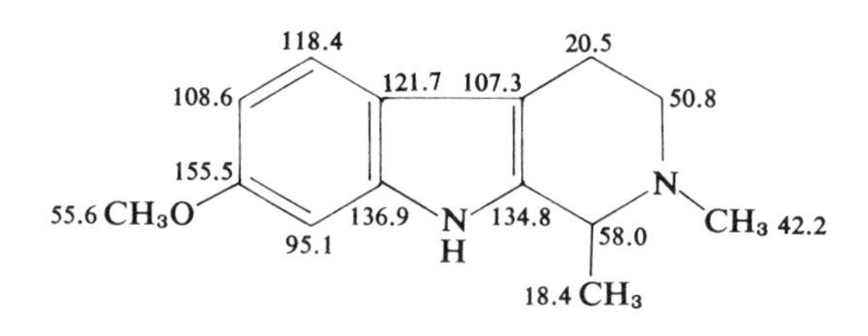

Ref. 21

504 ***Dracontomelum mangiferum* base**

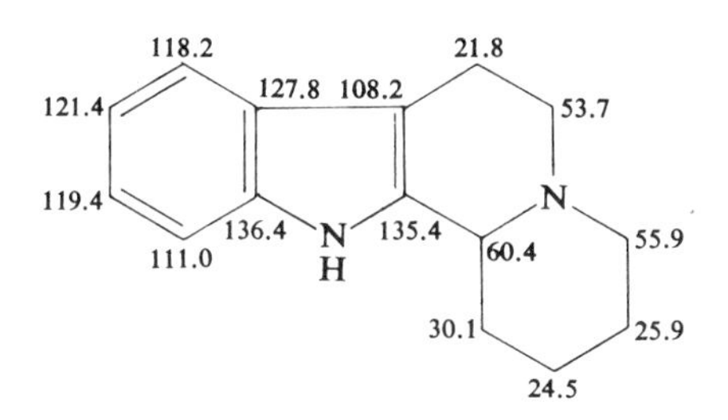

Ref. 21;
see also Ref. 87

505

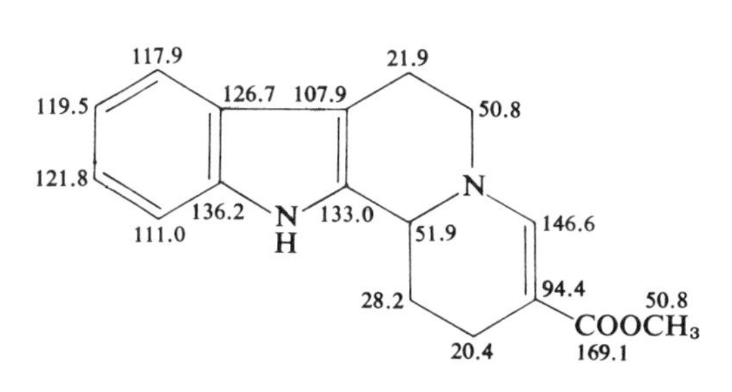

Ref. 88

b. Yohimbinoids

506 Yohimbane

(normal stereochemistry)

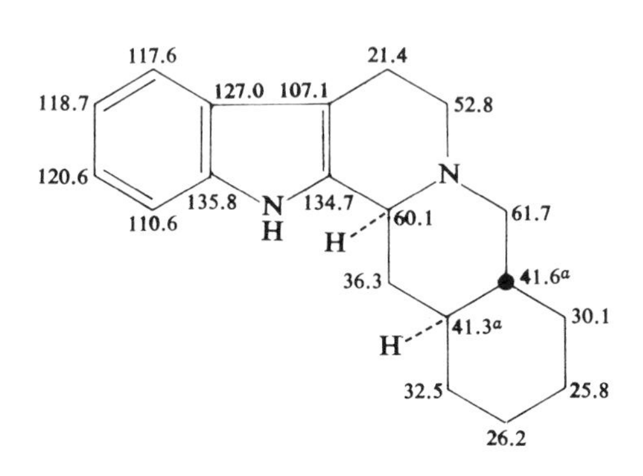

Ref. 88

507 Alloyohimbane

(allo stereochemistry)

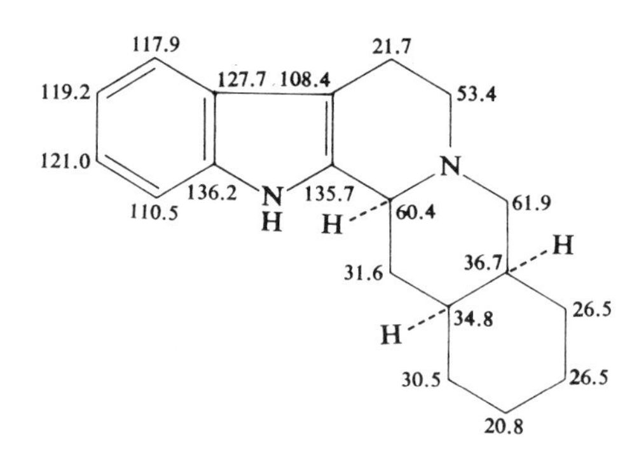

Ref. 88

508 Epialloyohimbane

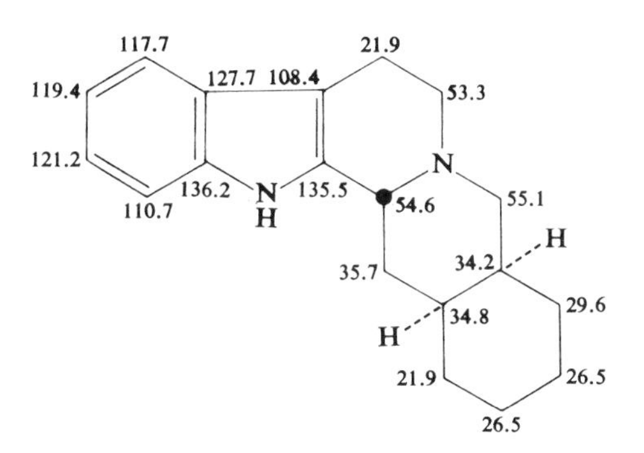

Ref. 88

509 Pseudoyohimbone

(pseudo stereochemistry)
(DMSO-d_6)

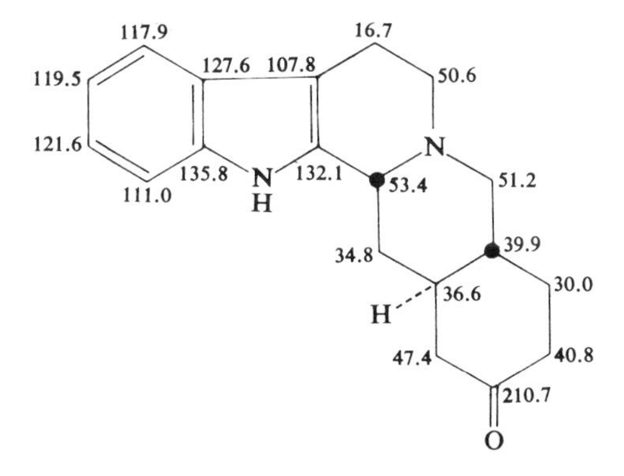

Ref. 88

510 **Yohimbinone**
(normal stereochemistry)
(DMSO-d_6)

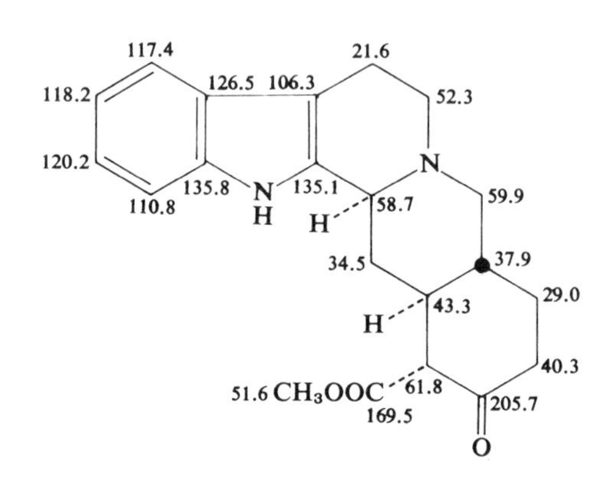

Ref. 88

511 **Yohimbine**
(normal stereochemistry)

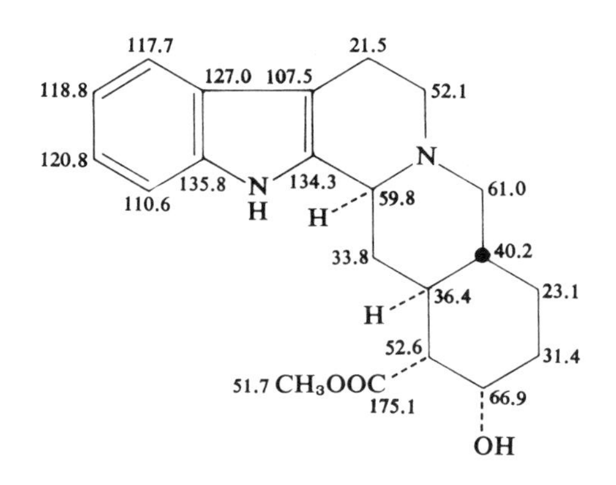

Ref. 88;
see also Ref. 89

512 **β-Yohimbine**
(normal stereochemistry)

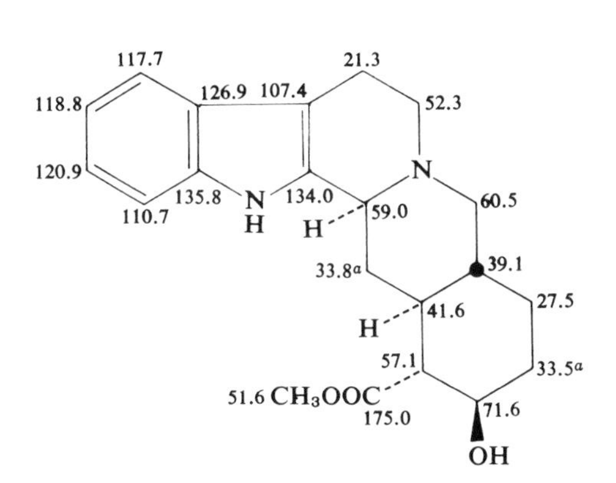

Ref. 88

513 **Corynanthine**
(normal stereochemistry)
(DMSO-d_6)

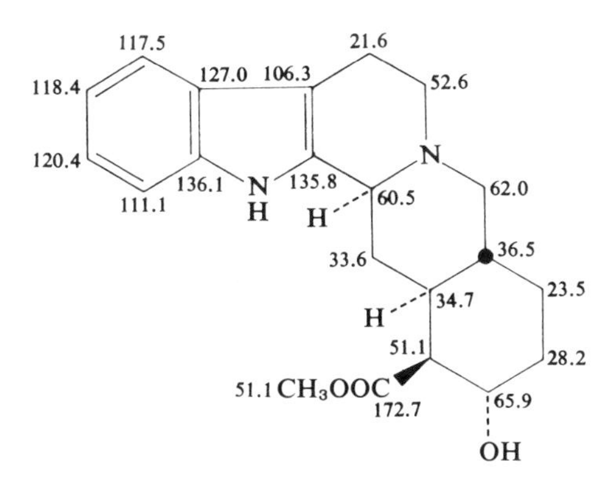

Ref. 88;
see also Ref. 89

514 Alloyohimbine
(allo stereochemistry)

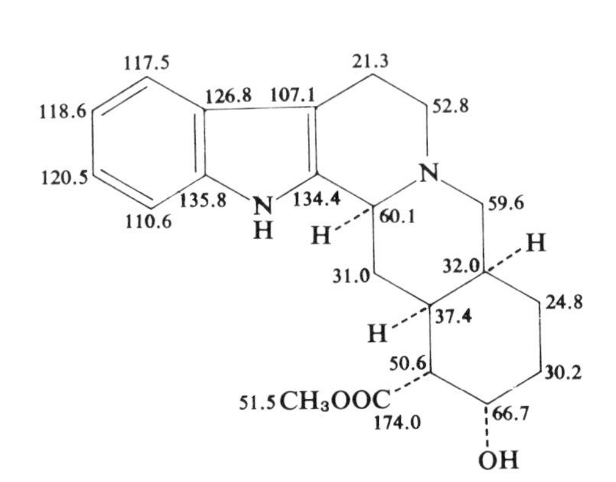

Ref. 88

515 α-Yohimbine
(allo stereochemistry)

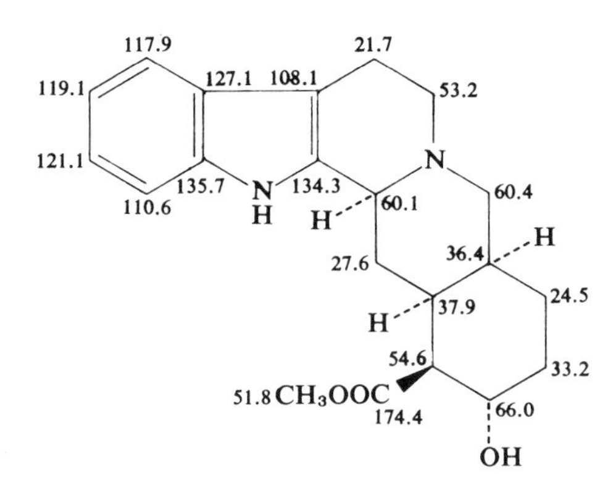

Ref. 88

516 **3-Epi-α-yohimbine**
(epiallo stereochemistry)

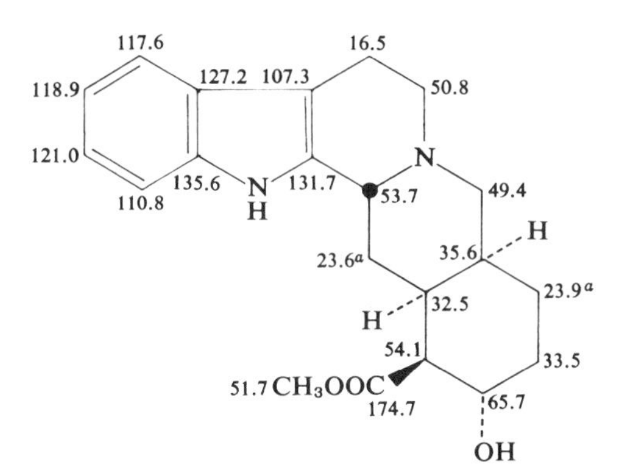

Ref. 88

517 **Pseudoyohimbine**
(pseudo stereochemistry)
(DMSO-d_6)

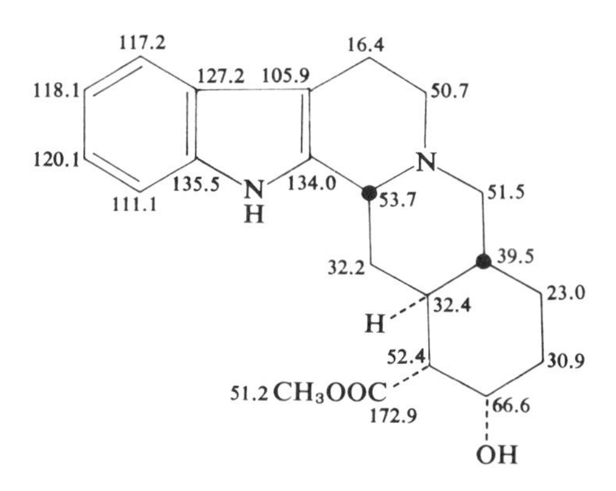

Ref. 88

518 Reserpine

(epiallo stereochemistry)

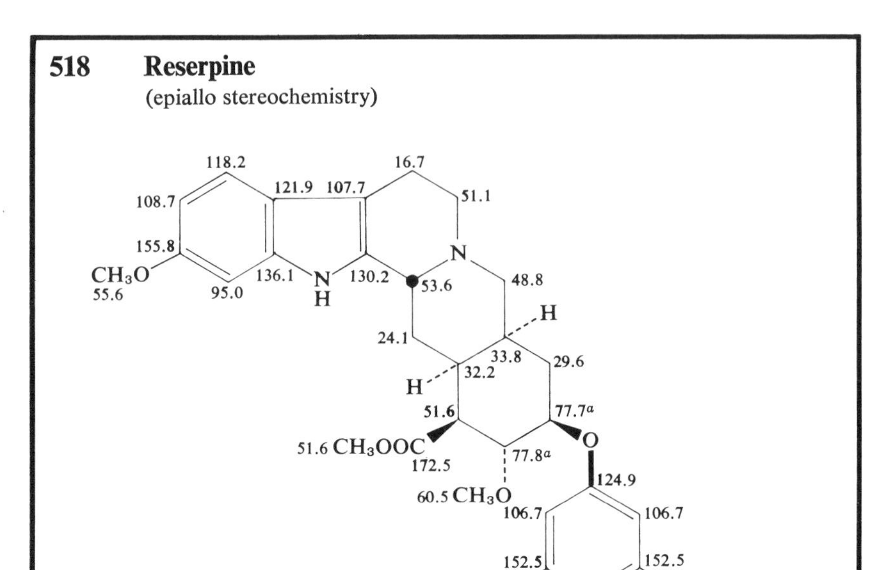

Ref. 88; see also Ref. 89

519 Ajmalicine

(normal stereochemistry)

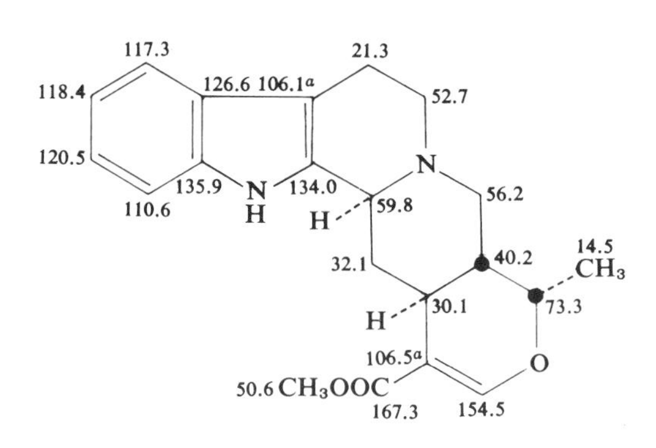

Ref. 88

520 **Tetrahydroalstonine**
(allo stereochemistry)

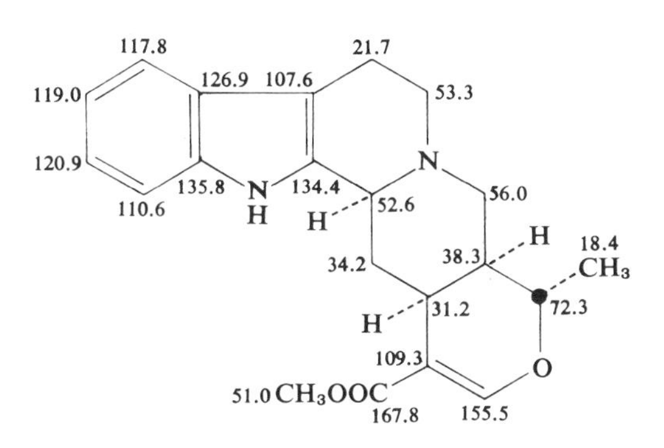

Ref. 88

521 **Rauniticine**
(allo stereochemistry)

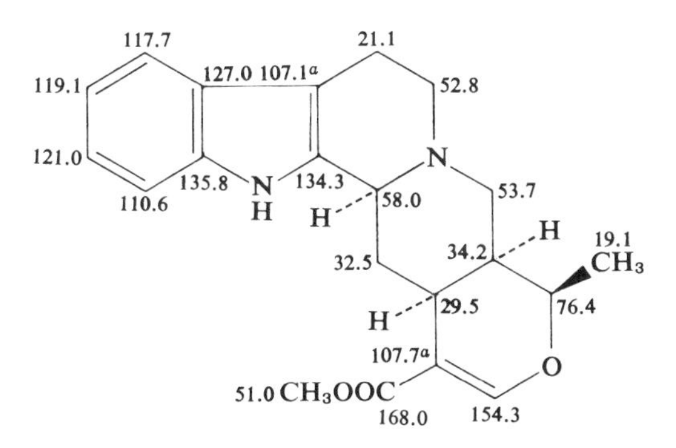

Ref. 88

522 **Akuammigine**
(epiallo stereochemistry)

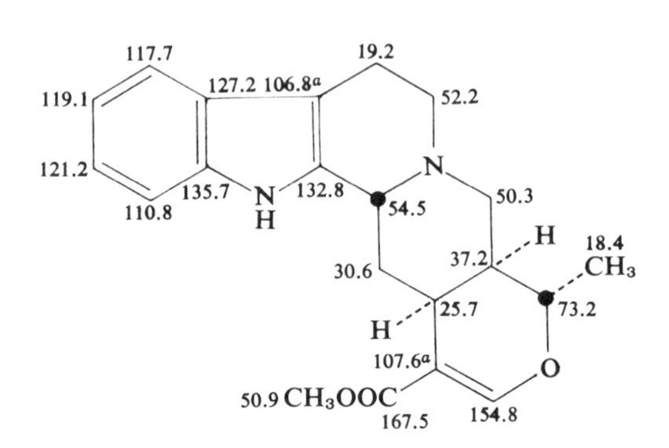

Ref. 88

523 **3-Iso-19-epiajmalicine**
(pseudo stereochemistry)

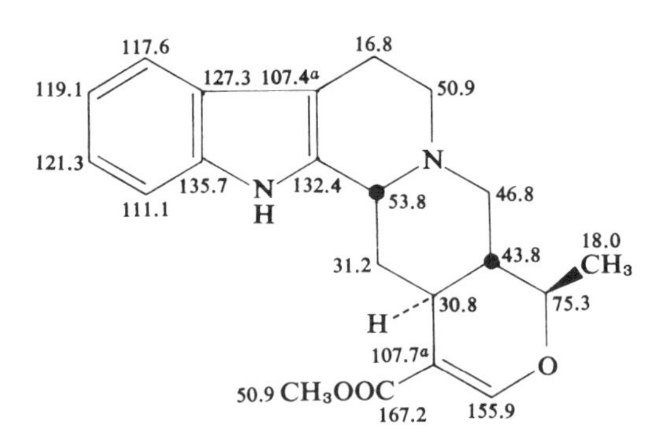

Ref. 88

524 Corynantheine

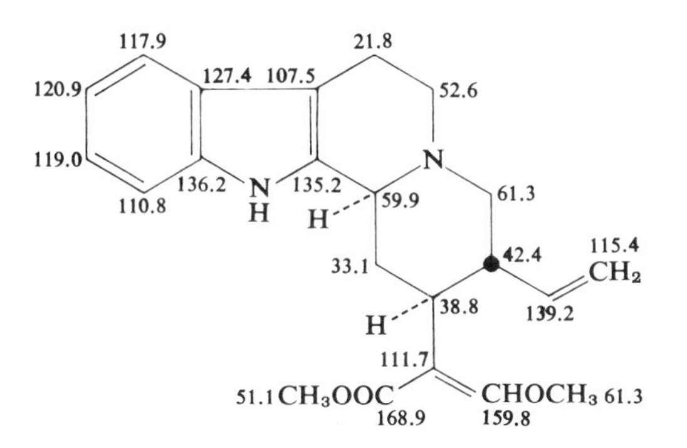

Ref. 21

525 Dihydrocorynantheine

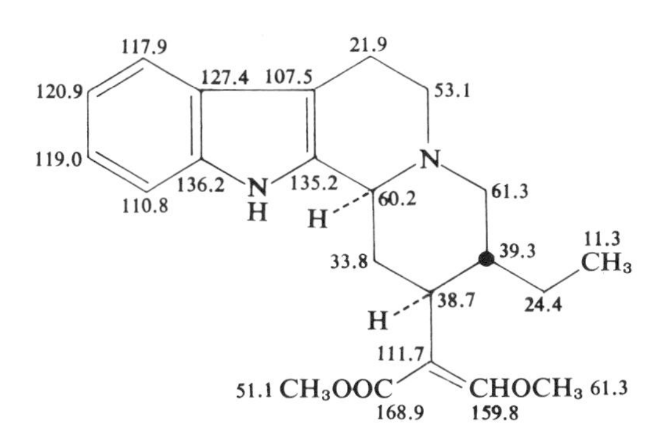

Ref. 21

526 Corynantheidine

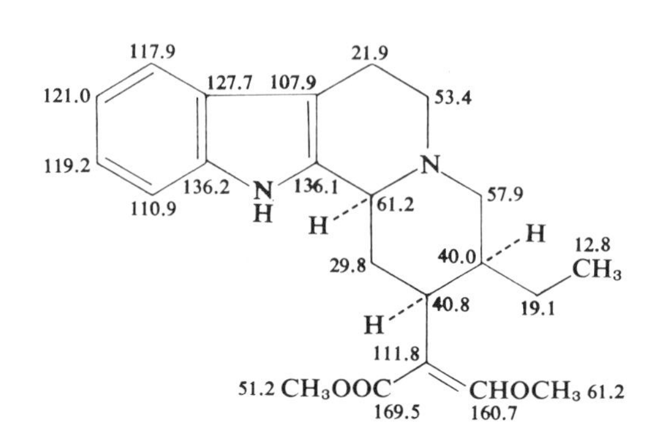

Ref. 21

527

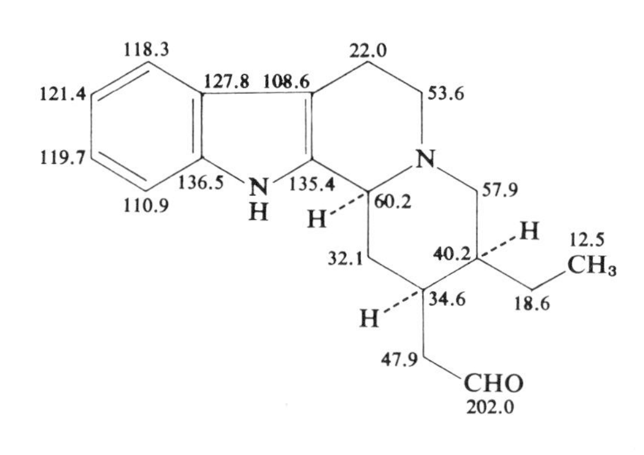

Ref. 21

528

(DMSO-d_6)

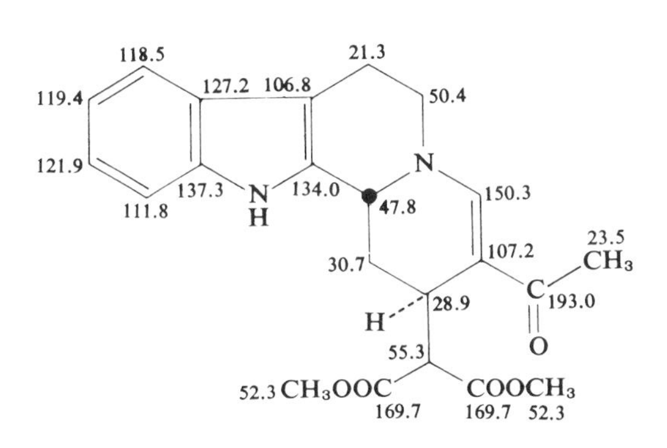

Ref. 88

529 Geissoschizine

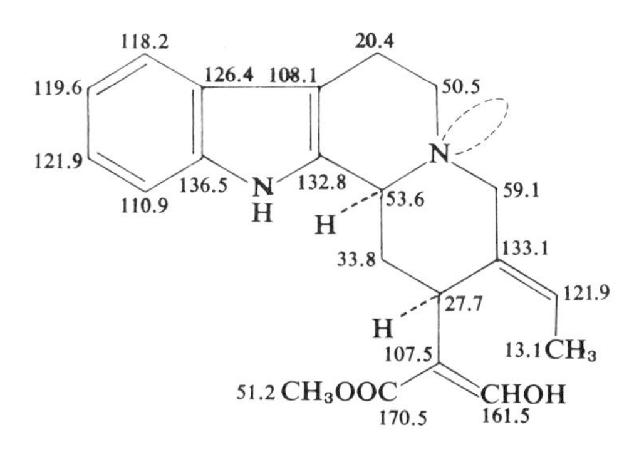

Ref. 90

530 *O*-Methylgeissoschizine

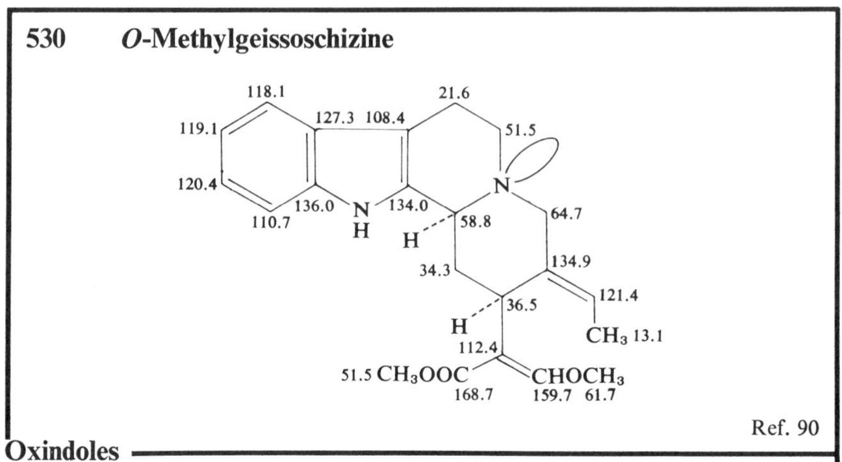

Ref. 90

c. Oxindoles

531

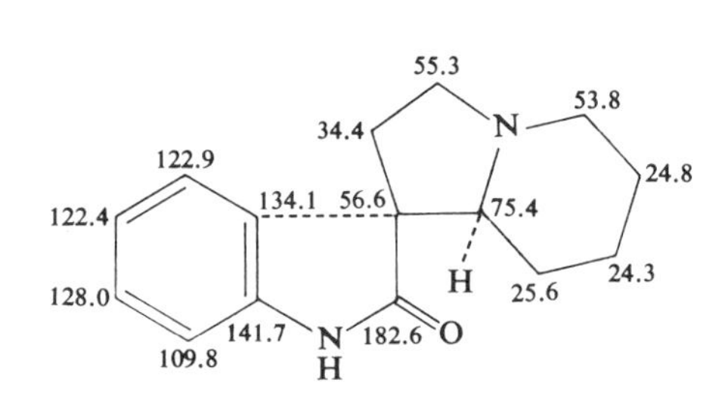

Ref. 21

532

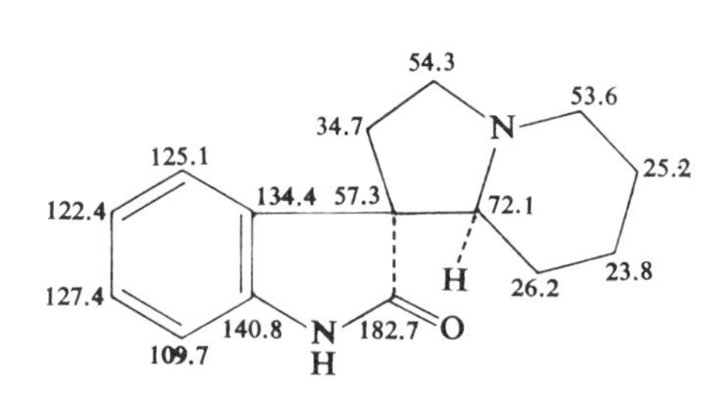

Ref. 21

533 Rhyncophyllal

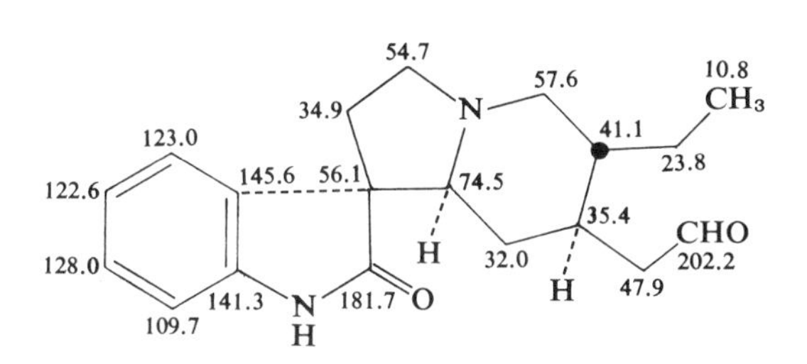

Ref. 21

534 Rhyncophylline

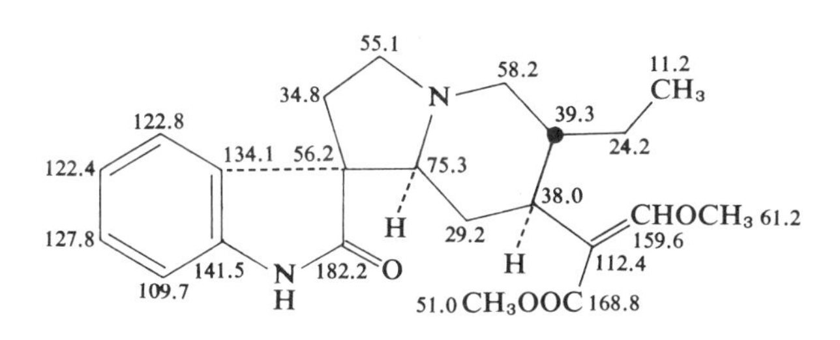

Ref. 21

535 Isorhyncophylline

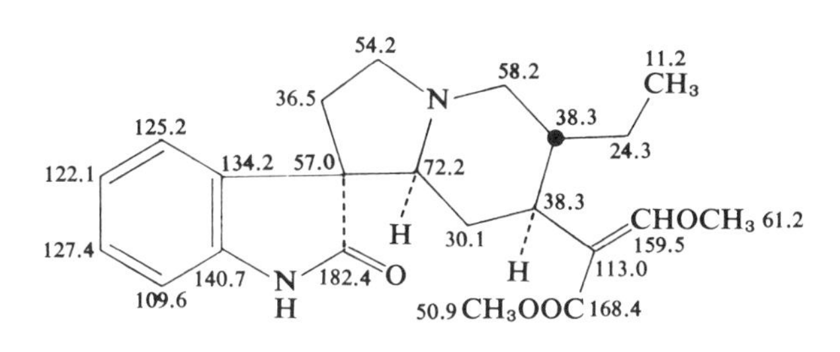

Ref. 21

d. Iboga Bases

536 Ibogaine

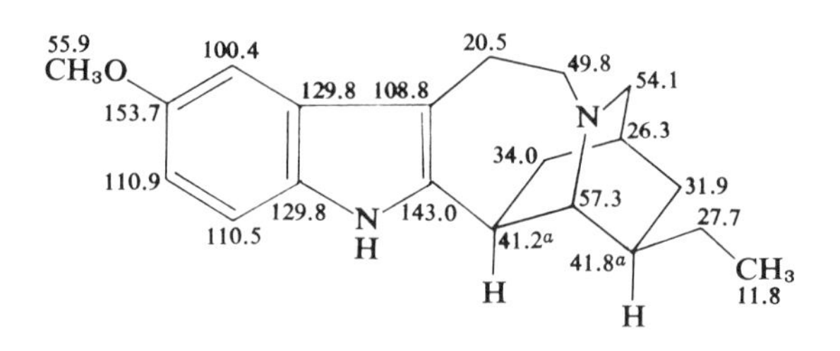

Ref. 34

537 Epiibogamine

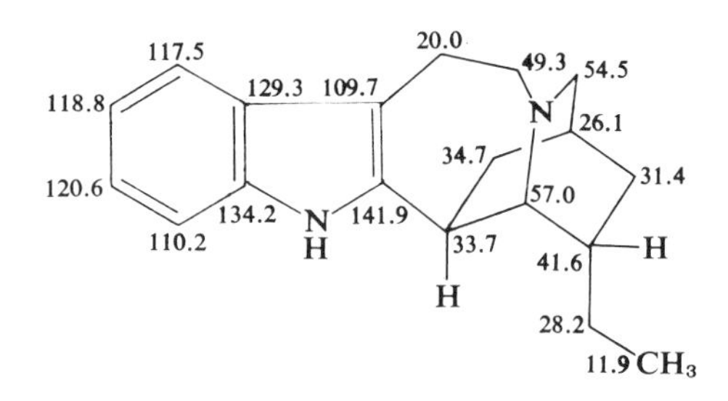

Ref. 34

538 Coronaridine

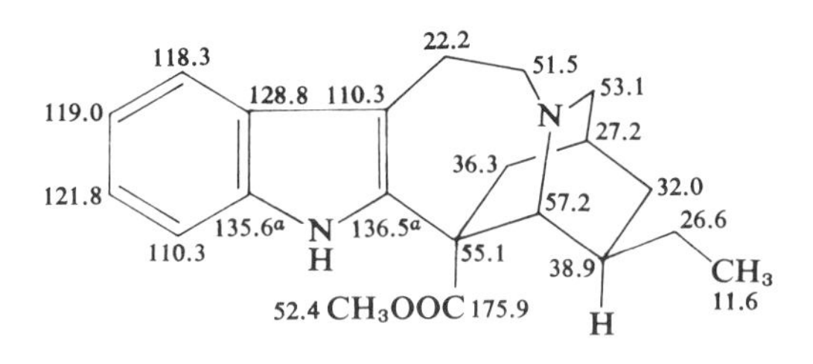

Ref. 34

539 Heyneanine

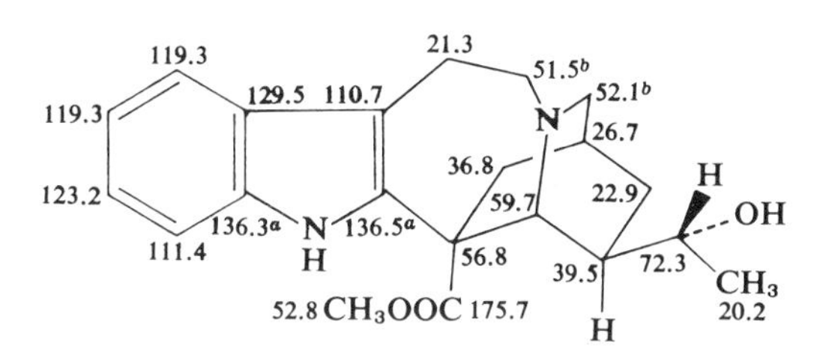

Ref. 34

540 Catharanthine

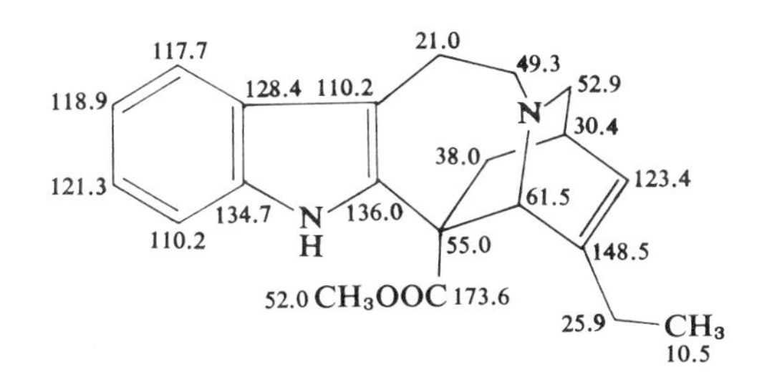

Ref. 34

541 Dehydrovoachalotine

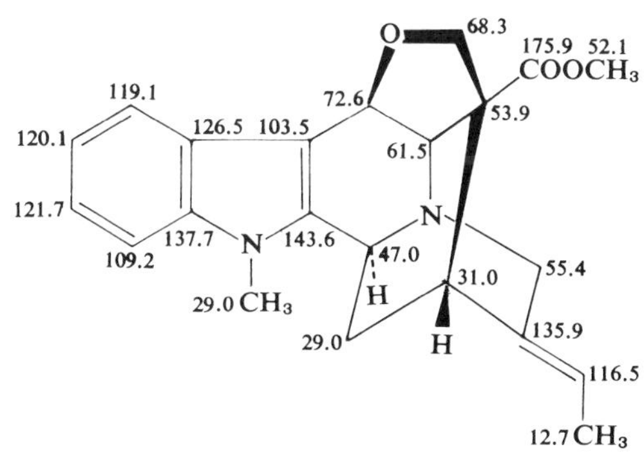

Ref. 91

542 Acetylvoachalotine

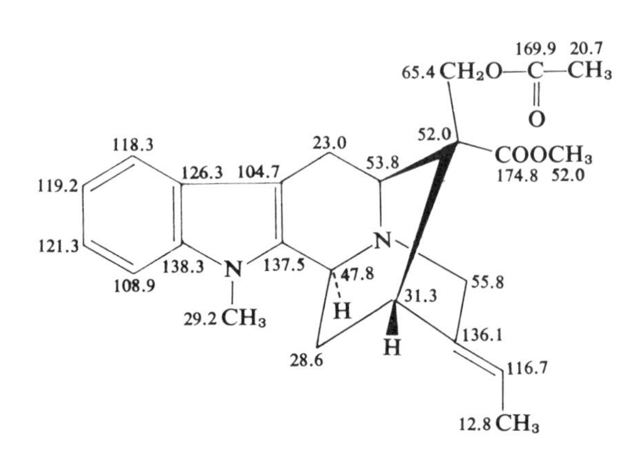

Ref. 91

543 17-*O*-Acetyl-19,20-dihydrovoachalotine

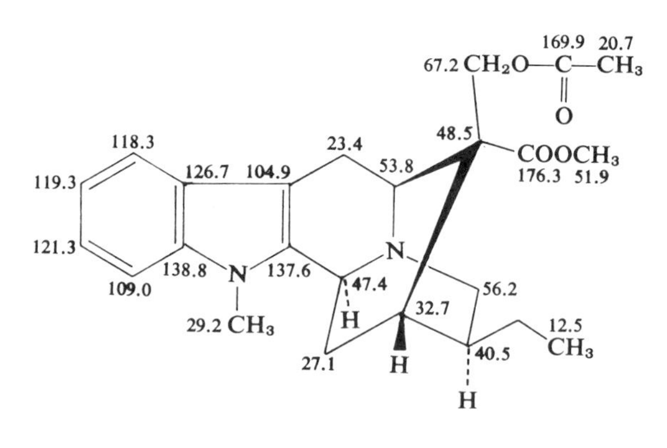

Ref. 91

544 Iboluteine

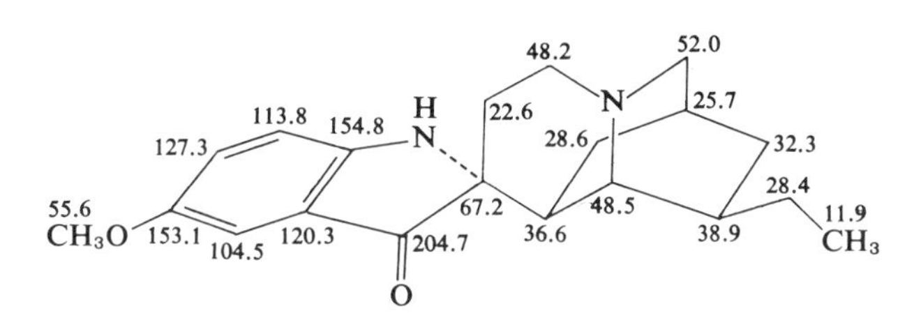

Ref. 92

545 Deoxydihydroiboluteine

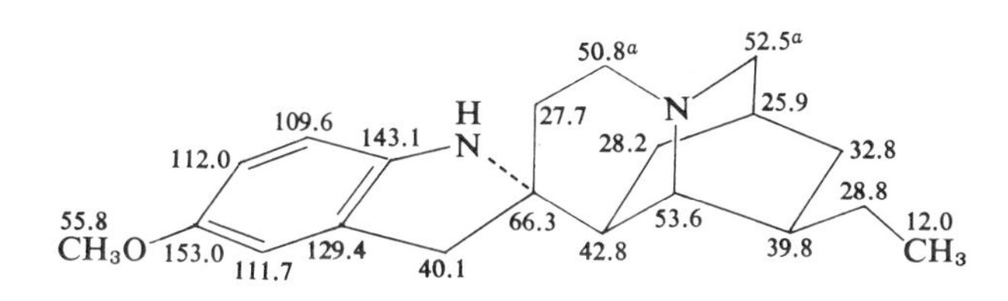

Ref. 92

e. Aspidosperma Alkaloids

546 Tabersonine
($CHCl_3$)

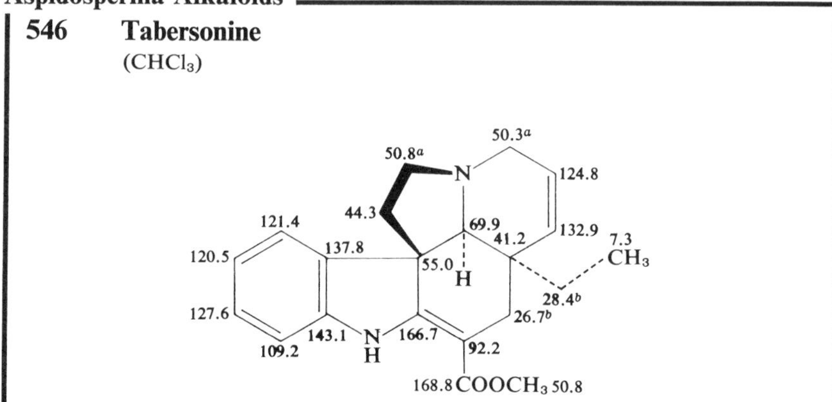

Ref. 24

547 Vincadifformine
($CHCl_3$)

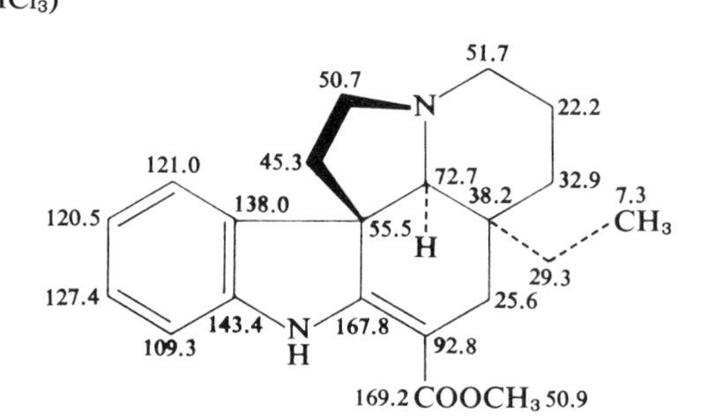

Ref. 24

548 Vindoline

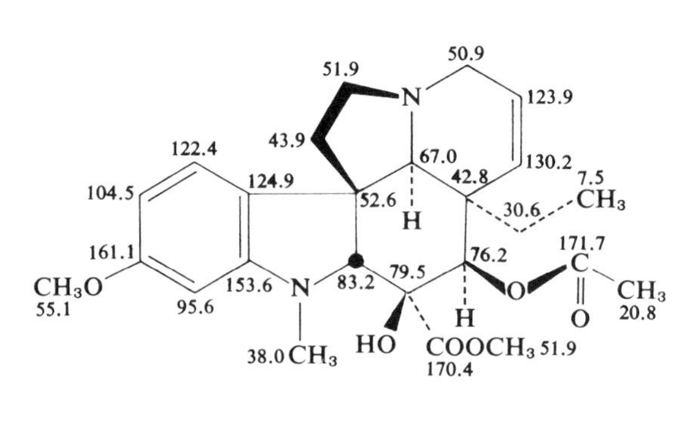

Ref. 24

549 Dihydrovindoline

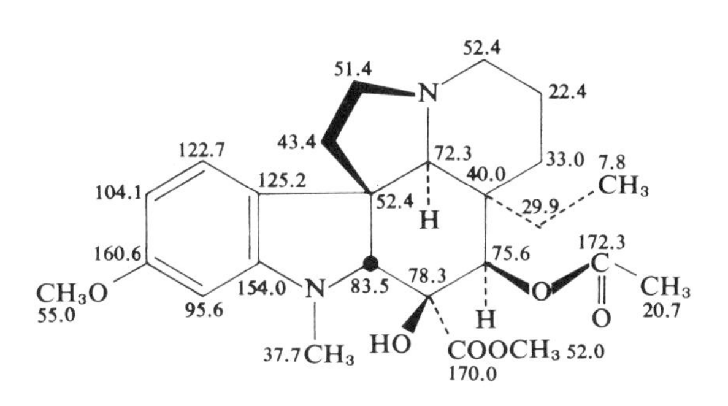

Ref. 24

550 (19S)-Vindolinine

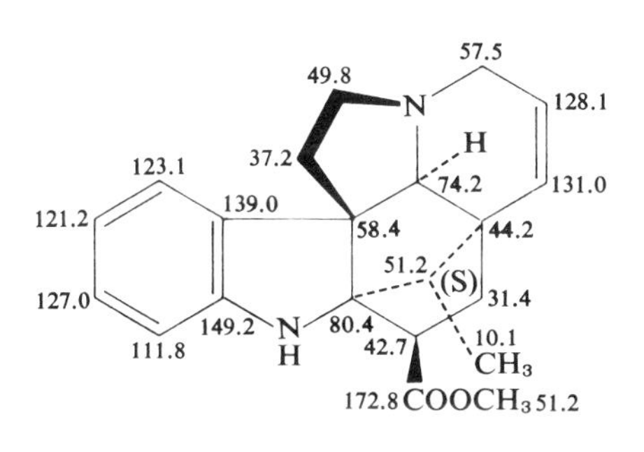

Ref. 93

551 (19R)-Vindolinine

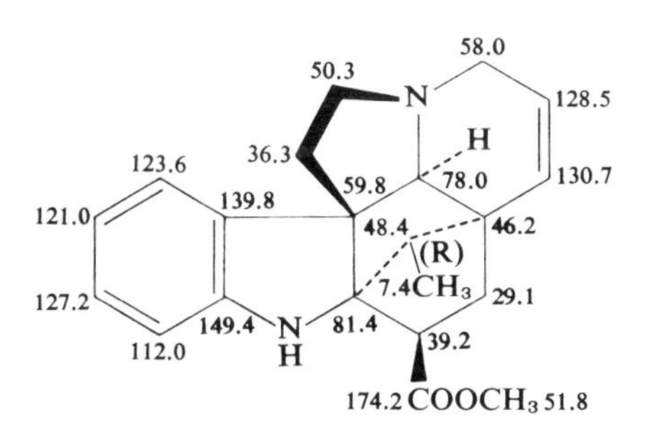

Ref. 94; see also Ref. 93

552 16-Epivindolinine

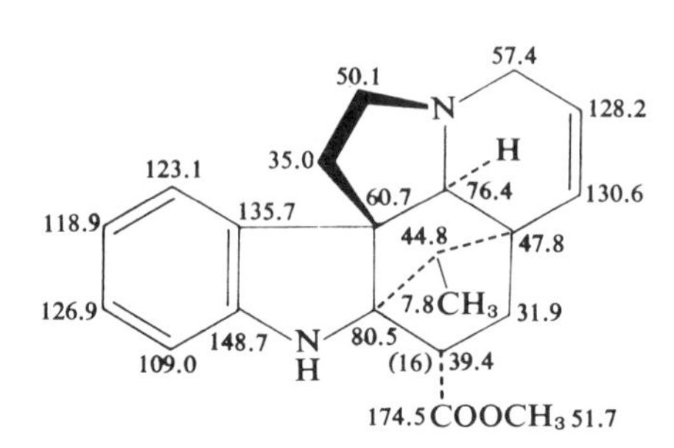

Ref. 94

553 Venalstonine

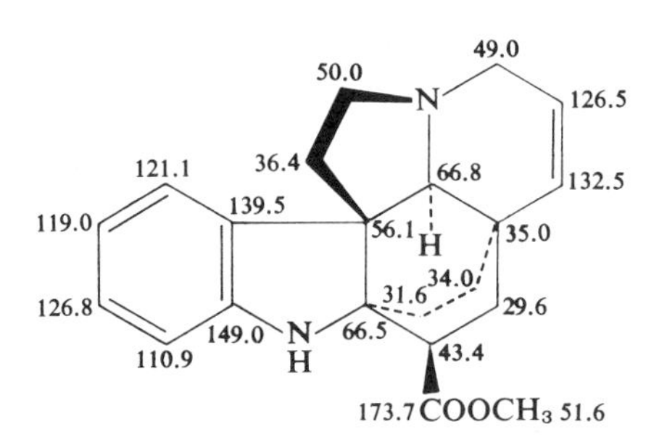

Ref. 94

554 Vandrikine

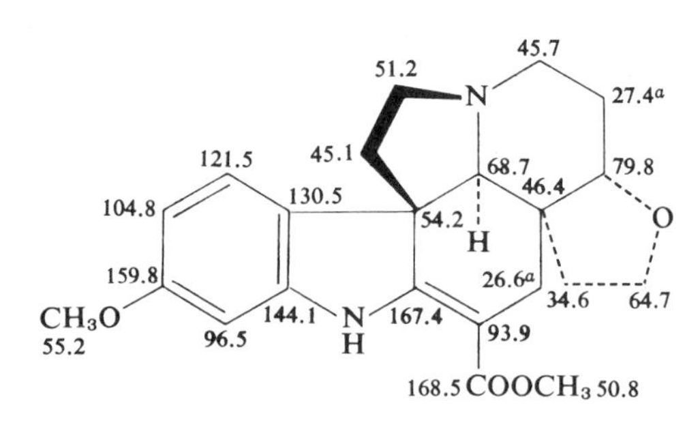

Ref. 24

555 Pandoline

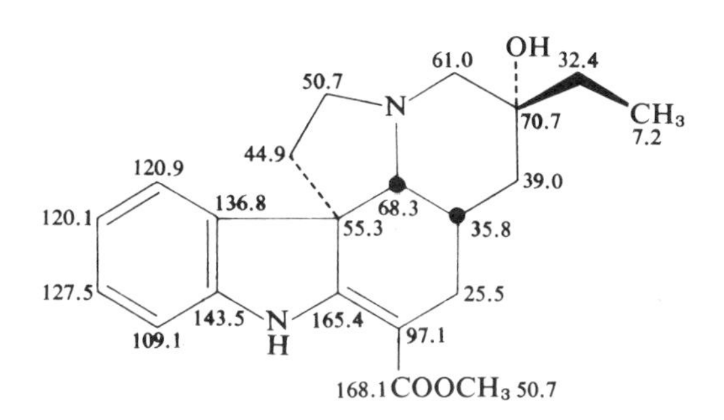

Ref. 95

556 Epipandoline

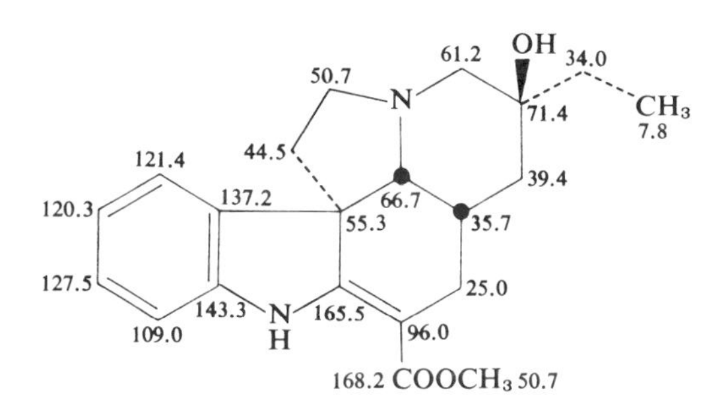

Ref. 95

557 Pandine

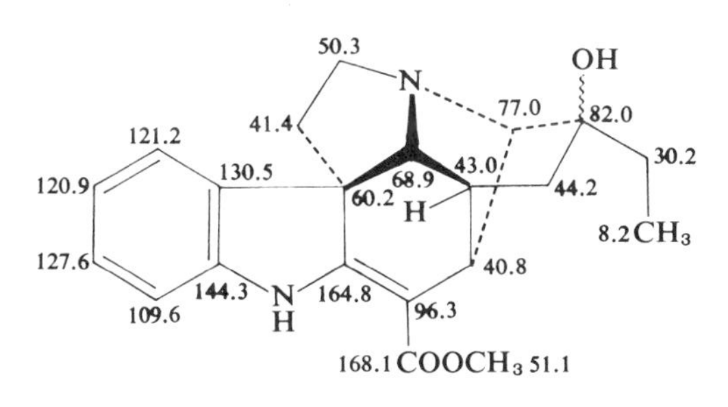

Ref. 96

558 Andrangine

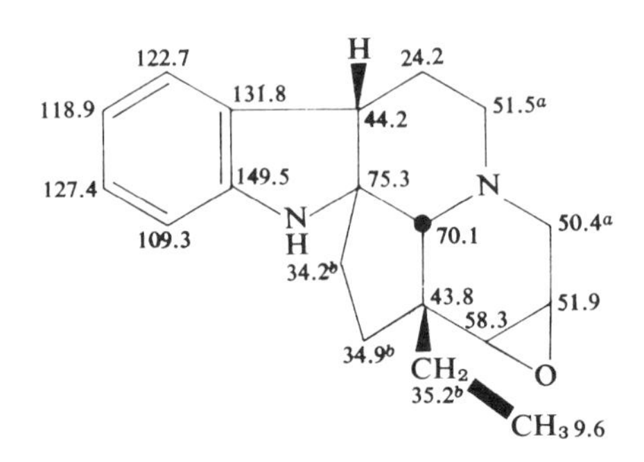

Ref. 97

559 Velbanamine

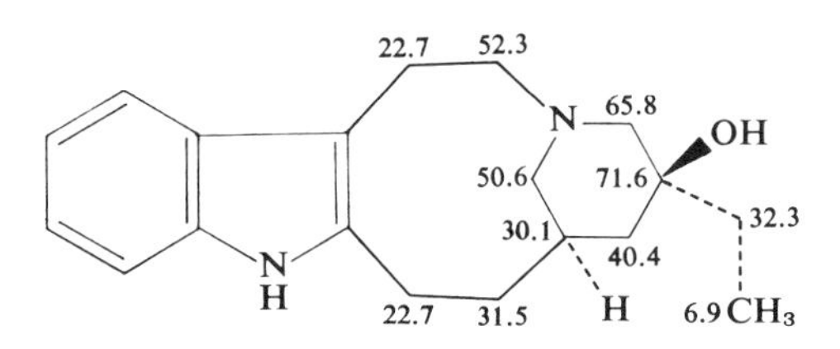

Ref. 95

560 3,4-Secopandoline A
(u = undetected signal)

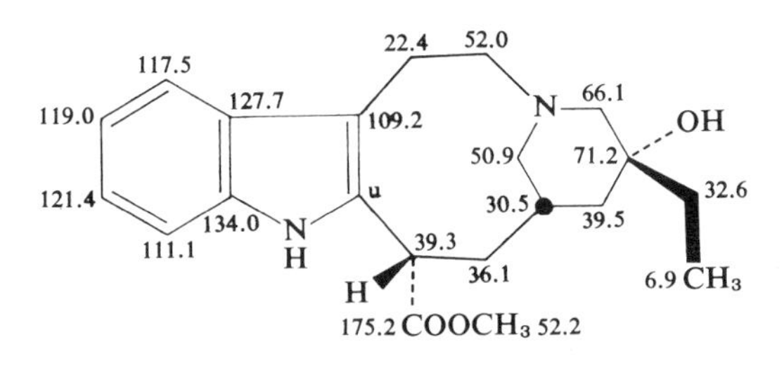

Ref. 95

561 3,4-Secopandoline B

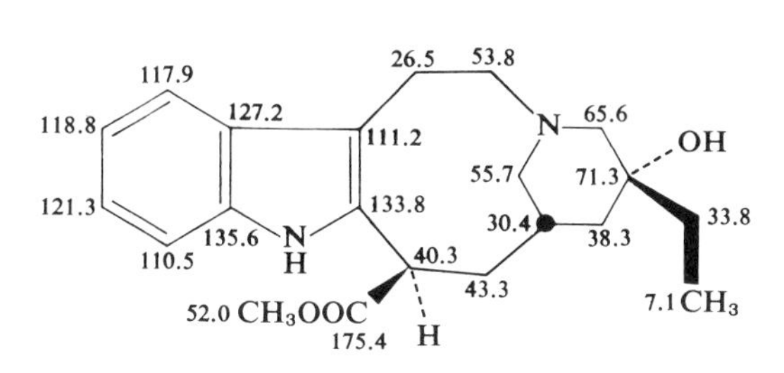

Ref. 95

f. Vincamines

562 Vincamine

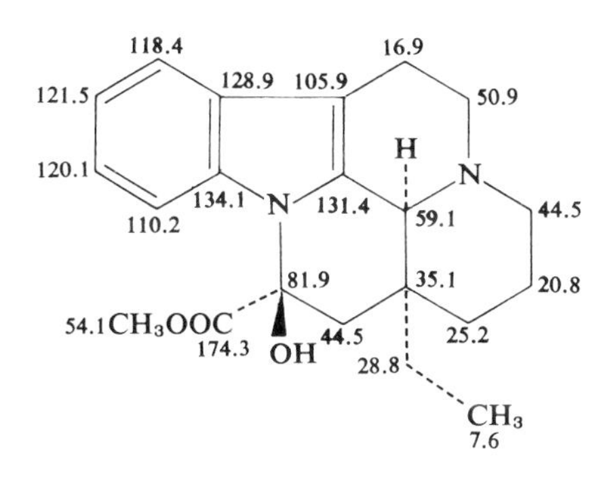

Ref. 98; see also Ref. 99

563 16-Epivincamine

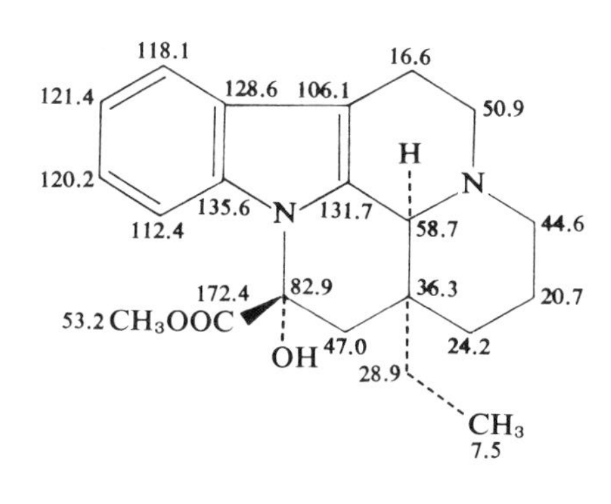

Ref. 99

564 Epivincine

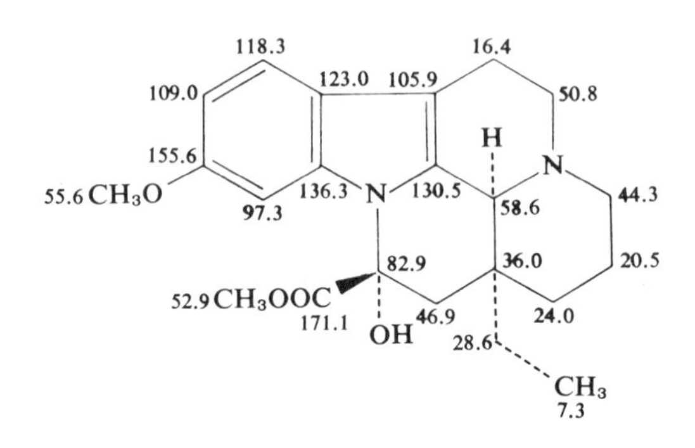

Ref. 98

565 14,15-Dehydroepivincine

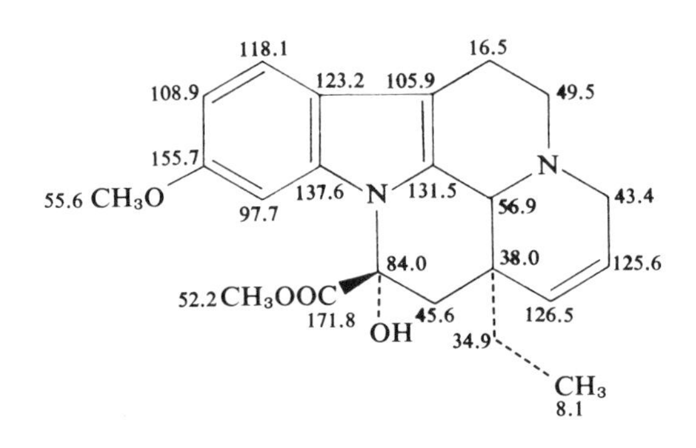

Ref. 98

566 14,15-Dehydrovincine

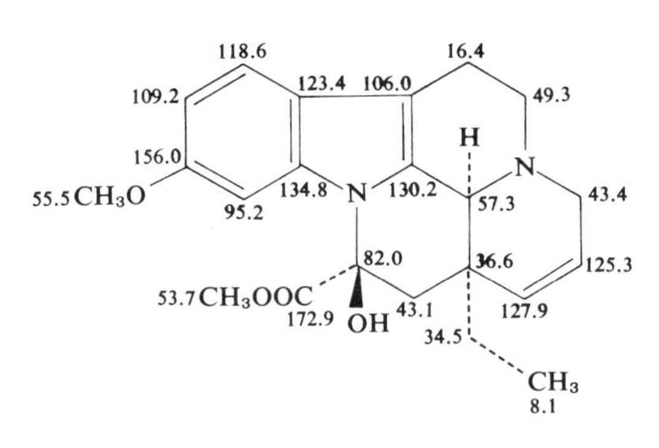

Ref. 98

567 Vincarodine

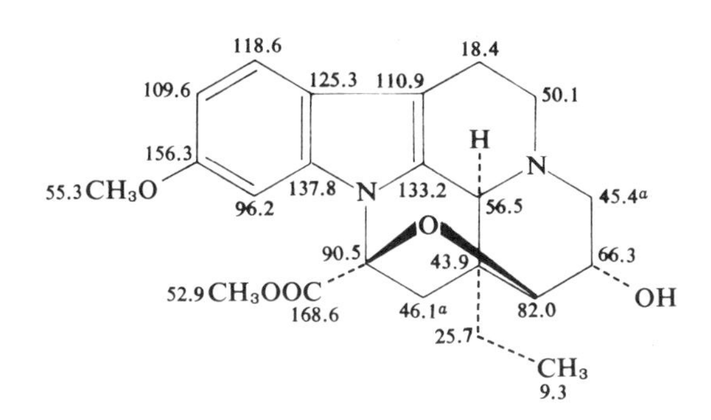

Ref. 98

568 Cuanzine

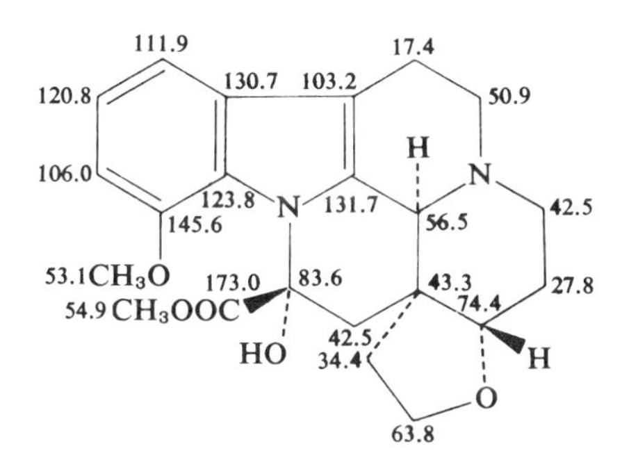

Ref. 99

g. 2-Acylindoles

569 Vobasine

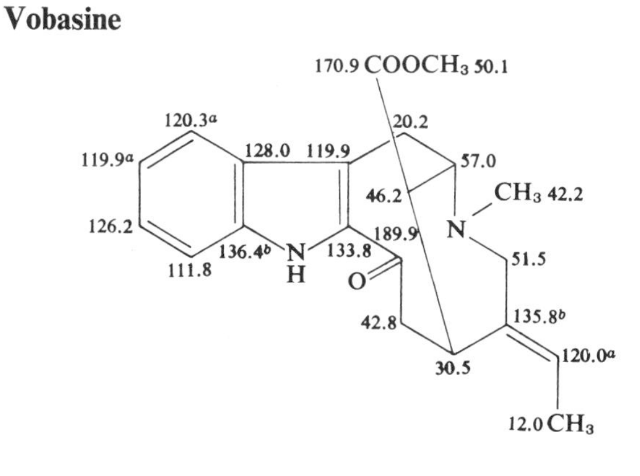

Ref. 100

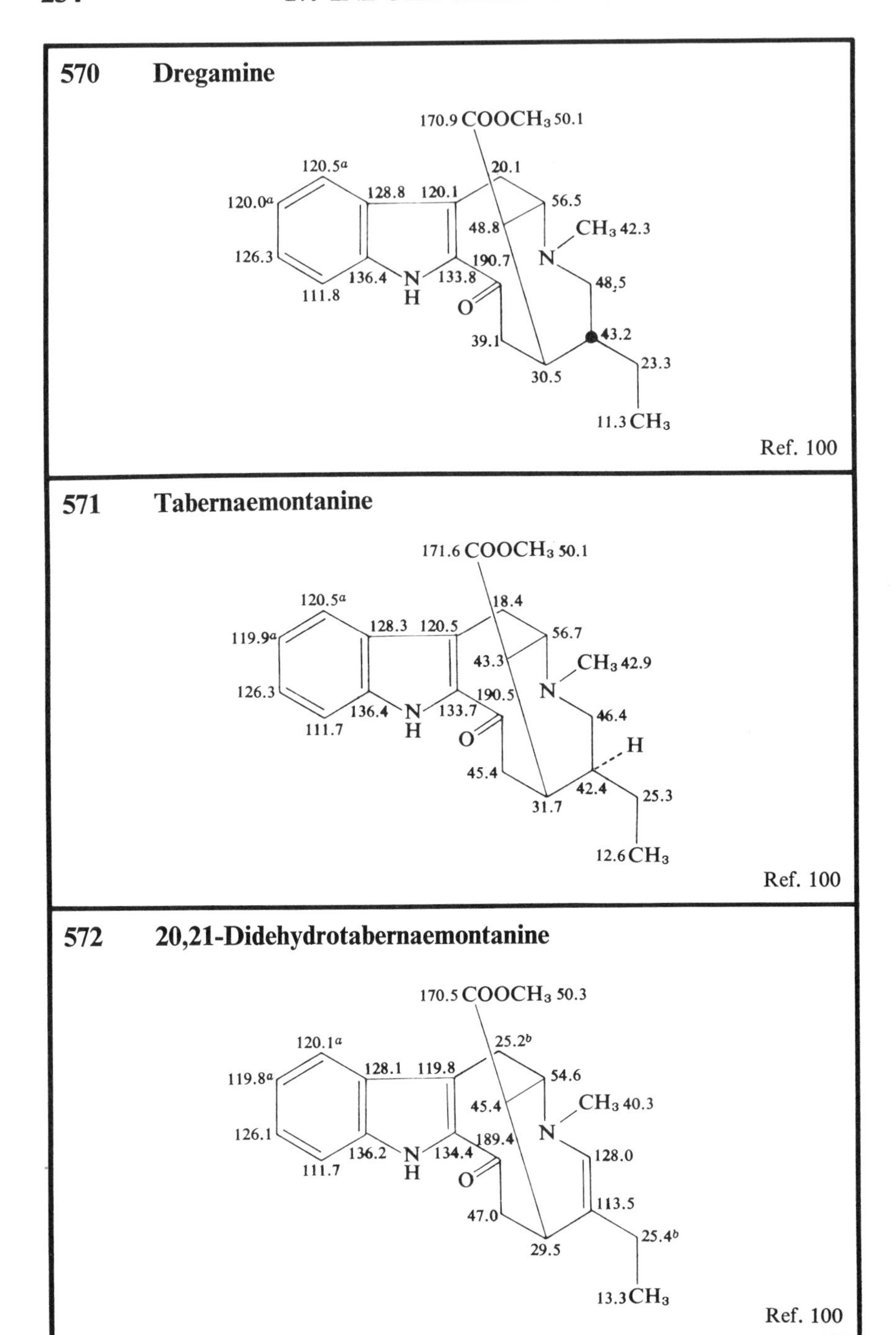
570 Dregamine
170.9 COOCH3 50.1
120.5a
20.1
120.0a
128.8
120.1
56.5
48.8
CH3 42.3
126.3
190.7
N
136.4
N
H
133.8
48.5
111.8
O
43.2
39.1
23.3
30.5
11.3 CH3
Ref. 100
571 Tabernaemontanine
171.6 COOCH3 50.1
120.5a
18.4
119.9a
128.3
120.5
56.7
43.3
CH3 42.9
126.3
190.5
N
136.4
N
H
133.7
46.4
111.7
O
H
45.4
42.4
25.3
31.7
12.6 CH3
Ref. 100
572 20,21-Didehydrotabernaemontanine
170.5 COOCH3 50.3
120.1a
25.2b
119.8a
128.1
119.8
54.6
45.4
CH3 40.3
126.1
189.4
N
136.2
N
H
134.4
128.0
111.7
O
113.5
47.0
25.4b
29.5
13.3 CH3
Ref. 100

573 Ochropamine

170.9 $COOCH_3$ 49.8
120.2[a] 21.0
119.8[a] 126.6 120.7 57.0
46.5 CH_3 42.2
125.8 190.7 N
138.7 N 133.3 51.8
109.5 O
CH_3
32.8 45.4 135.7
119.8
30.6
12.1 CH_3

Ref. 100

Miscellaneous Indoles

574 Vincorine

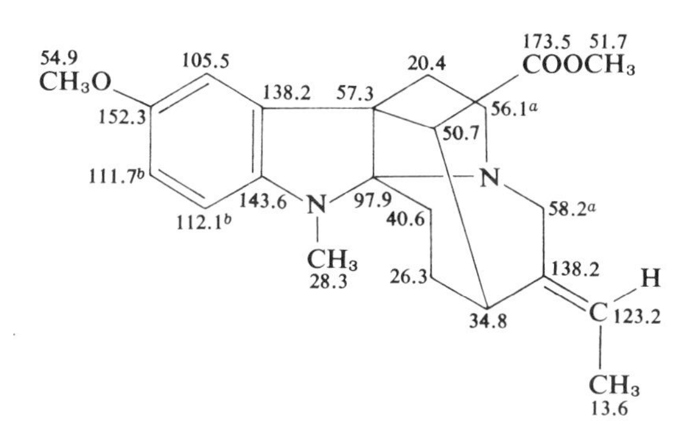

Ref. 101

575 Vinoxine
(C_6D_6)

119.3
120.0 127.6 101.5 OH
57.7
118.9 133.3 H
136.0 N 59.9 54.7
108.5 N
31.1
51.7 CH_3OOC 51.5 56.2
169.4 31.1
H 132.7
121.9
12.4 CH_3

Ref. 102

576 Strychnine

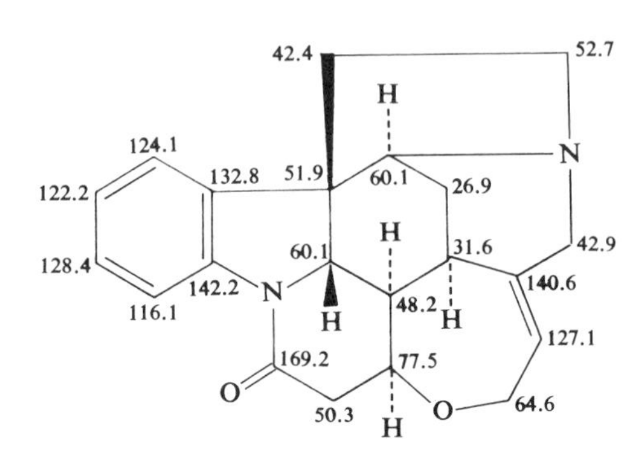

Ref. 30; see also Ref. 103

577 Gelsemine

(values adjusted from CS_2 reference)

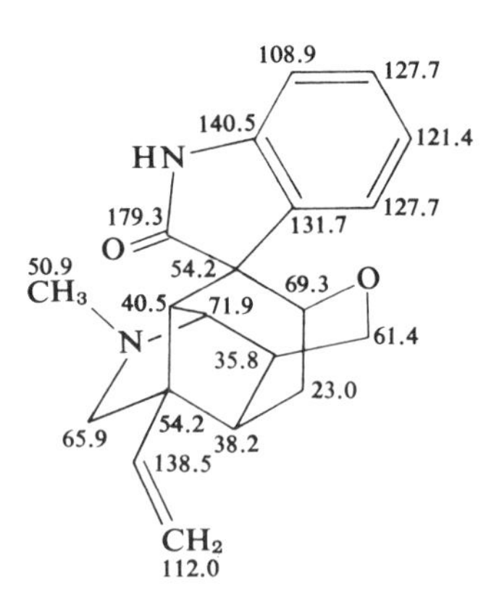

Ref. 56

578 Gelsedine

(values adjusted from CS_2 reference)

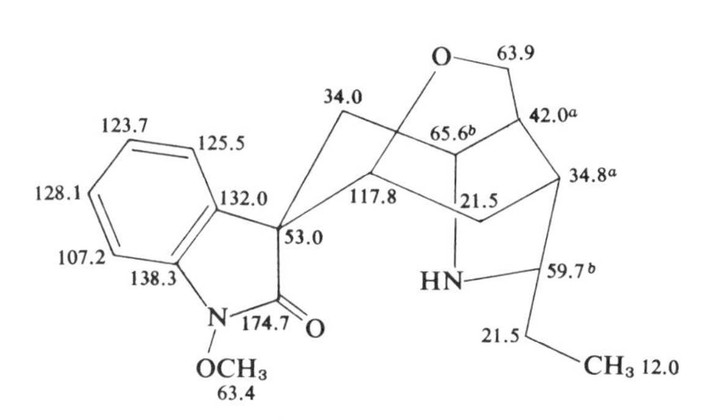

Ref. 104

579 Rutaecarpine

(DMSO-d_6)

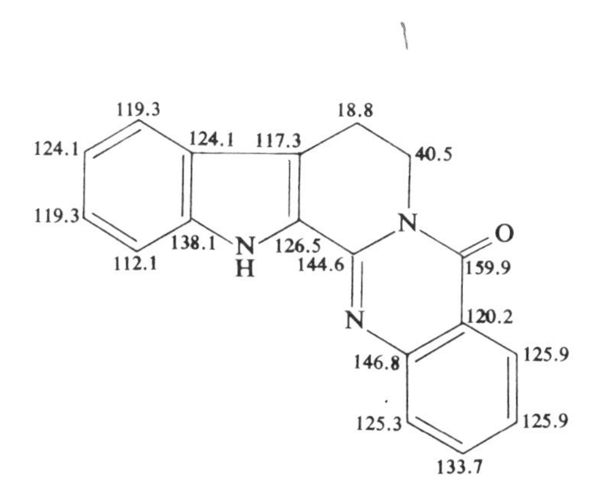

Ref. 105

i. Ergot Bases

580 1,3,4,5-Tetrahydrobenz[cd]indole

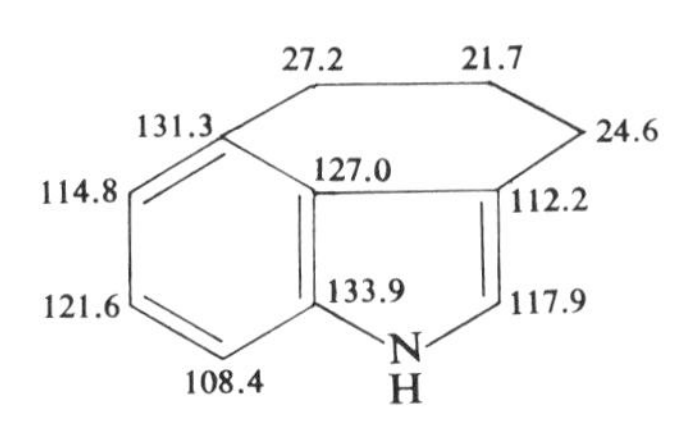

Ref. 25

581 1-Methyl-1,3,4,5-tetrahydrobenz[cd]indole

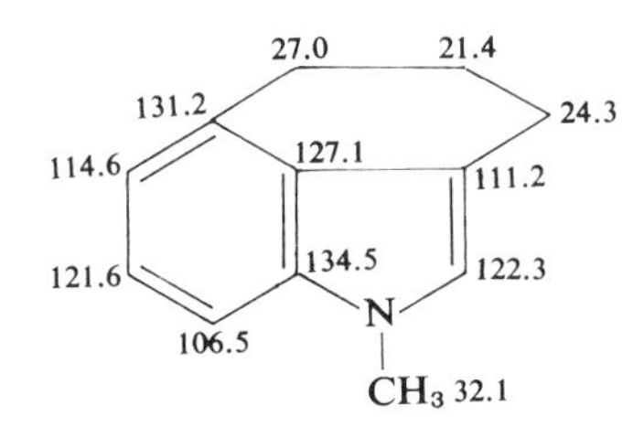

Ref. 25

582 Agroclavine

(pyridine-d_5)

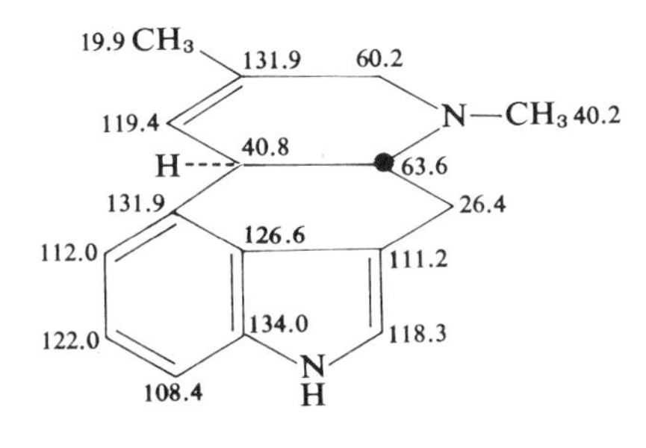

Ref. 25

583 Elymoclavine acetate

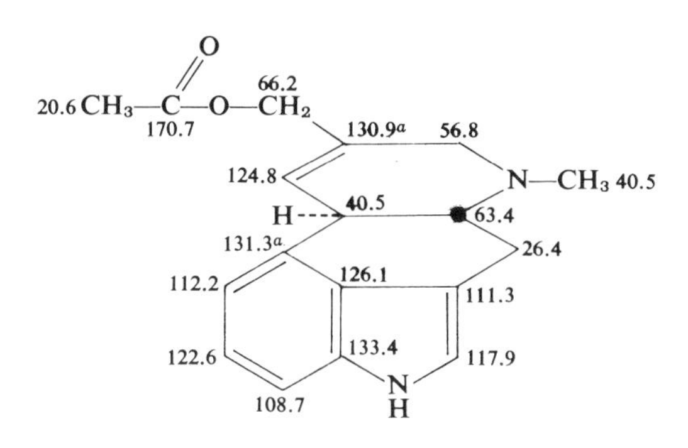

Ref. 25

584 Fumigaclavine B
(pyridine-d_5)

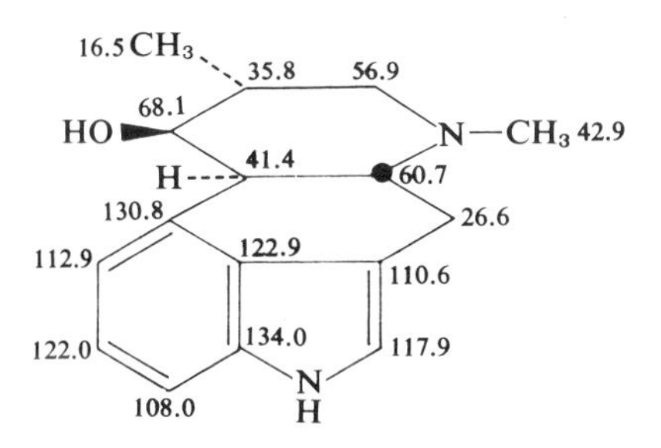

Ref. 25

585 Methyl 9,10-dihydrolysergate
(DMSO-d_6)

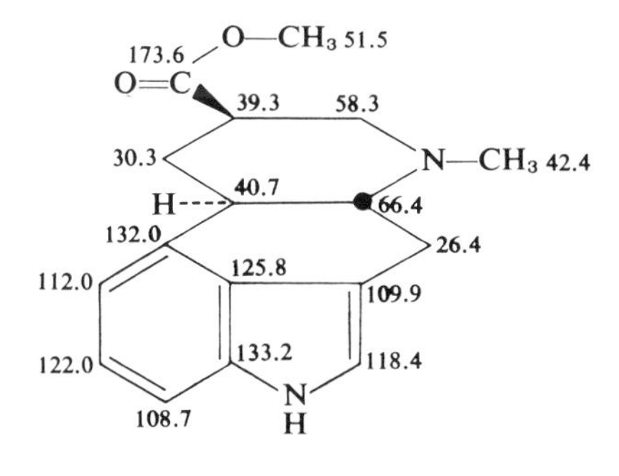

Ref. 25

586 1-Methyldihydrolysergamide (*trans* C/D, 8α isomer)

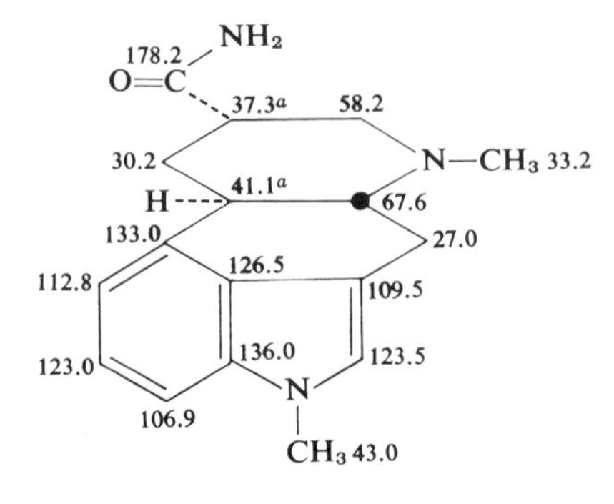

Ref. 106

587 1-Methyldihydrolysergamide (*trans* C/D, 8β isomer)

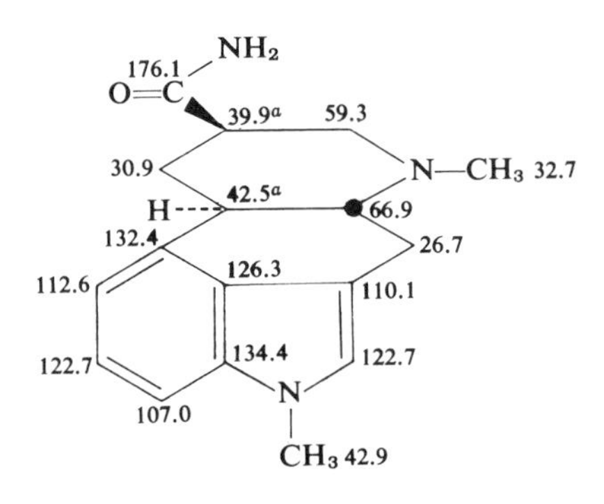

Ref. 106

588 1-Methyldihydrolysergamide (*cis* C/D, 8α isomer)

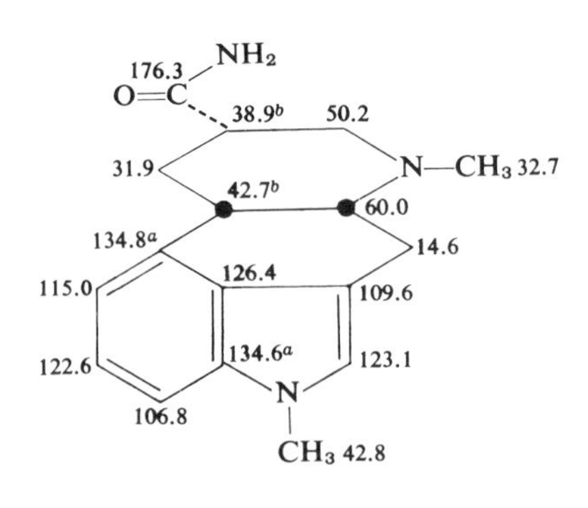

Ref. 106

589 1-Methyldihydrolysergamide (*cis* C/D, 8β isomer)

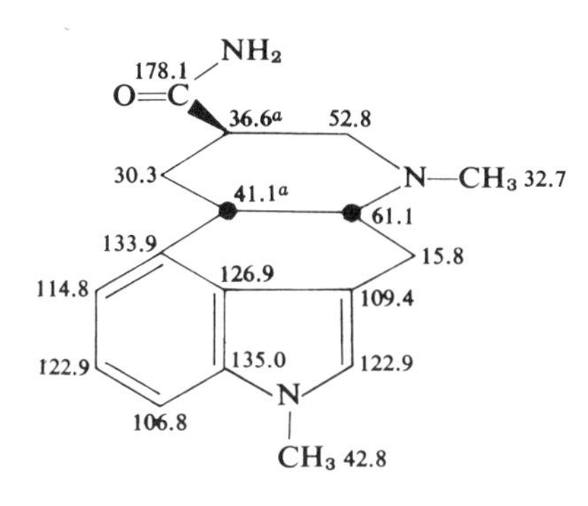

Ref. 106

590 10-Methoxydihydrolysergamide (8α,10α isomer)

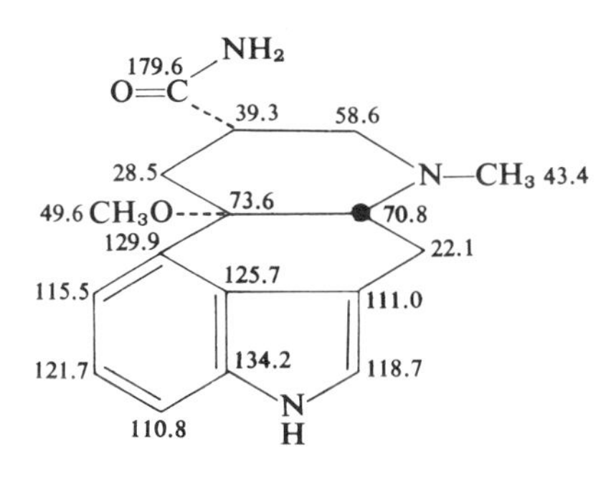

Ref. 106

591 Methyl 10-methoxydihydrolysergate (8α,10α isomer)

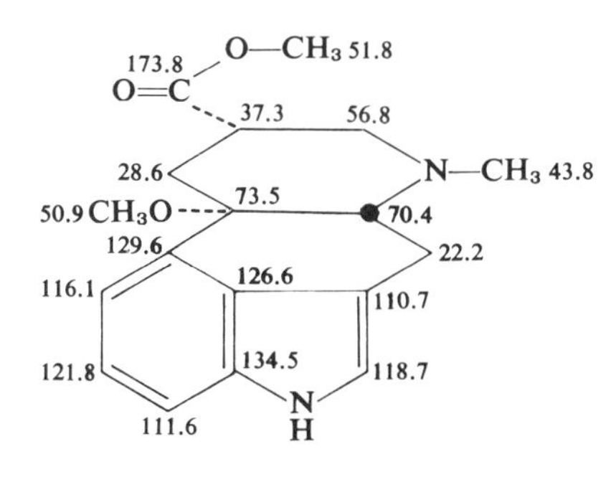

Ref. 107

592 **10-Methoxydihydrolysergamide (8β,10α isomer)**

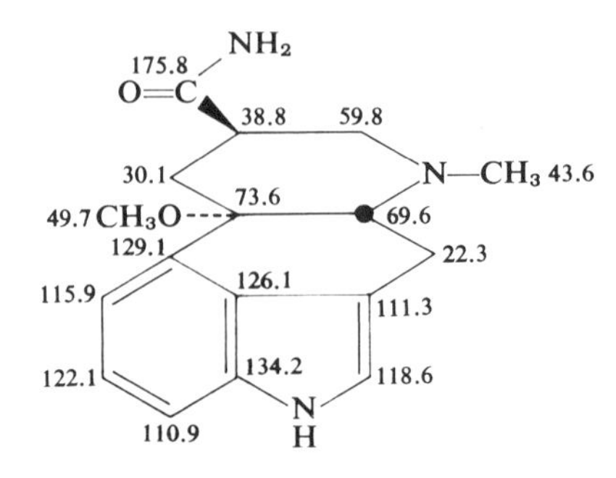

Ref. 106

593 **Methyl 10-methoxydihydrolysergate (8β,10α isomer)**

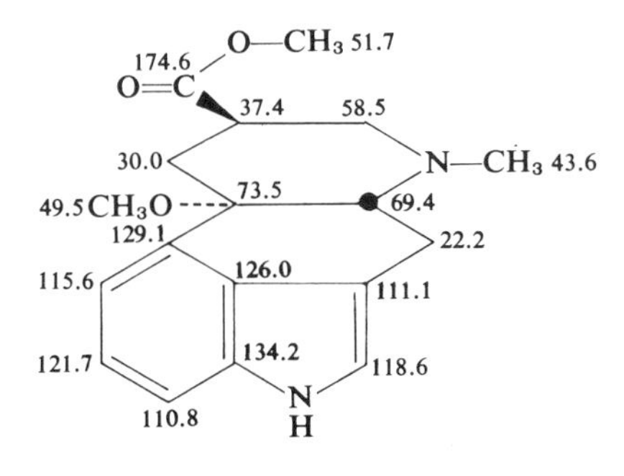

Ref. 107

594 **10-Methoxydihydrolysergamide (8α,10β isomer)**

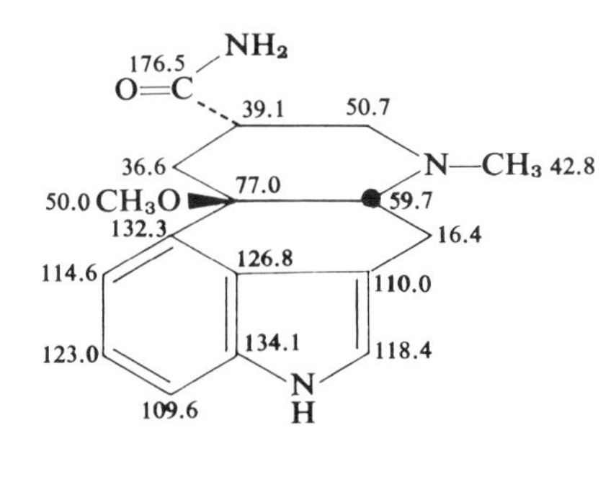

Ref. 106

595 **Methyl 10-methoxydihydrolysergate (8α,10β isomer)**

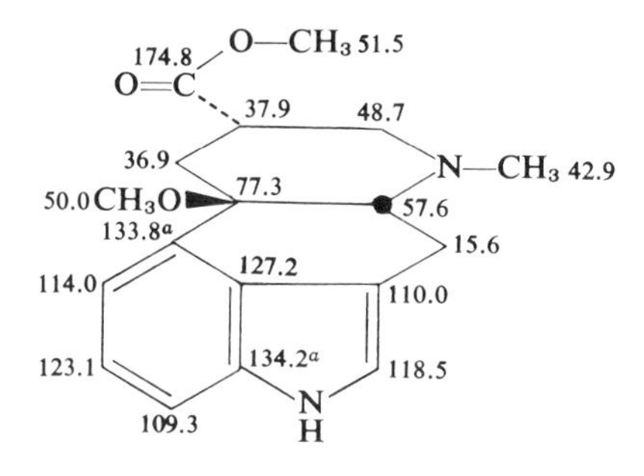

Ref. 107

596 **10-Methoxydihydrolysergamide (8β,10β isomer)**

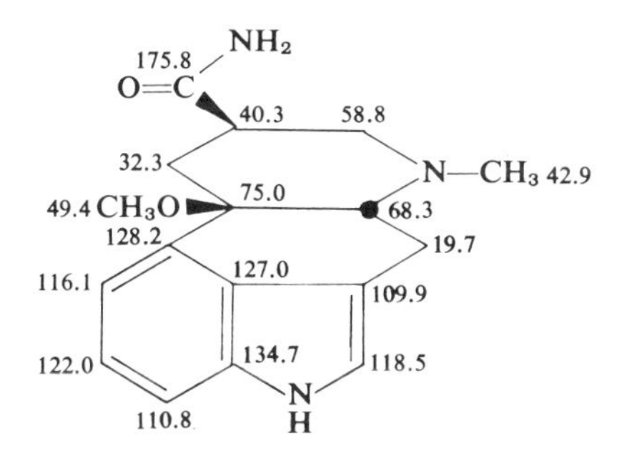

Ref. 106

597 **Methyl 10-methoxydihydrolysergate (8β,10β isomer)**

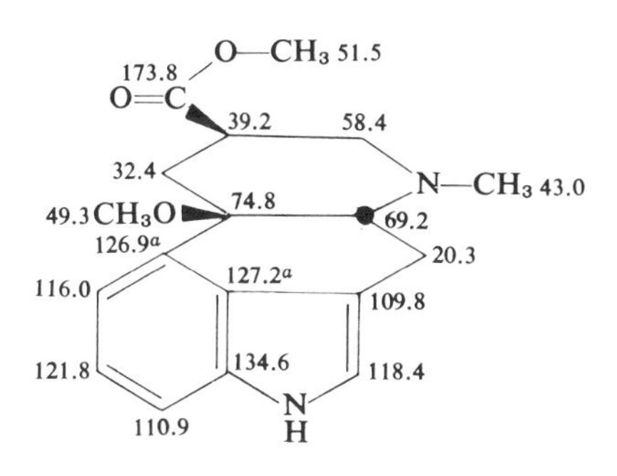

Ref. 107

598 **Ergonovine**
(DMSO-d_6)

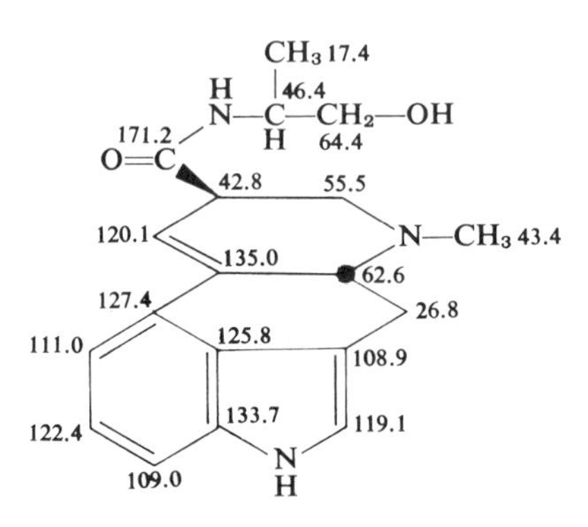

Ref. 25

599 **Ergonovinine**
(DMSO-d_6)

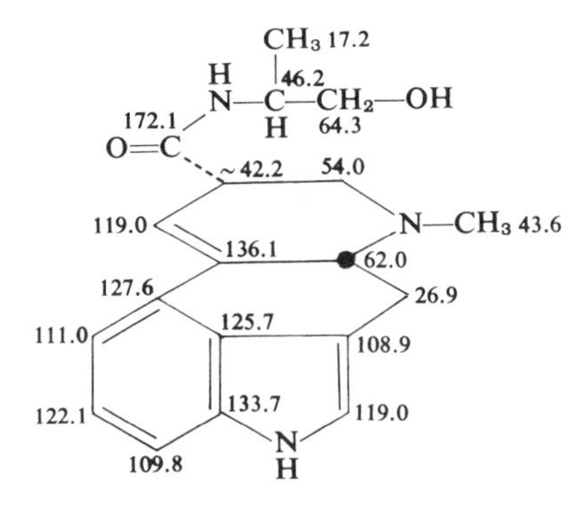

Ref. 25

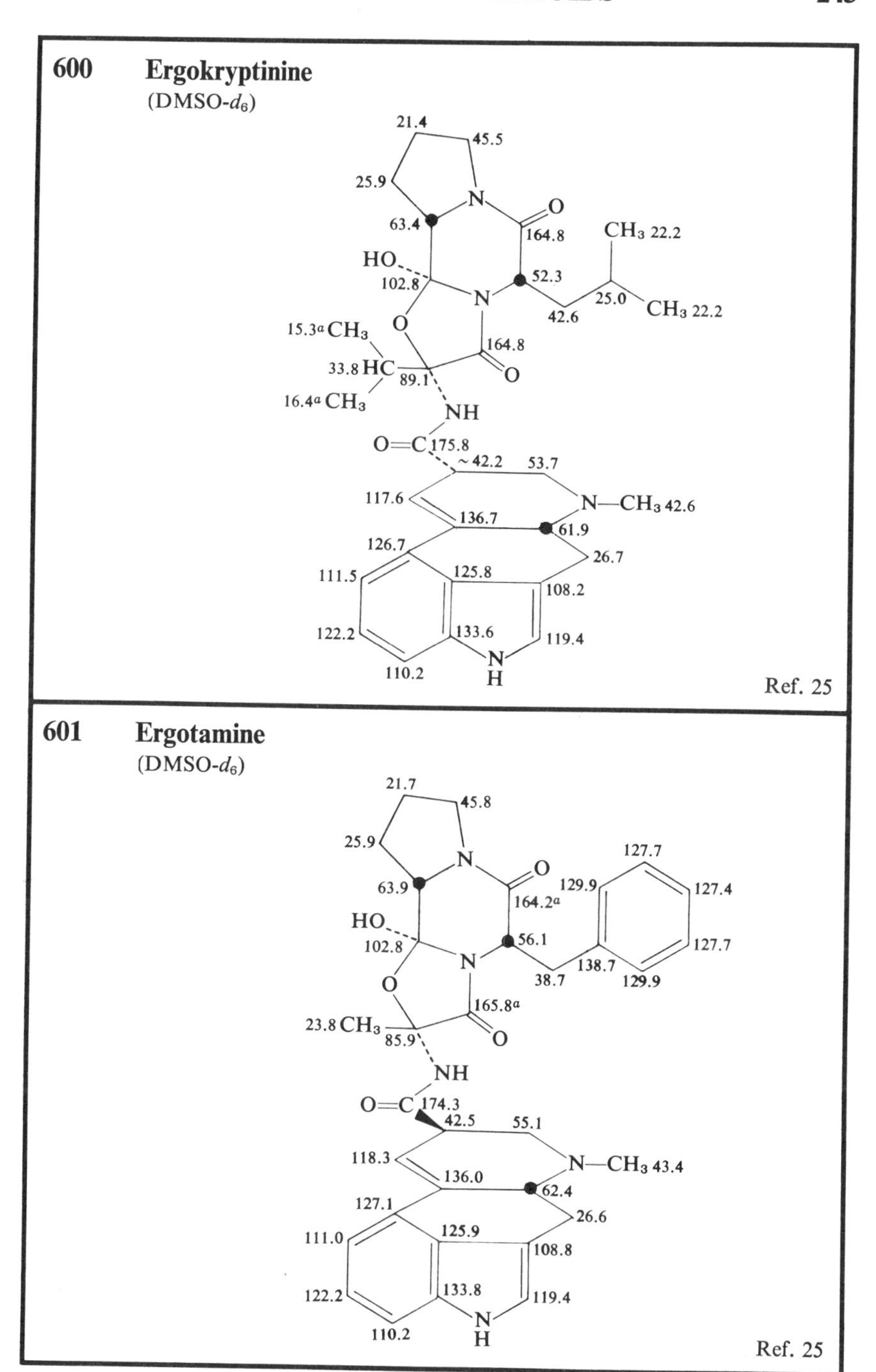
600 Ergokryptinine
(DMSO-d6)
21.4
45.5
25.9
N
O
63.4
164.8
CH3 22.2
HO
102.8
52.3
N
25.0
CH3 22.2
42.6
15.3a CH3
O
164.8
33.8 HC
89.1
O
16.4a CH3
NH
O=C 175.8
~42.2
53.7
117.6
N—CH3 42.6
136.7
61.9
126.7
26.7
111.5
125.8
108.2
122.2
133.6
119.4
N
H
110.2
Ref. 25
601 Ergotamine
(DMSO-d6)
21.7
45.8
25.9
127.7
N
O
63.9
129.9
127.4
164.2a
HO
102.8
56.1
127.7
N
138.7
38.7
129.9
O
165.8a
23.8 CH3
85.9
O
NH
O=C 174.3
42.5
55.1
118.3
N—CH3 43.4
136.0
62.4
127.1
26.6
125.9
111.0
108.8
122.2
133.8
119.4
N
H
110.2
Ref. 25

602 Ergotaminine

(DMSO-d_6)

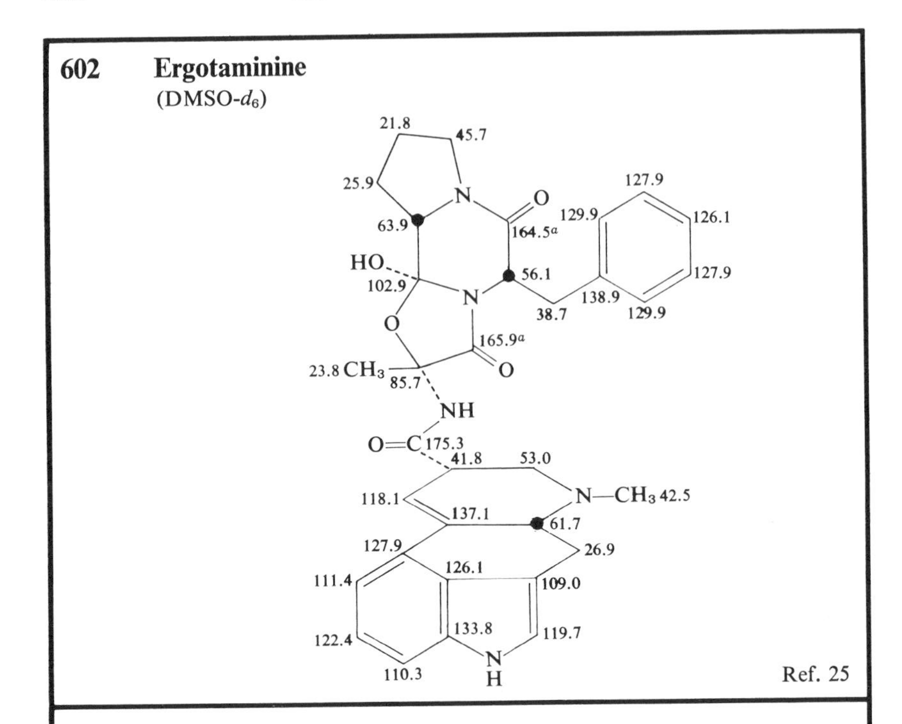

Ref. 25

18. REARRANGED (QUINOLINIC) INDOLE ALKALOIDS

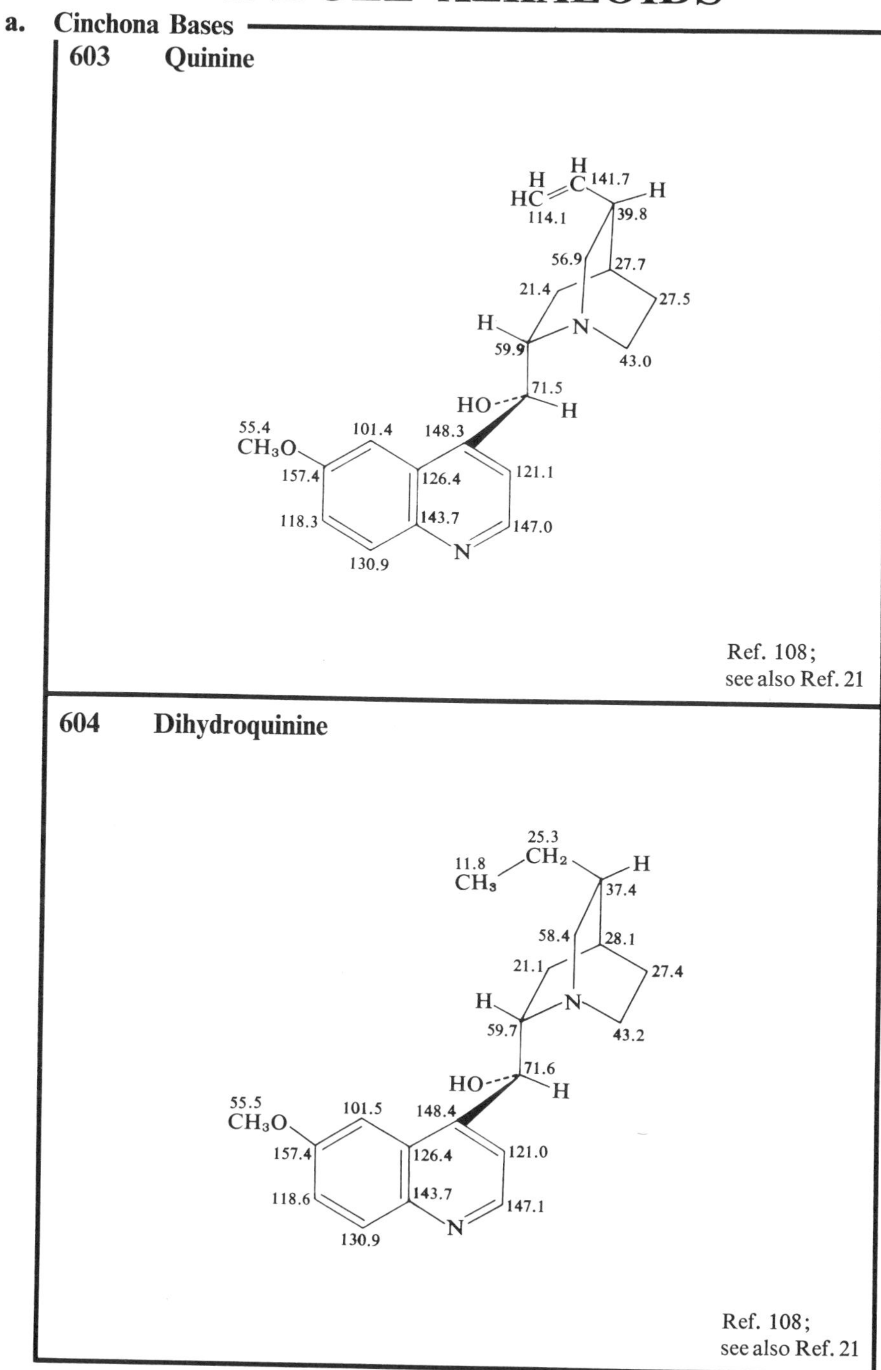

605 Epiquinine

H
H C 141.2 H
HC 39.6
114.3
55.3 27.76
24.9 27.1
H N
61.3 40.6
71.2
H OH
55.8 102.5 144.6[a]
CH_3O
157.3 128.0 121.0
119.9 144.6[a] 147.3
N
131.3

Ref. 108

606 Cinchonidine

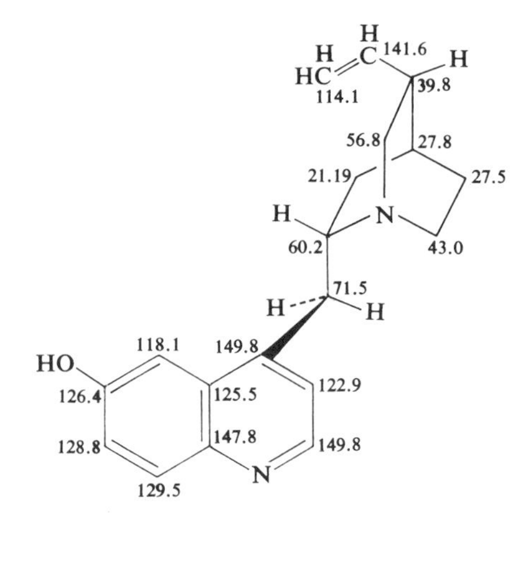

Ref. 108

607 Quinidine

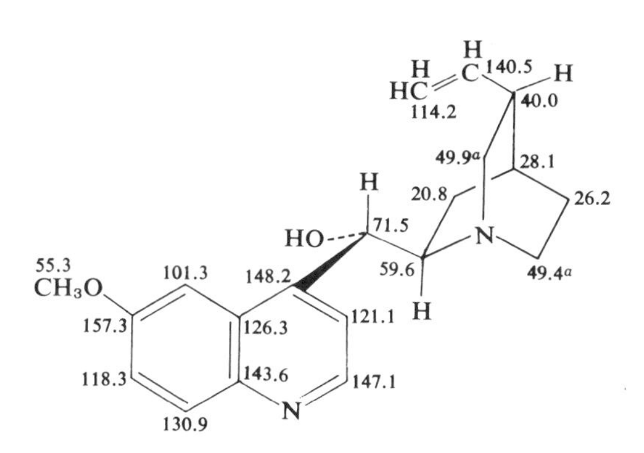

Ref. 108;
see also Ref. 21

608 Dihydroquinidine

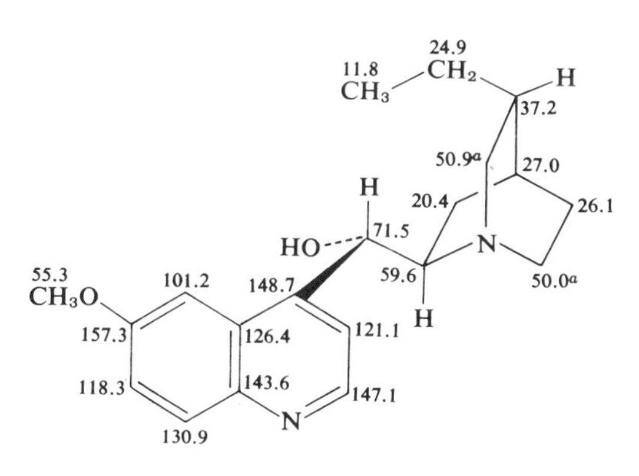

Ref. 108;
see also Ref. 21

609 Epiquinidine

Ref. 108

610 Dihydroepiquinidine

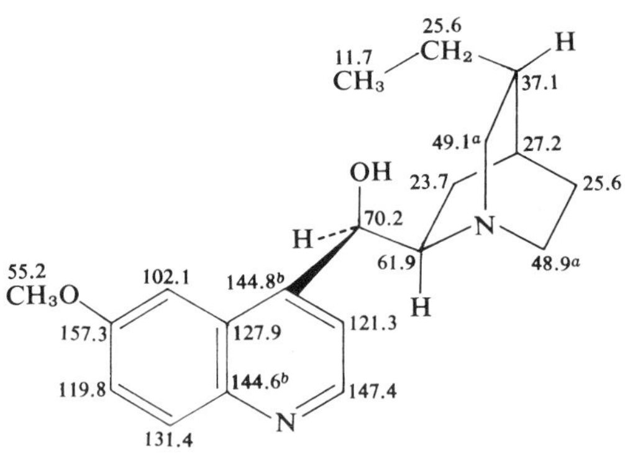

Ref. 108

611 (3S)-3-Hydroxyquinidine

(DMSO-d_6)

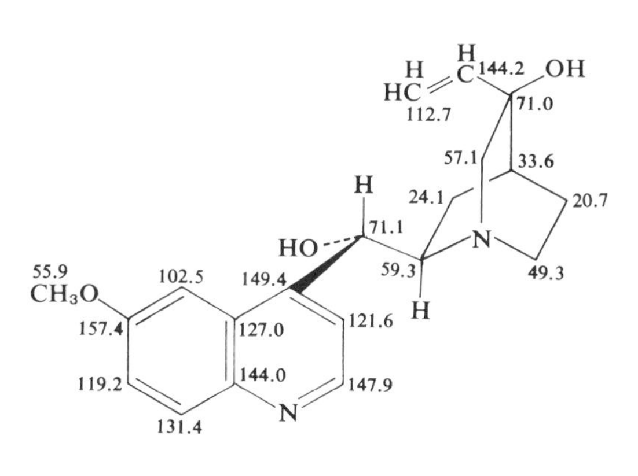

Refs. 109, 110

612 Metabolite of Quinidine

(DMSO-d_6)

H
H
C
141.4
H
HC
114.6
39.9
49.4[a]
28.0
H
22.7
26.5
71.5
N
HO
55.7
107.0
153.7
59.8
48.6[a]
CH3O
154.0
117.2
118.7[b] H
119.1[b]
133.6
161.7
O
N
H
119.1[b]

Ref. 109

b. Melodinus Bases

613 Scandine

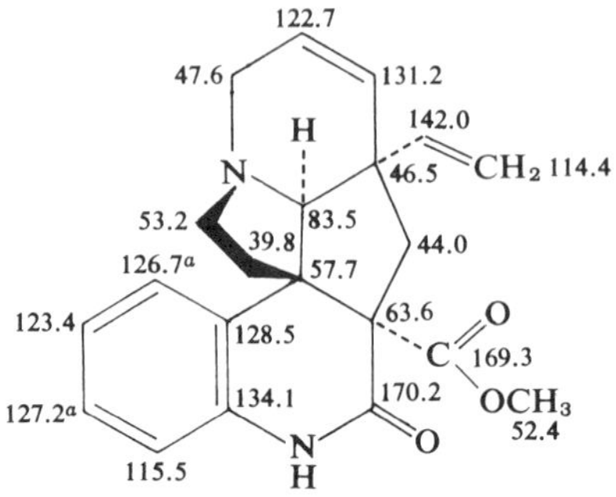

Ref. 111

614 Meloscine

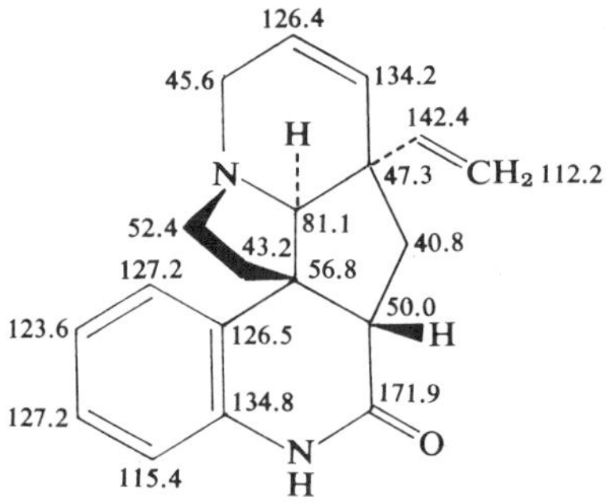

Ref. 111

615 Epimeloscine

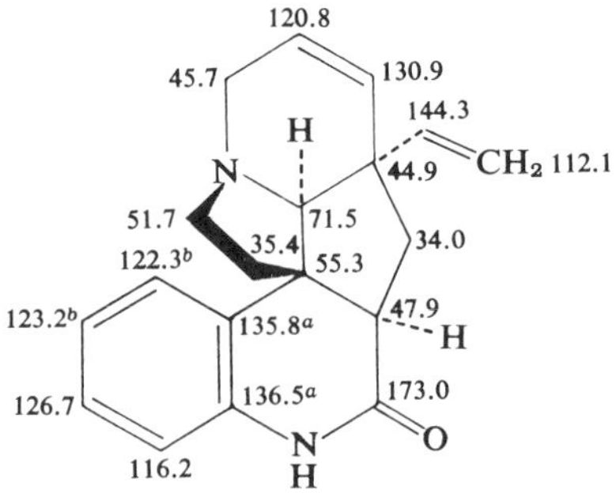

Ref. 111

616 Meloscandonine

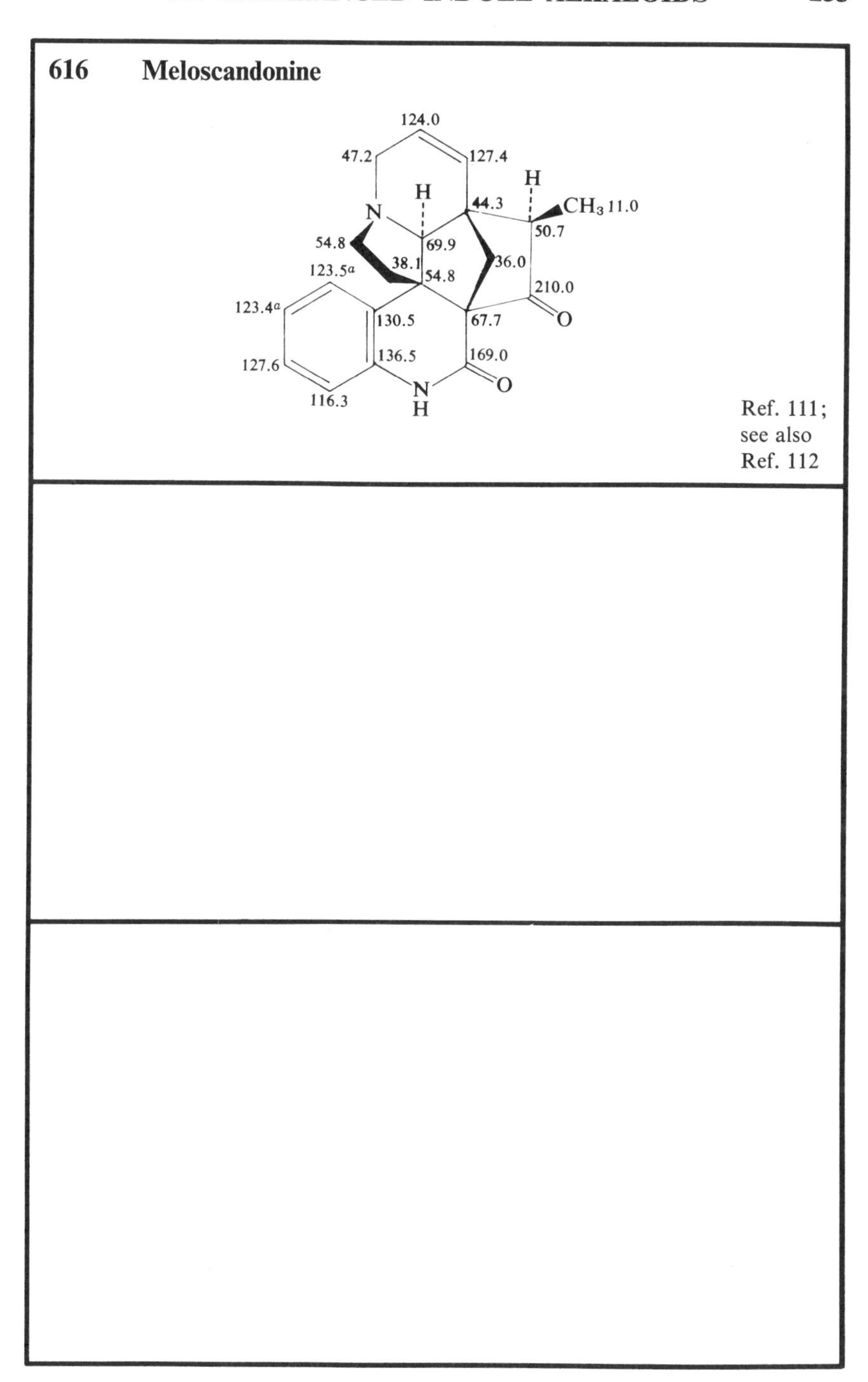

Ref. 111; see also Ref. 112

19. DIMERIC INDOLE ALKALOIDS[113]

617 **Ochrolifuanine A**

(the shifts of like carbons of the two indole units are undifferentiated)

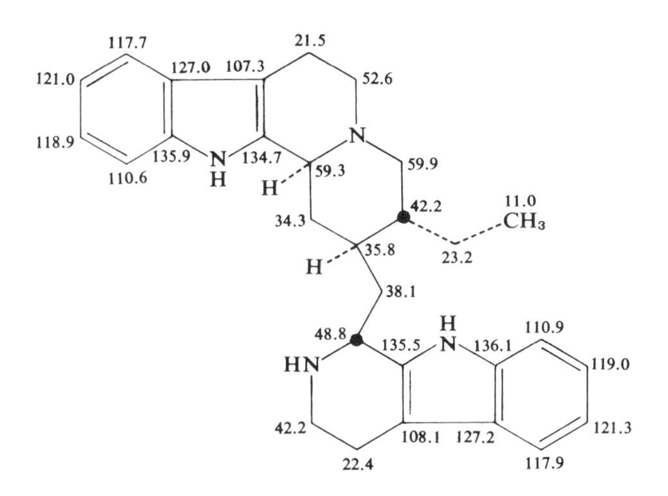

Ref. 114

618 **Ochrolifuanine B**

(the shifts of like carbons of the two indole units are undifferentiated)

117.7
21.6
120.9
127.0
107.3
52.9
N
118.9
135.8
N
134.6
60.1
110.8
H
H
59.5
42.5
11.2
CH3
36.4
37.8
23.8
H
38.4
H
H
110.8
135.5
N
135.9
HN
51.9
119.3
42.0
108.6
127.2
121.6
22.4
117.9

Ref. 114

619

(the shifts of like carbons of the two indole units are undifferentiated)

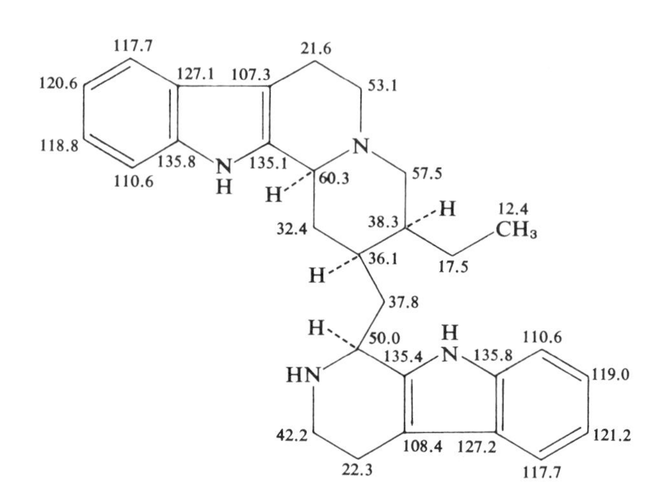

Ref. 114

620

(the shifts of like carbons of the two indole units are undifferentiated)

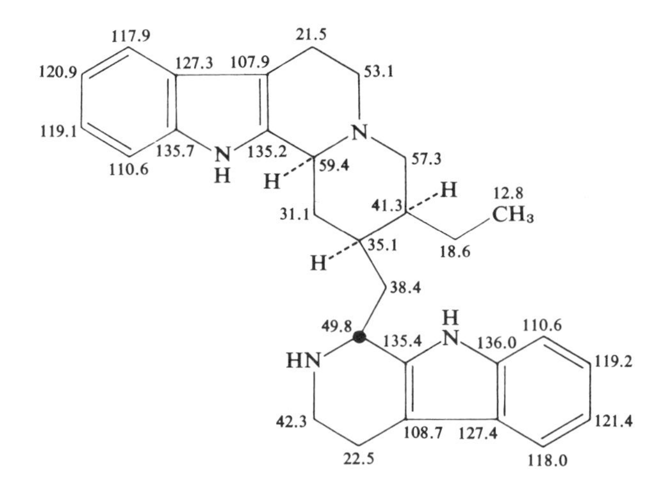

Ref. 114

621 Vincaleucoblastine

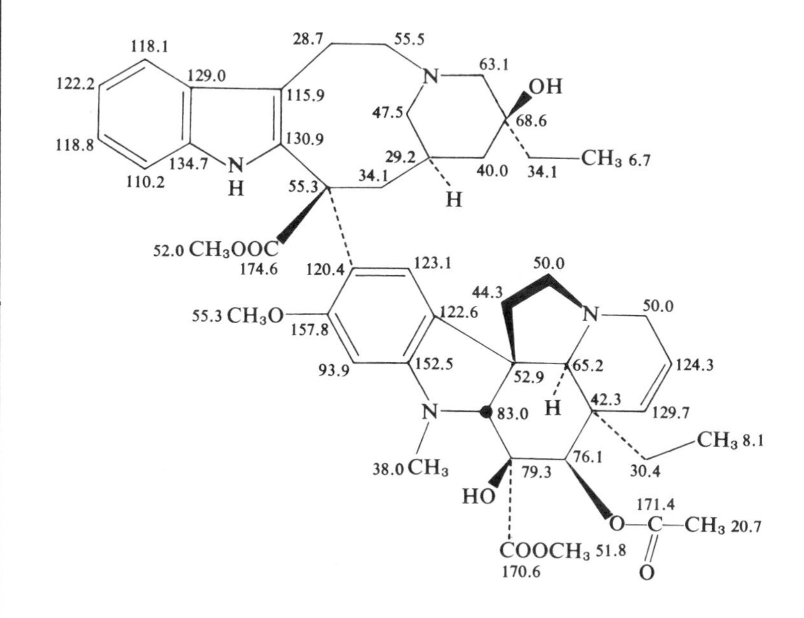
118.1
28.7
55.5
63.1
122.2
129.0
N
OH
115.9
47.5
68.6
118.8
130.9
29.2
CH3 6.7
134.7
N
40.0
34.1
110.2
H
55.3
34.1
H
52.0 CH3OOC
174.6
120.4
123.1
50.0
44.3
55.3 CH3O
157.8
122.6
N
50.0
93.9
152.5
52.9
65.2
124.3
N
83.0
H
42.3
129.7
CH3 8.1
38.0 CH3
79.3
76.1
30.4
HO
171.4
O—C—CH3 20.7
COOCH3 51.8
170.6
O

Ref. 113f

622 Villalstonine

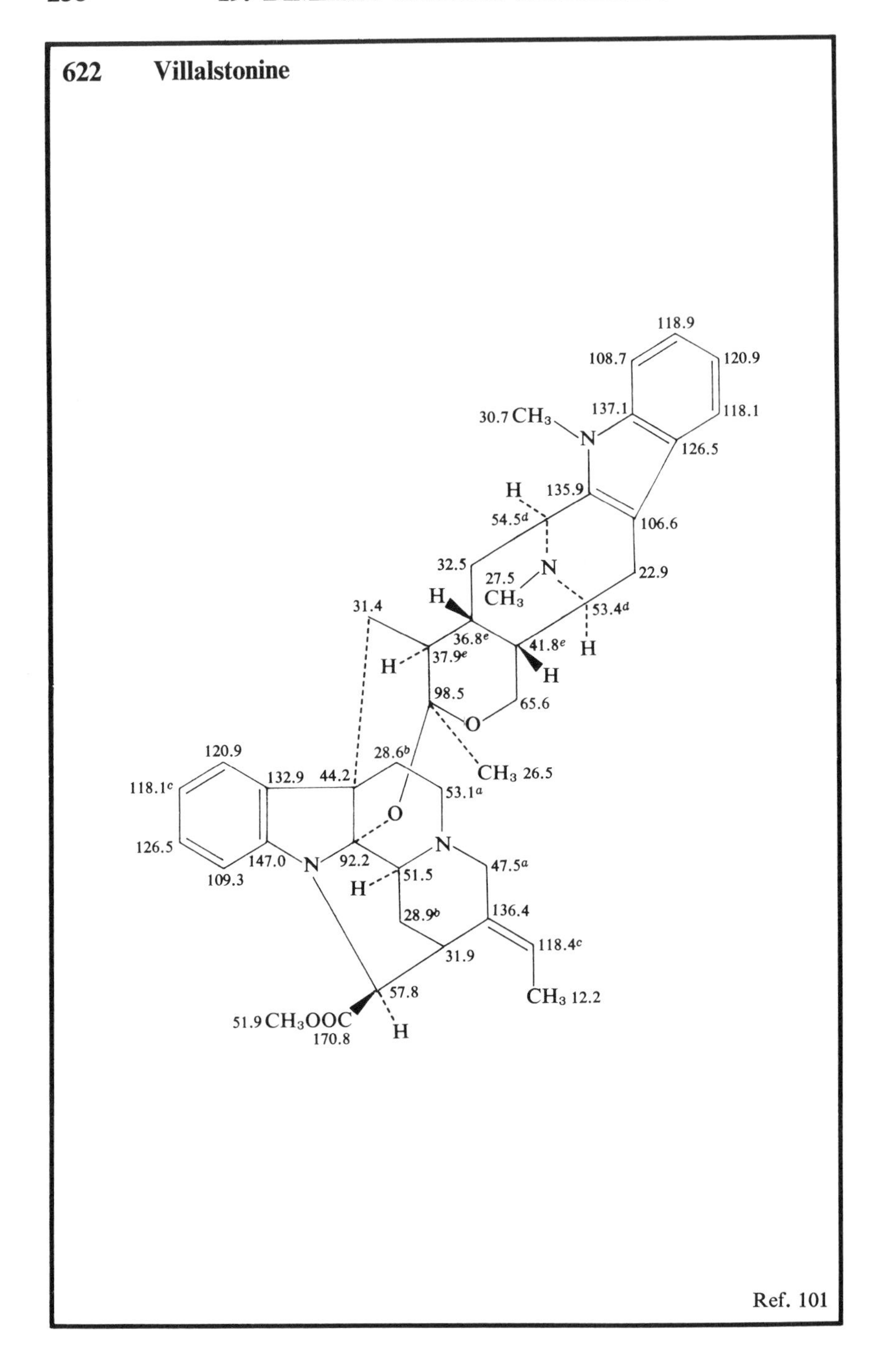

Ref. 101

623 Macralstonidine

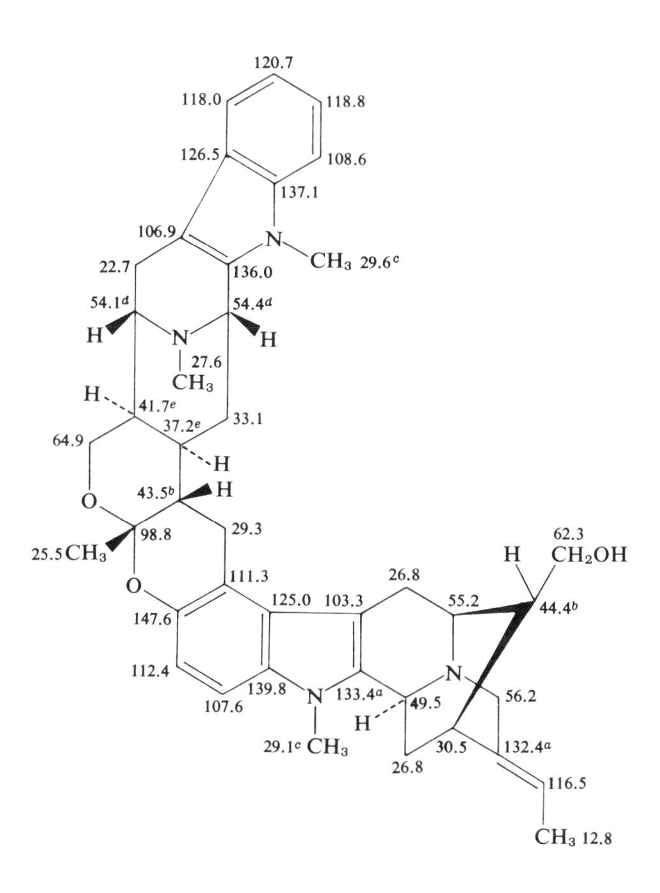

Ref. 101

624 Anhydrovobtusine

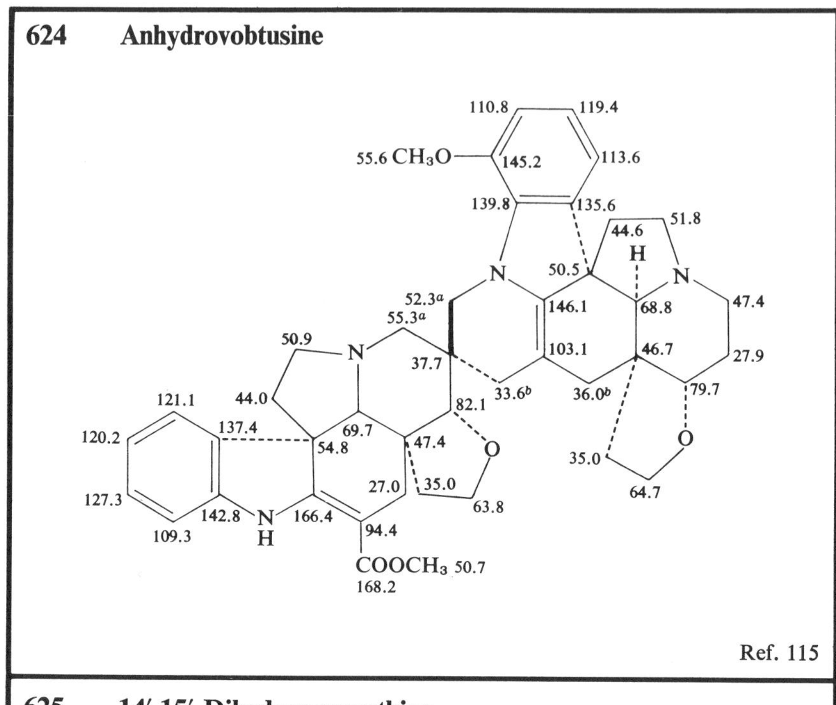

Ref. 115

625 14′,15′-Dihydropycnanthine

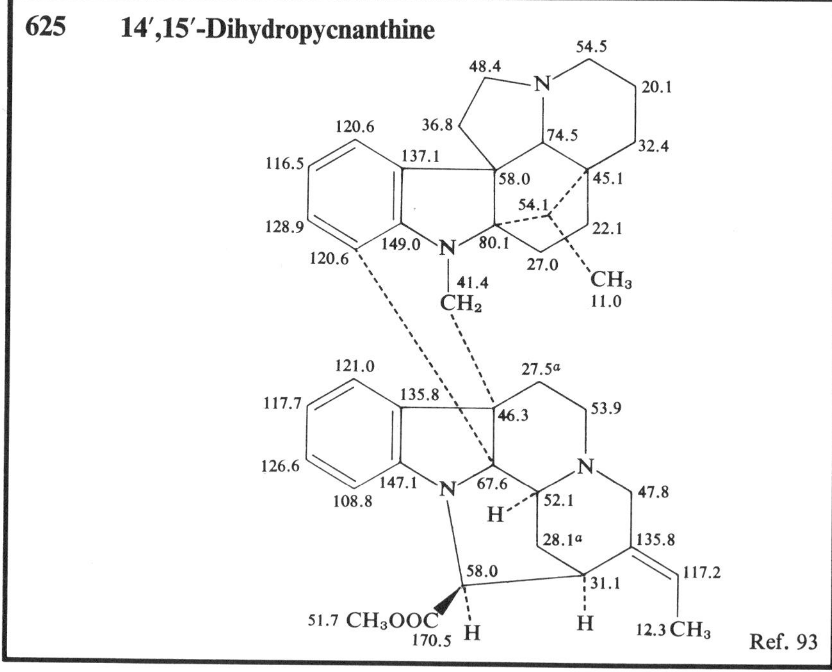

Ref. 93

626 Pleiocorine

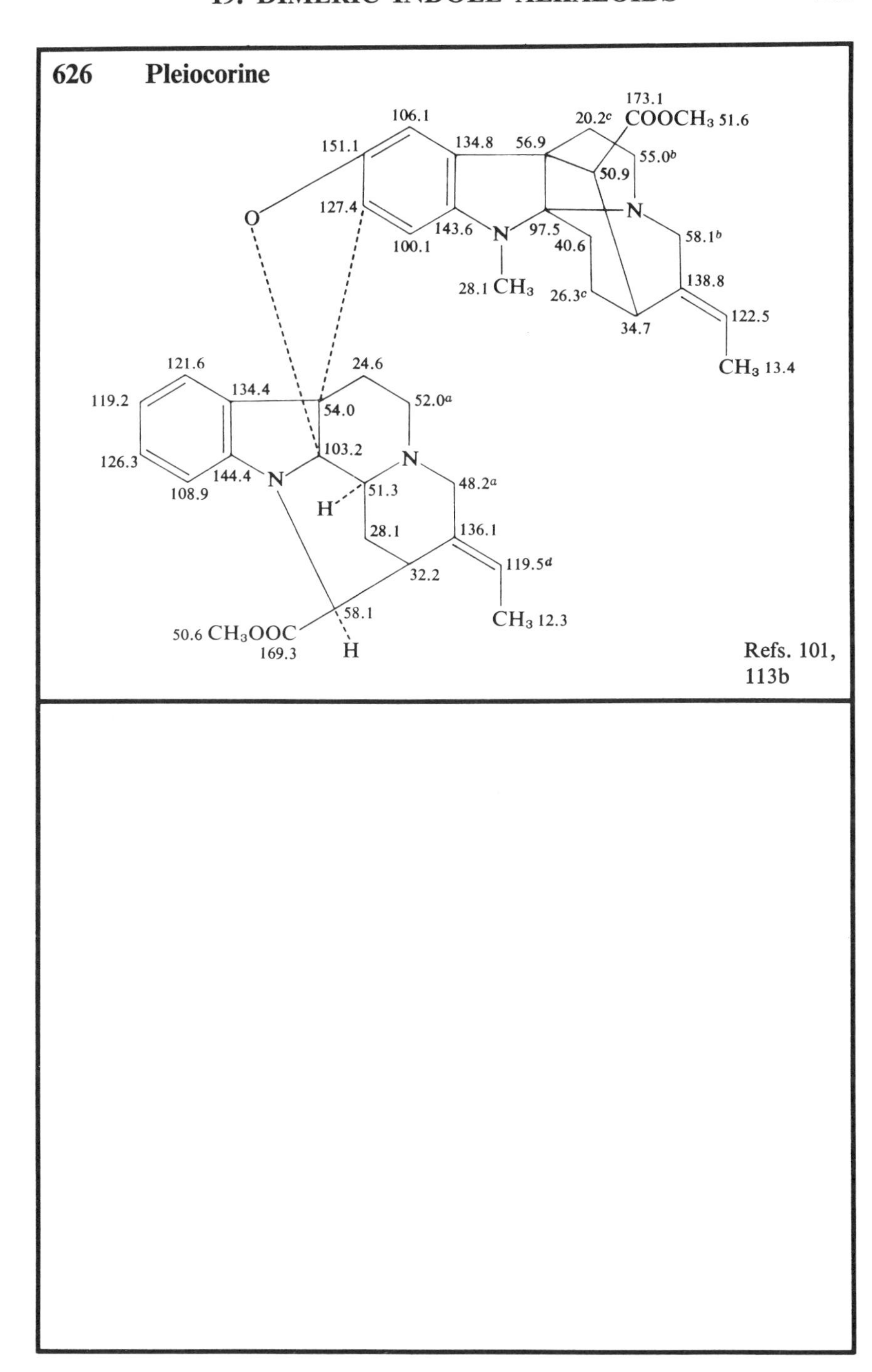

Refs. 101, 113b

20. YUZURIMINE ALKALOIDS

627 Yuzurimine

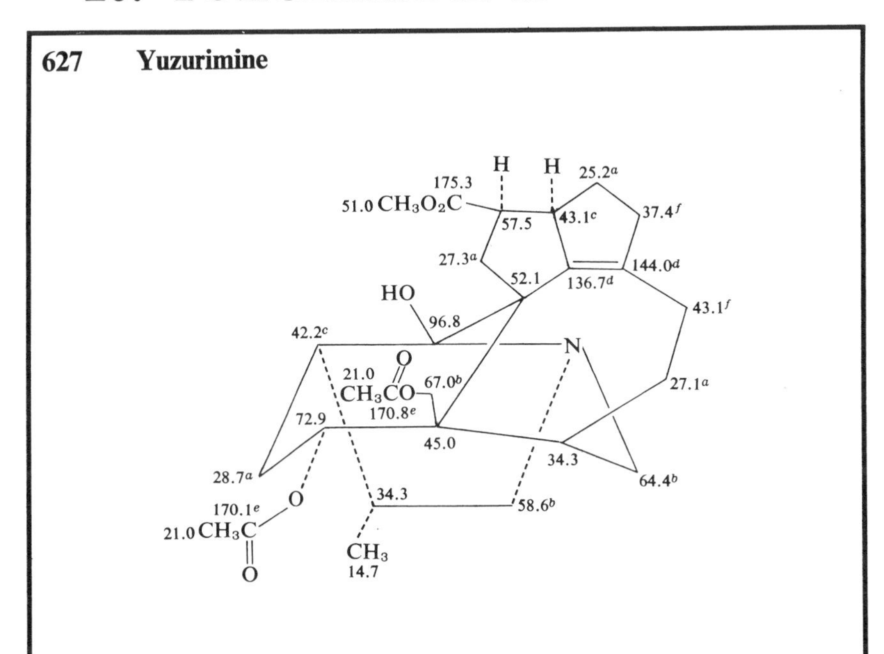

Ref. 116

628

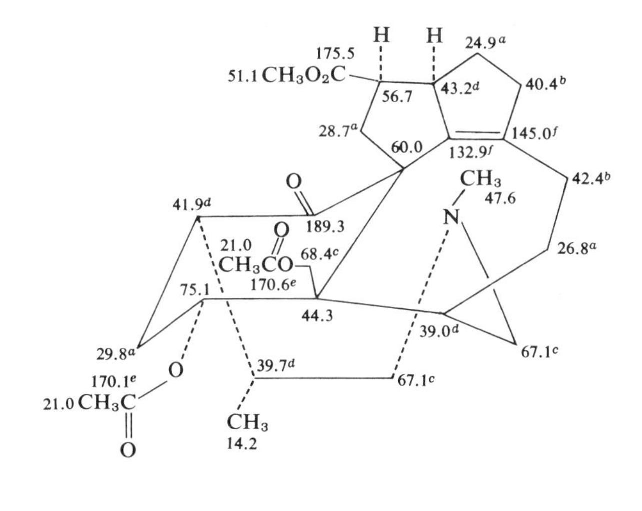

Ref. 116

629

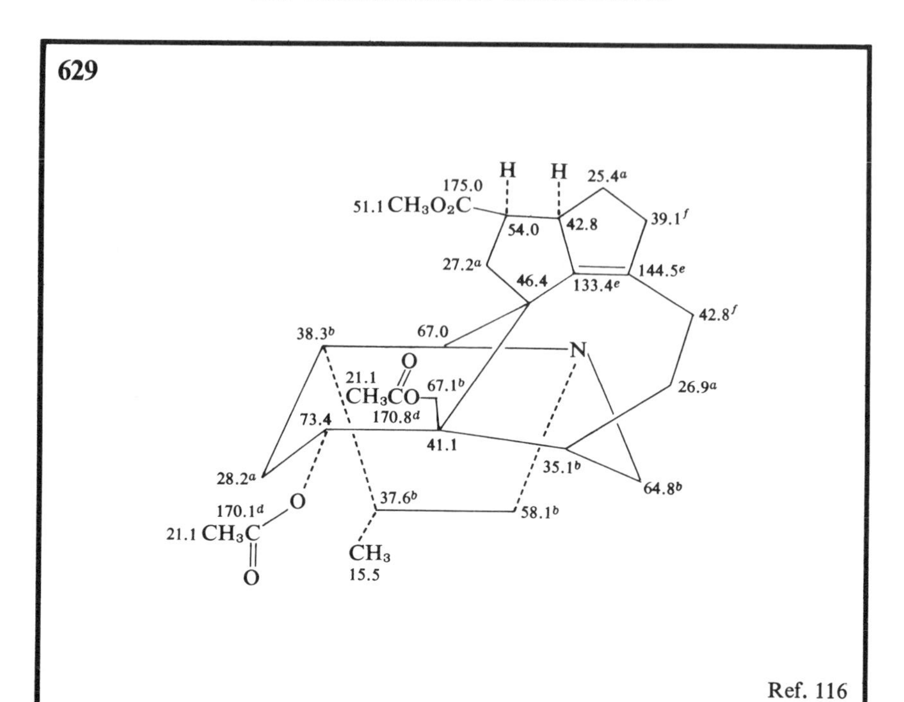

Ref. 116

630

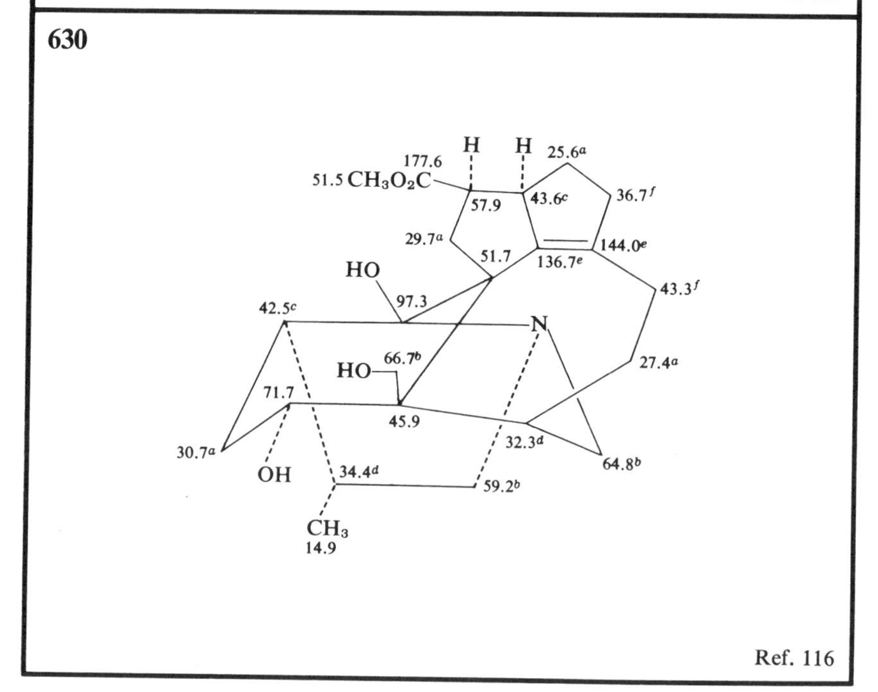

Ref. 116

631 Yuzurimine-C

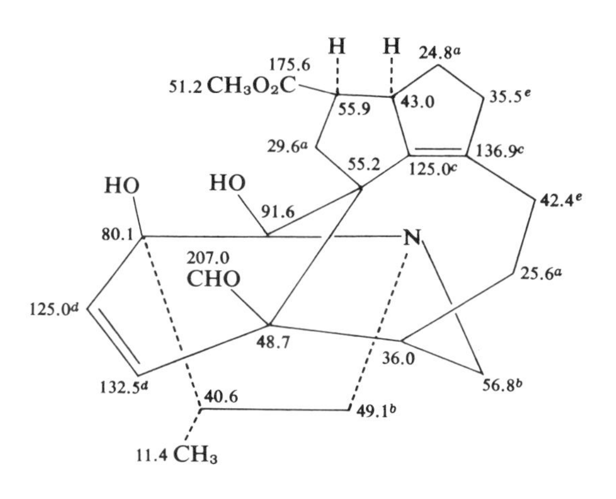

Ref. 116

632

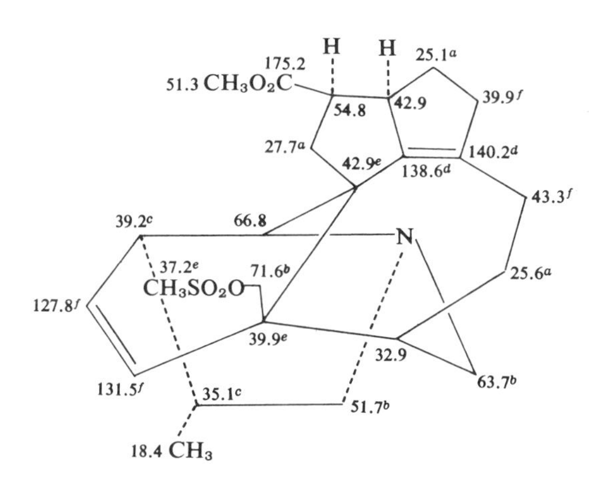

Ref. 116

21. DITERPENOID ALKALOIDS[117]

633 Atisine

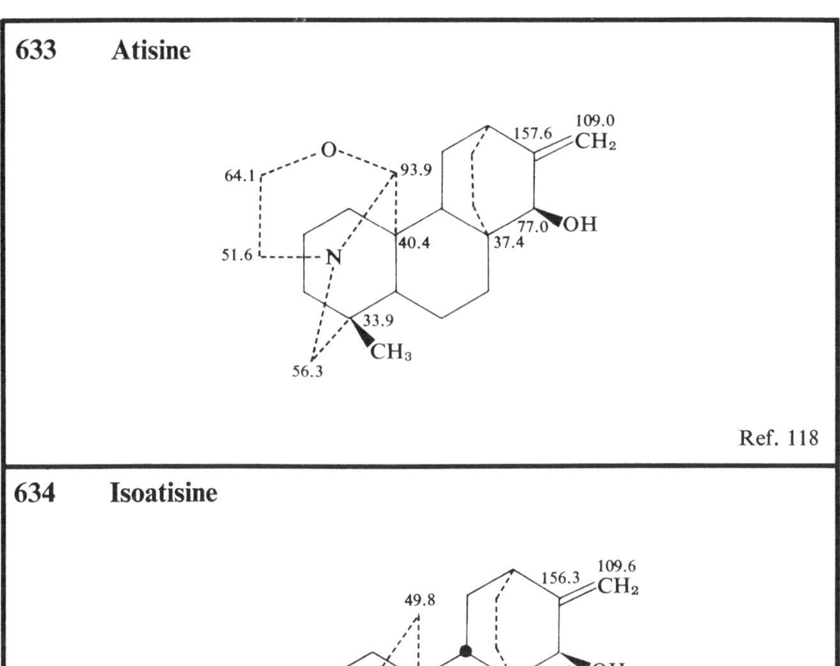

Ref. 118

634 Isoatisine

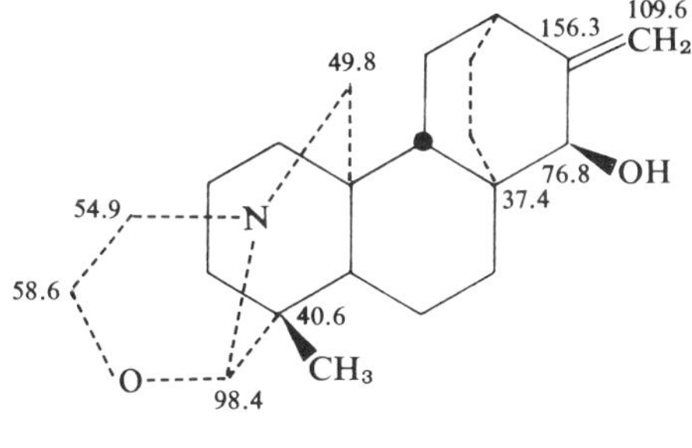

Ref. 118

635 Veatchine

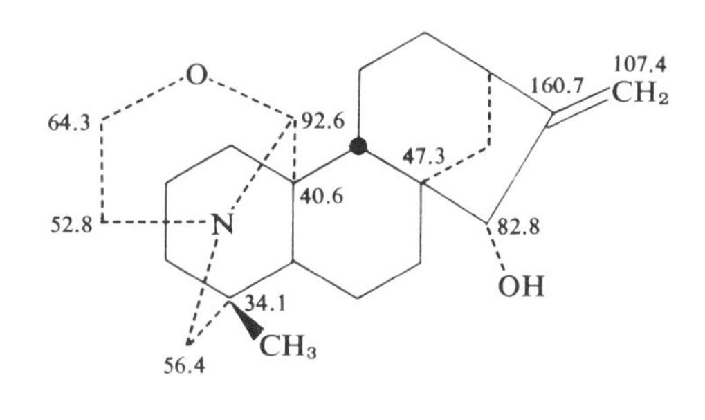

Ref. 118

636 Garryine

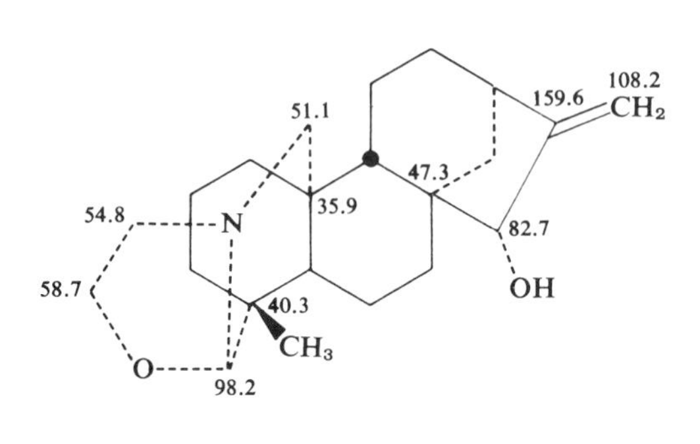

Ref. 118

637 Heterasine

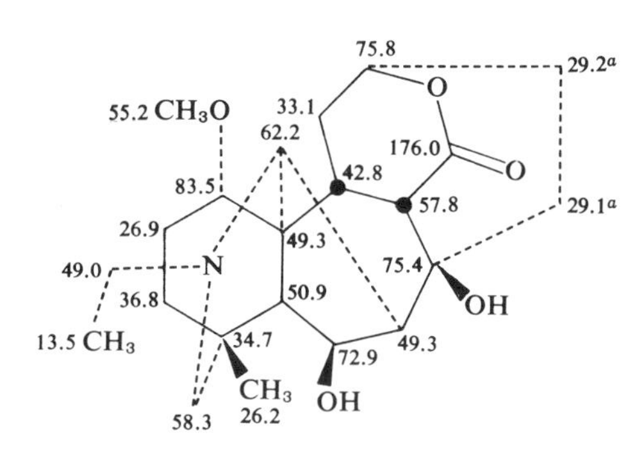

Ref. 119

638 Neoline

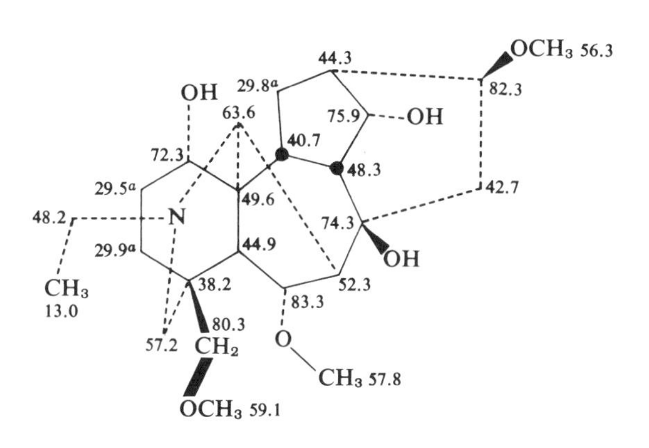

Ref. 119

639 Chasmanine

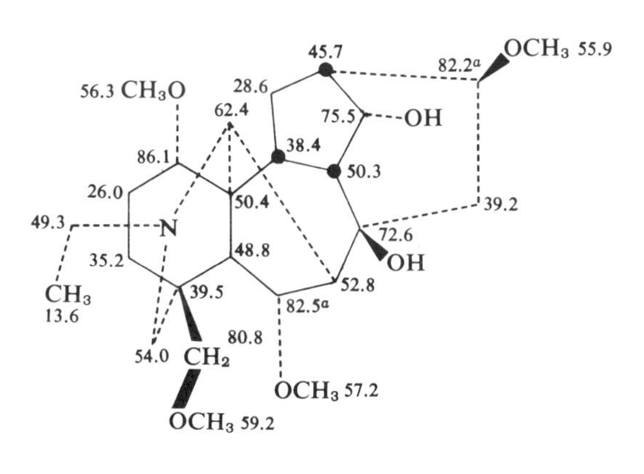

Ref. 120

640 **Alkaloid A from *Delphinium bicolor***

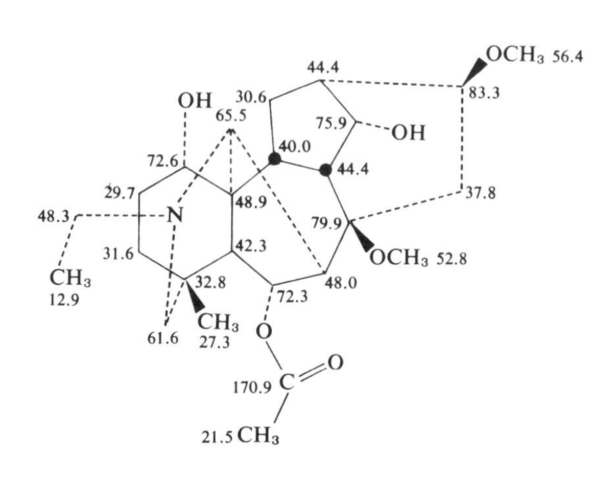

Ref. 119

641 **Delphisine**

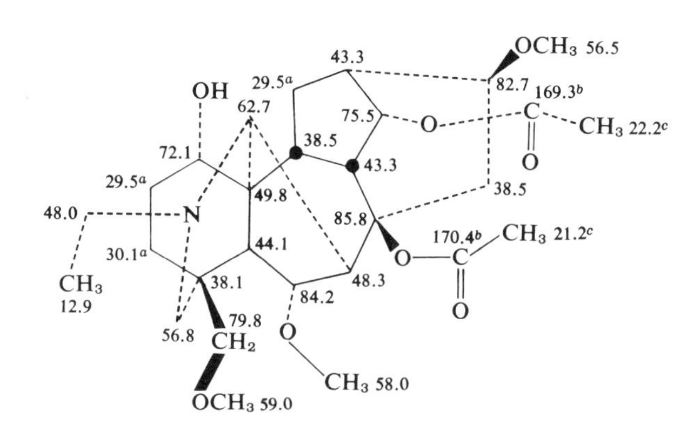

Ref. 119

642 Aconitine

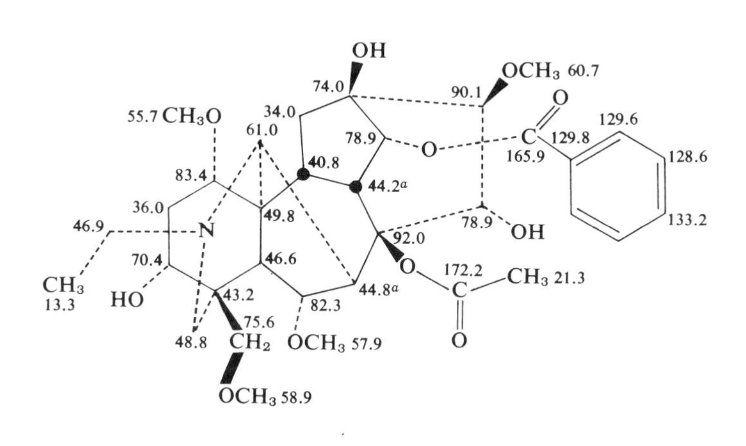

Ref. 120

643 Anhydroaconitine

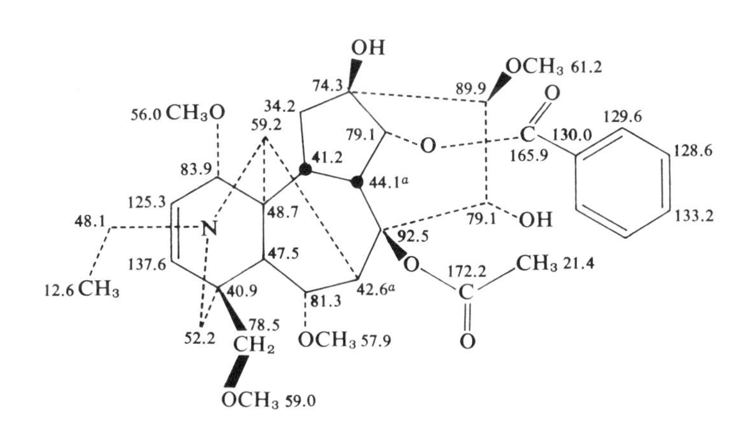

Ref. 120

644 Delcosine

645 Browniine

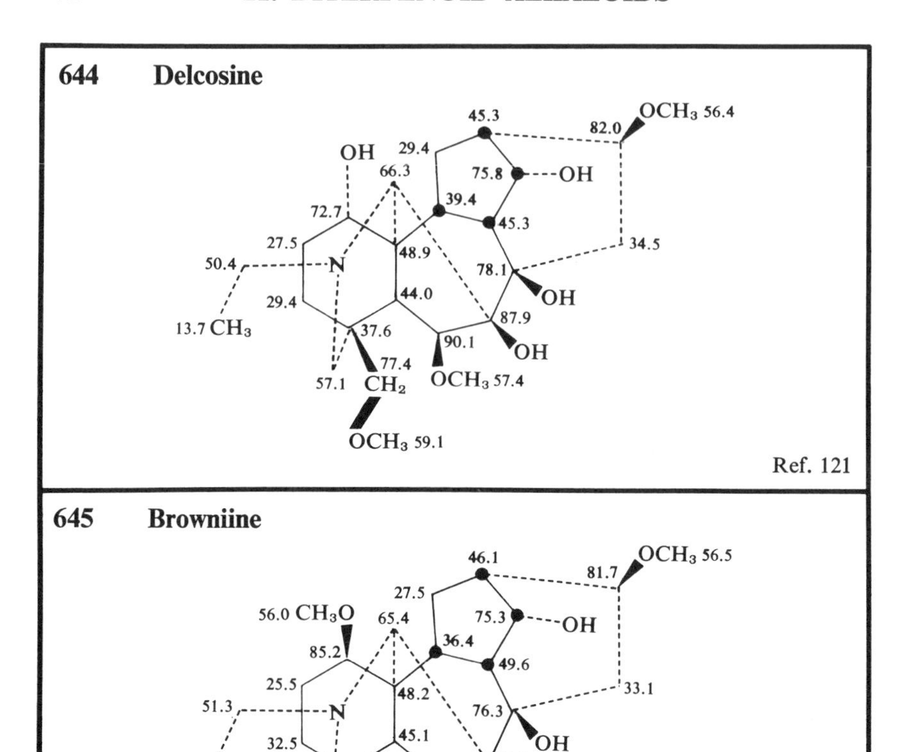

Ref. 121

Ref. 121

646 Lycoctonine

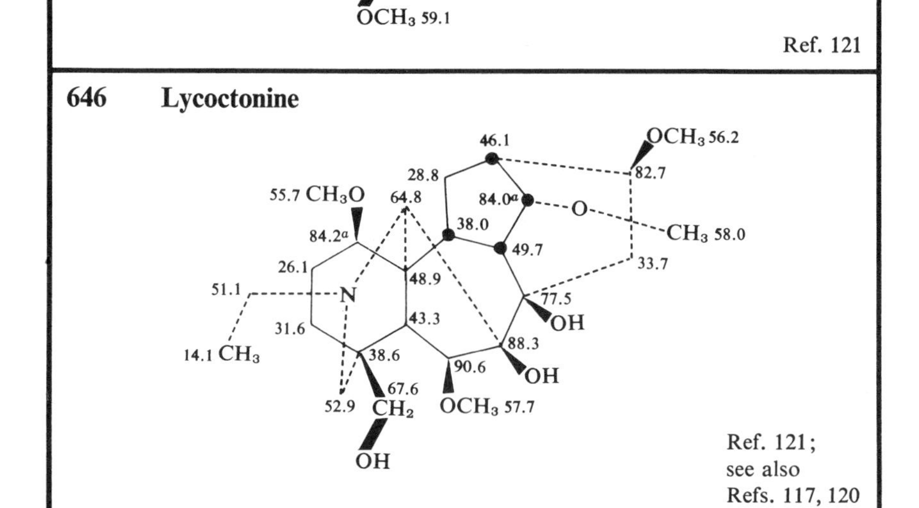

Ref. 121; see also Refs. 117, 120

22. STEROID ALKALOIDS

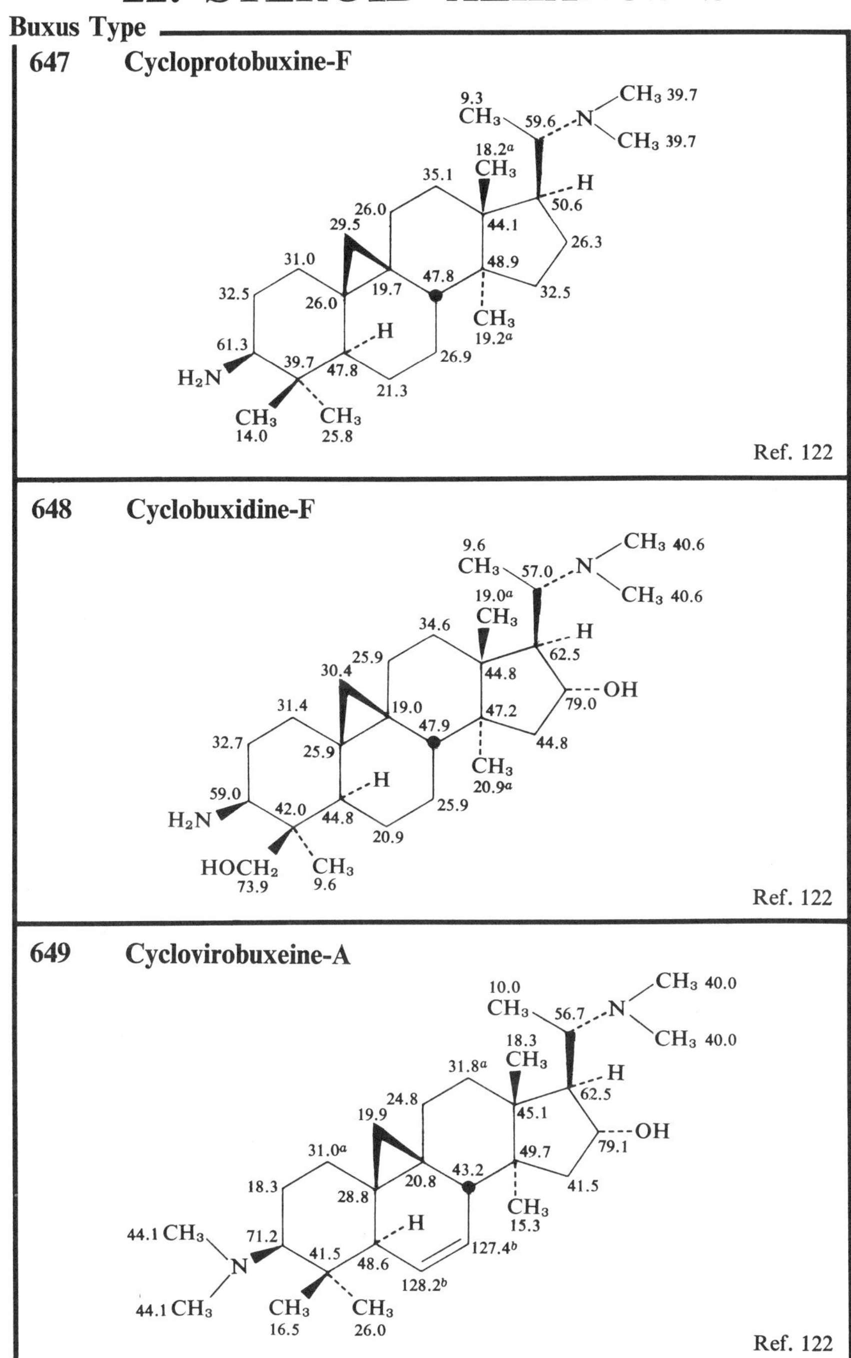

650 Cyclomicrophylline-B

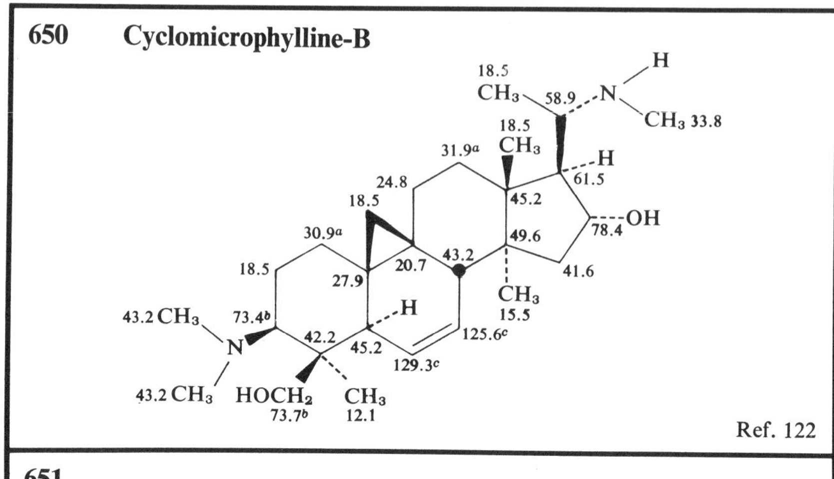

Ref. 122

651

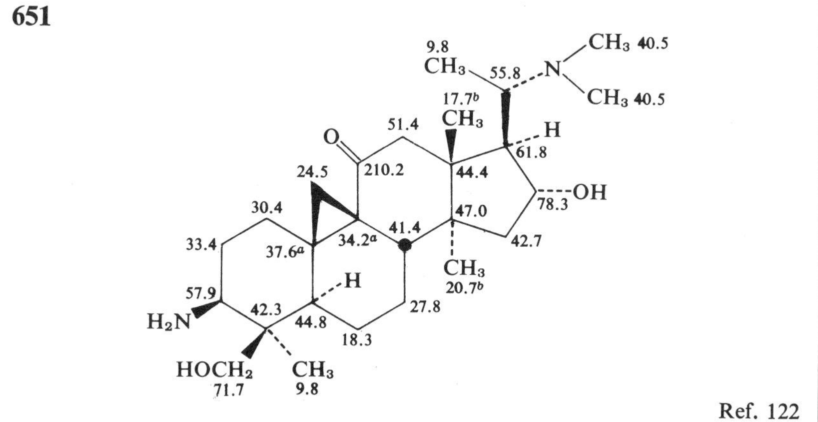

Ref. 122

652

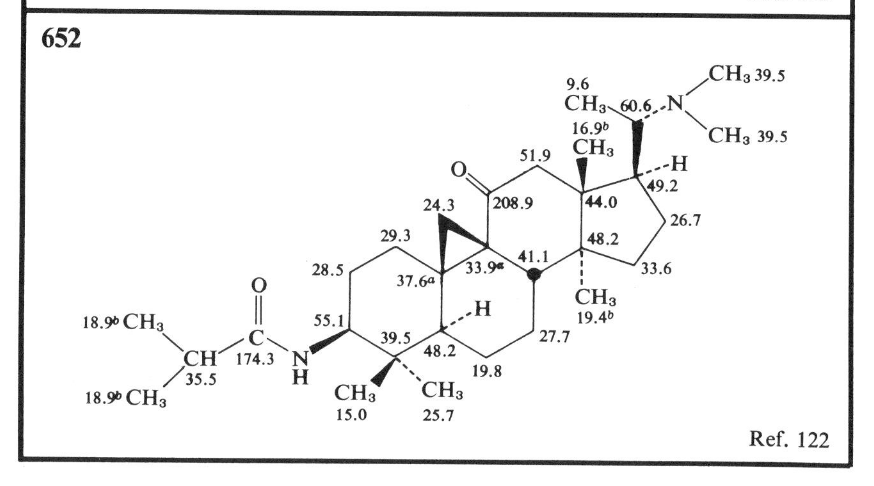

Ref. 122

653

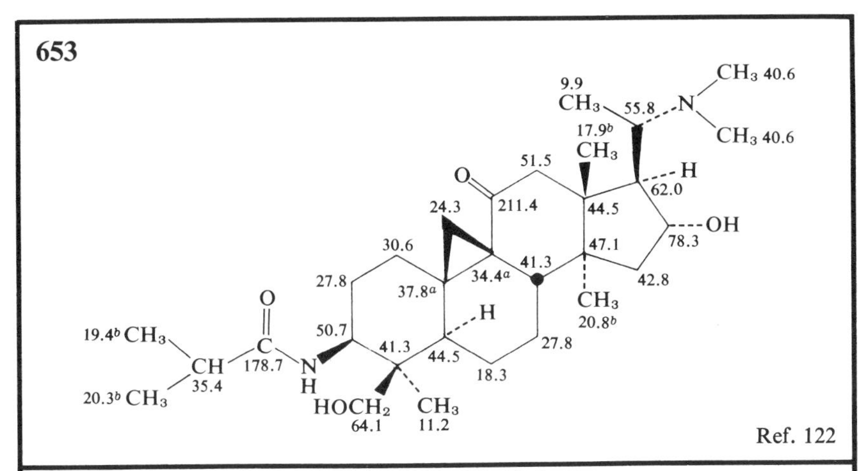

Ref. 122

654

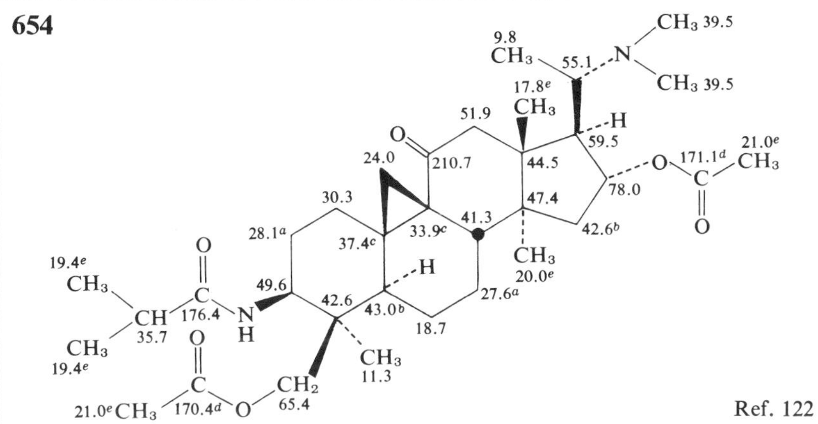

Ref. 122

655

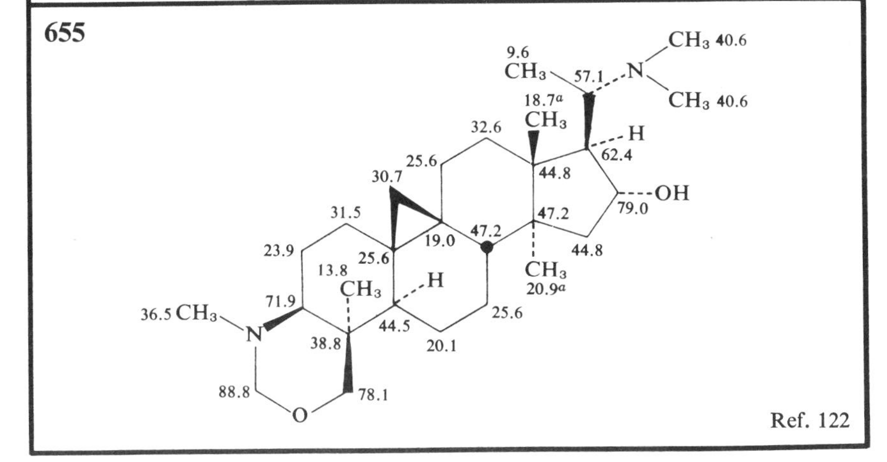

Ref. 122

656

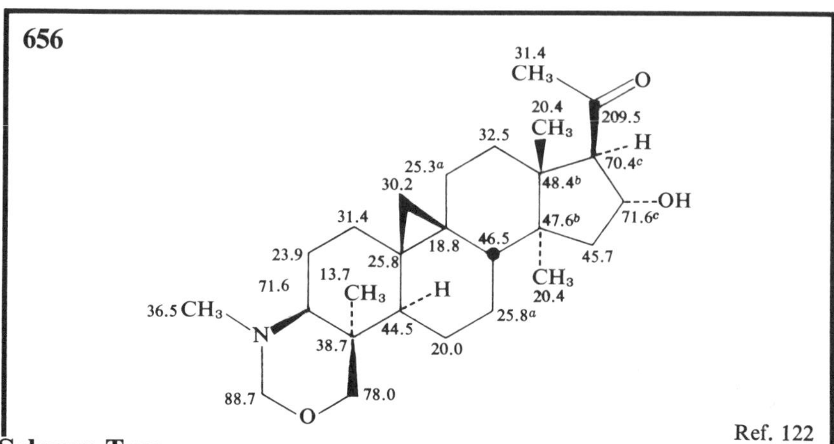

Ref. 122

b. Solanum Type

657 Solanidine

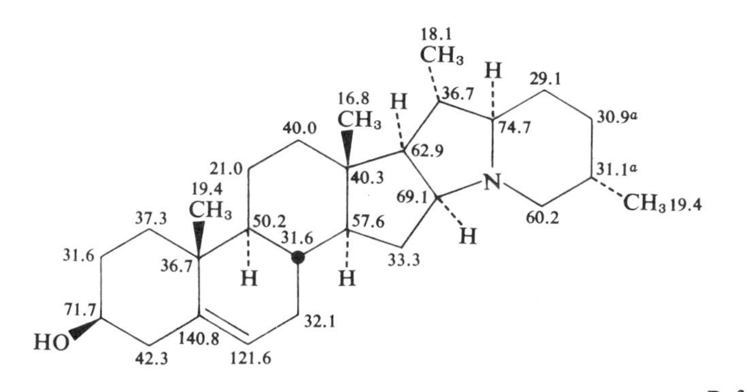

Ref. 123

658 Demissidine

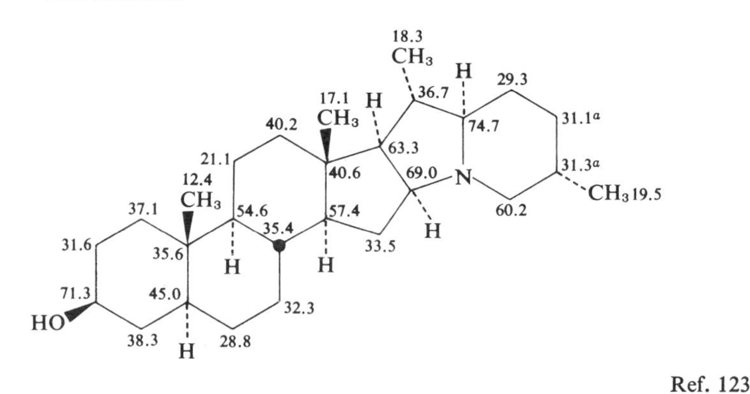

Ref. 123

659 Soladulcidine

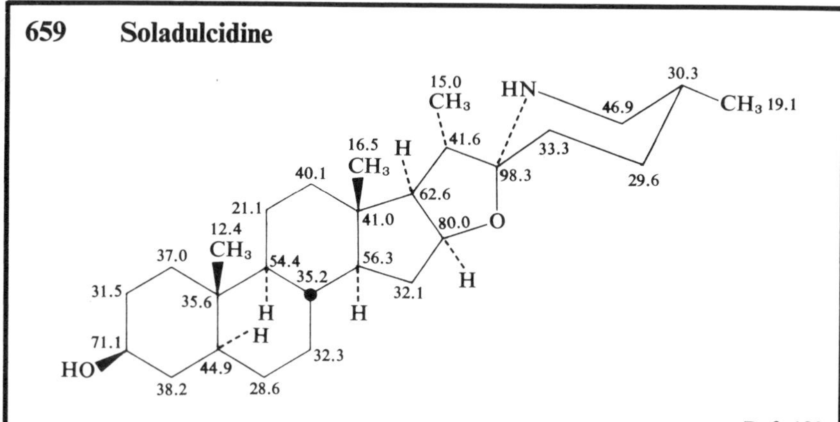

Ref. 123

660

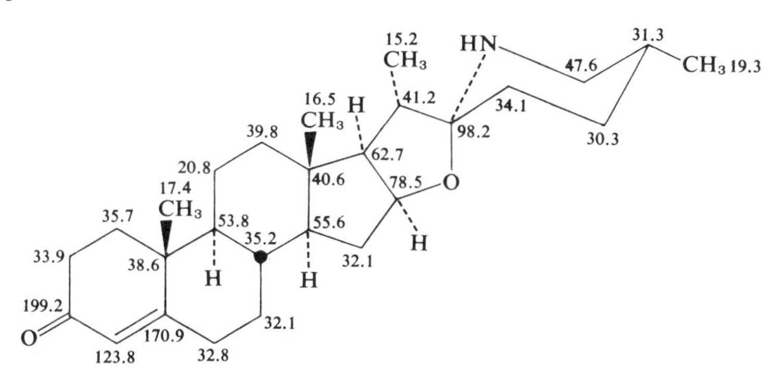

Ref. 123

661 Solasodine

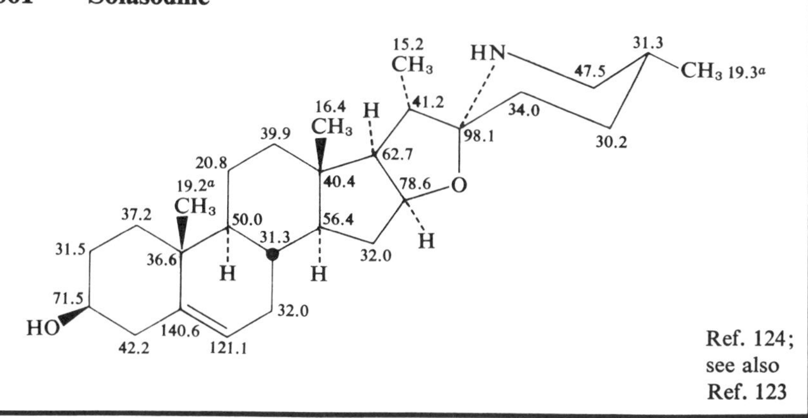

Ref. 124; see also Ref. 123

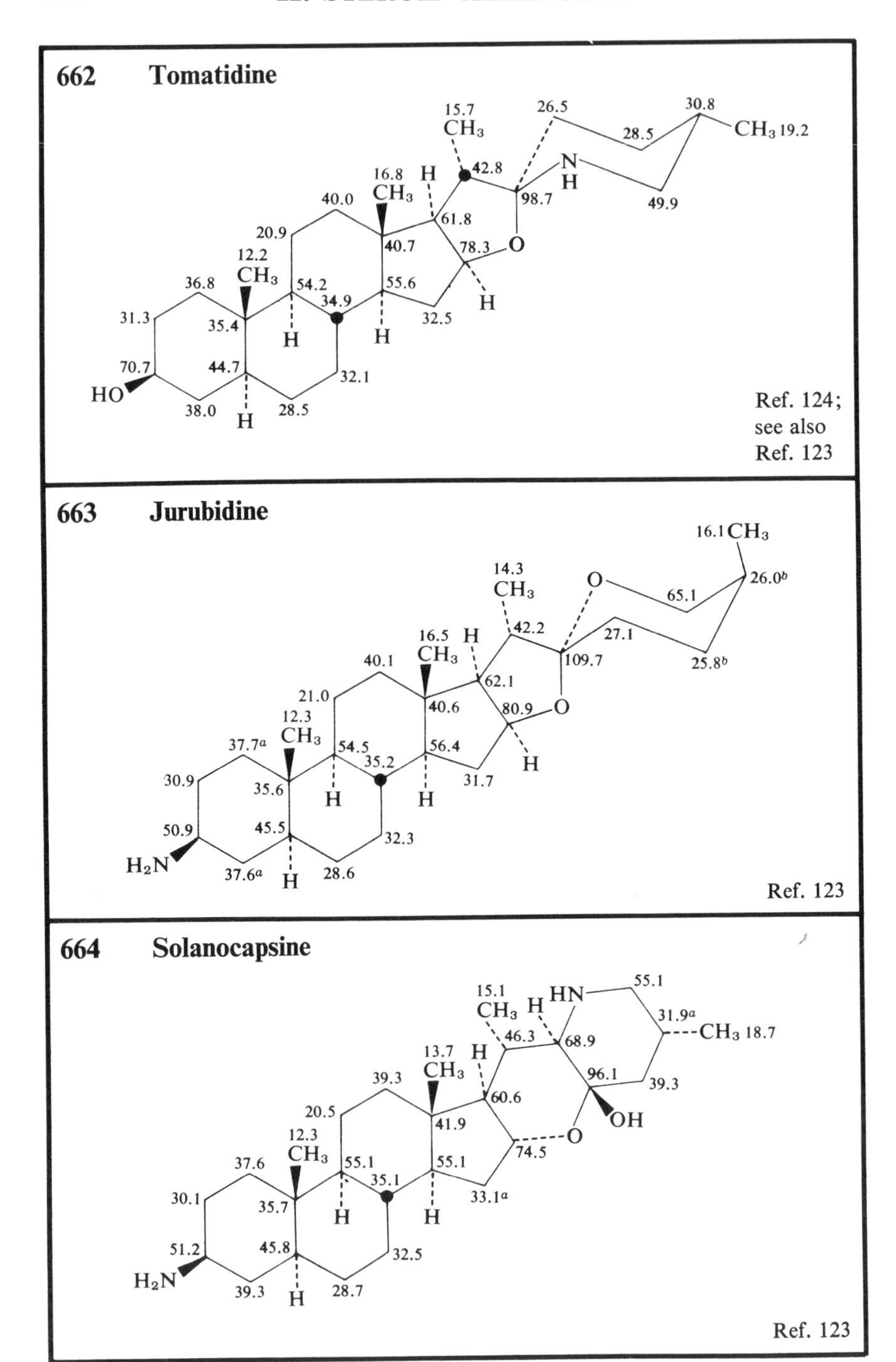
662 Tomatidine
15.7
CH_3
26.5
30.8
28.5
CH_3 19.2
42.8
16.8 H
CH_3
N
H
98.7
49.9
40.0
61.8
20.9
40.7
78.3 O
12.2
CH_3
36.8
54.2
55.6
34.9
H
31.3
35.4
32.5
H H
70.7
44.7
32.1
HO
38.0
H
28.5
Ref. 124;
see also
Ref. 123
663 Jurubidine
16.1 CH_3
14.3
CH_3
O
26.0[b]
65.1
42.2
27.1
16.5 H
CH_3
109.7
40.1
25.8[b]
62.1
21.0
40.6
80.9 O
12.3
CH_3
37.7[a]
54.5
56.4
35.2
H
30.9
35.6
31.7
H H
50.9
45.5
32.3
H_2N
37.6[a]
H
28.6
Ref. 123
664 Solanocapsine
15.1
CH_3 H
HN
55.1
31.9[a]
CH_3 18.7
46.3
68.9
13.7 H
CH_3
96.1
39.3
39.3
60.6
OH
20.5
41.9
O
12.3
CH_3
74.5
37.6
55.1
55.1
35.1
33.1[a]
30.1
35.7
H H
51.2
45.8
32.5
H_2N
39.3
H
28.7
Ref. 123

23. ADDENDA

665 **Europine *N*-oxide**
(D_2O)

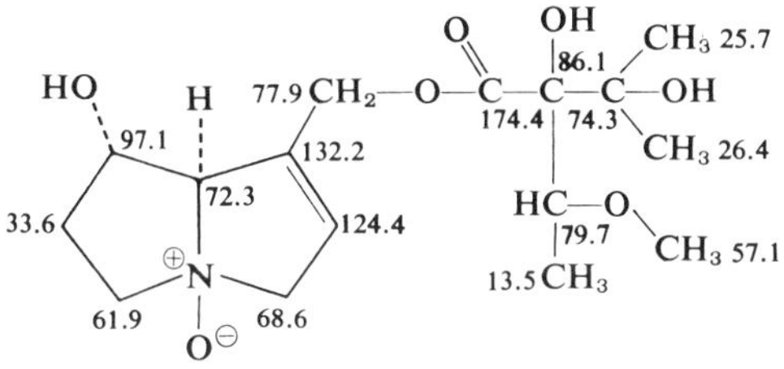

Ref. 125

666
(DMSO-d_6)

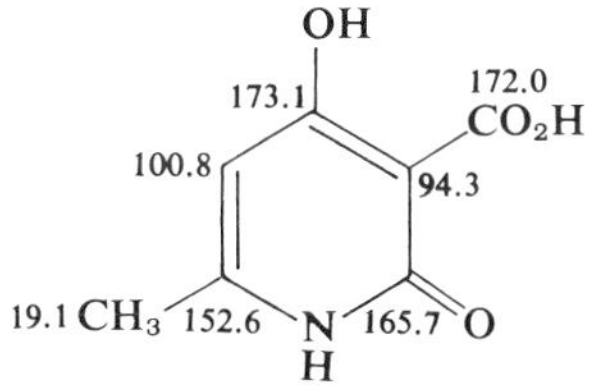

Ref. 126

667
(DMSO-d_6)

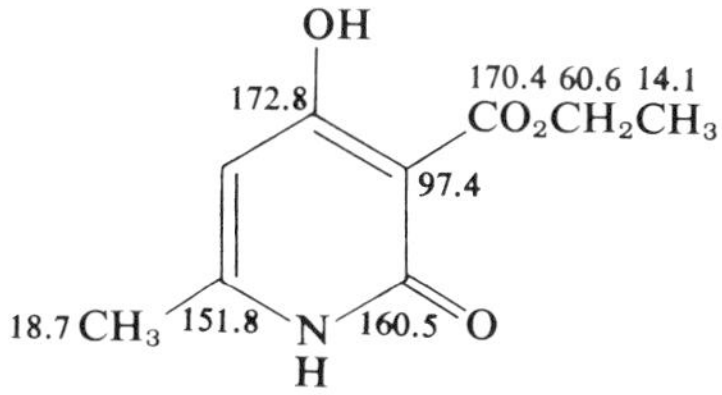

Ref. 126

668

(DMSO-d_6)

OH, O, 175.4, 95.5, C, 99.6, 172.4, NH_2, 18.9 CH_3, 150.0, N, H, 163.1, O

Ref. 126

669 **Tenellin**

(DMSO-d_6)

115.0, HO, 130.2, 156.9, OH, O, 173.0, 122.8, 149.7, 150.8, 29.4, 115.0, 105.9, C, 132.6, 34.6, 193.8, CH_3, 130.2, 110.9, 123.1, 11.7, CH_3, CH_3, 140.0, 12.3, 19.8, N, 157.5, O, OH

Ref. 127

670

(DMSO-d_6)

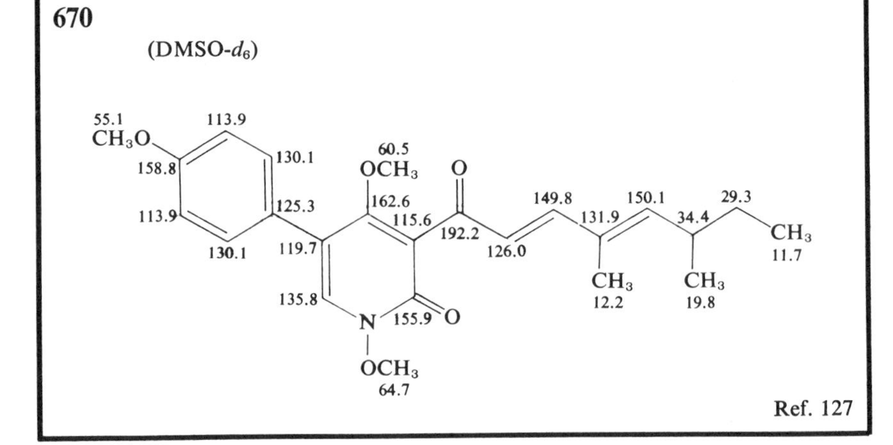

Ref. 127

671 ***N*-Benzylidenebenzylamine**
($CHCl_3$–$CDCl_3$, 7:5)

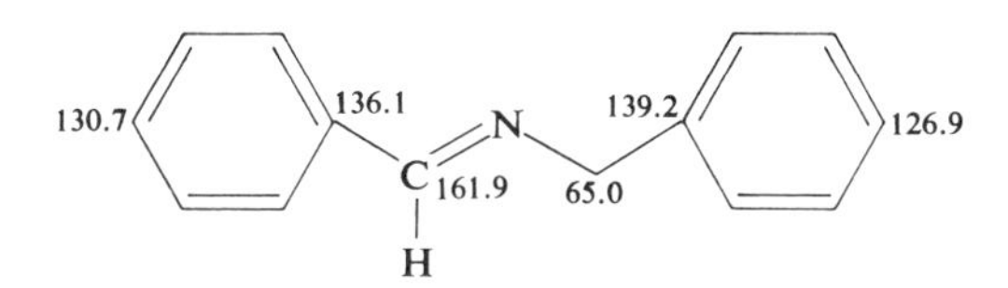

Ref. 128

672 **Catuabine A**

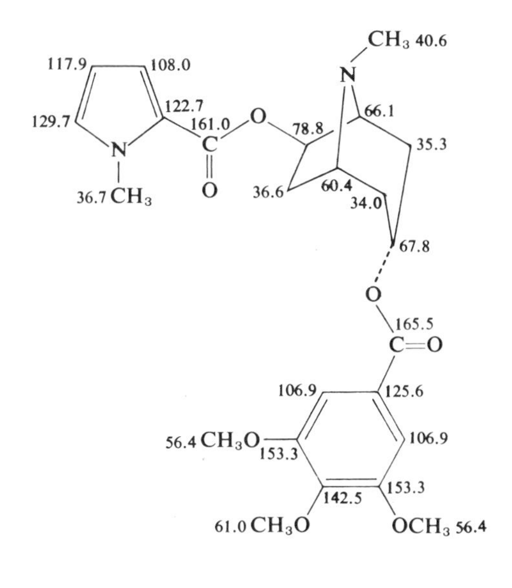

Ref. 129

673 Acridine

135.9 130.3
126.5
125.5
128.5
N 149.1
129.5

Ref. 130

674

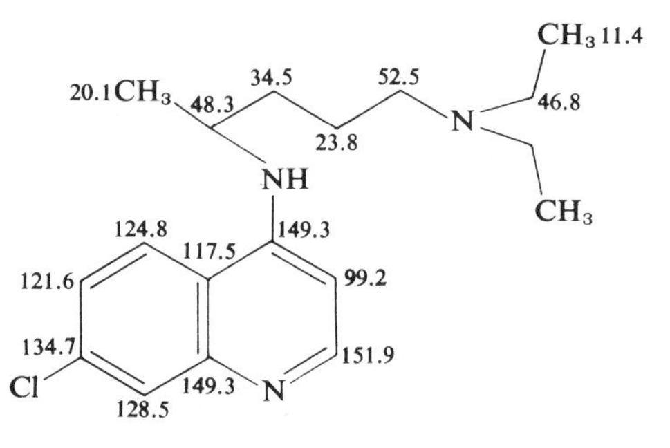

Ref. 130

675 Quinacrine

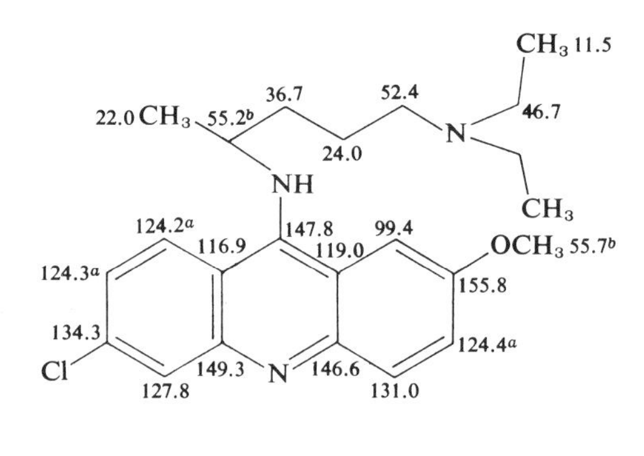

Ref. 130

676 Streptonigrin
(DMSO-d_6)

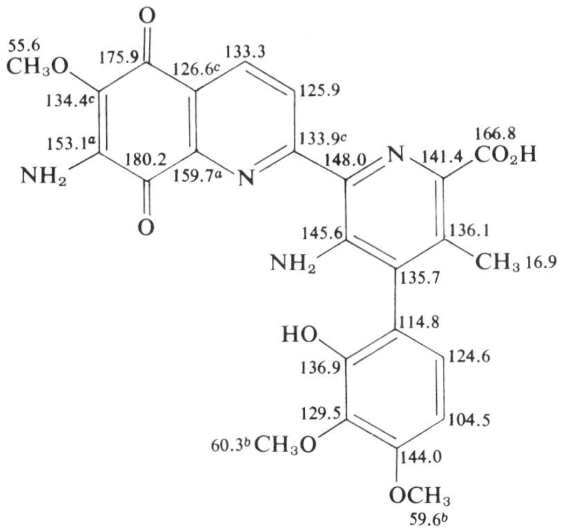

55.6 CH_3O
175.9
O
126.6[c]
133.3
125.9
134.4[c]
153.1[a]
180.2
NH_2
O
159.7[a] N
133.9[c]
148.0 N 141.4
166.8 CO_2H
145.6
NH_2
136.1
CH_3 16.9
135.7
114.8
HO
136.9
124.6
129.5
60.3[b] CH_3O
104.5
144.0
OCH_3
59.6[b]

Ref. 131

677 Berberine pseudobase
(pyridine-d_5)

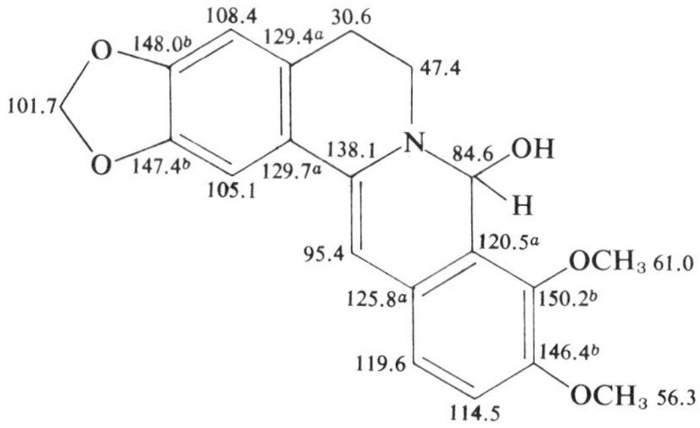

108.4
30.6
O 148.0[b]
129.4[a]
47.4
101.7
O 147.4[b]
105.1
129.7[a]
138.1 N
84.6 OH
H
95.4
120.5[a]
OCH_3 61.0
125.8[a]
150.2[b]
146.4[b]
119.6
114.5
OCH_3 56.3

Ref. 59

678 16β-Methyltubilfolidine

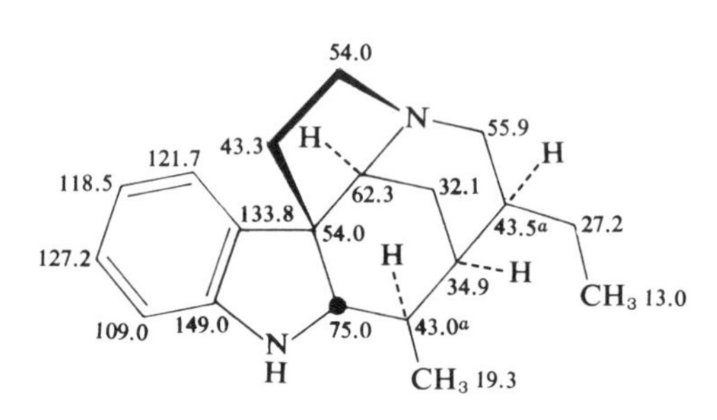

Ref. 132

679 16-Isoretuline

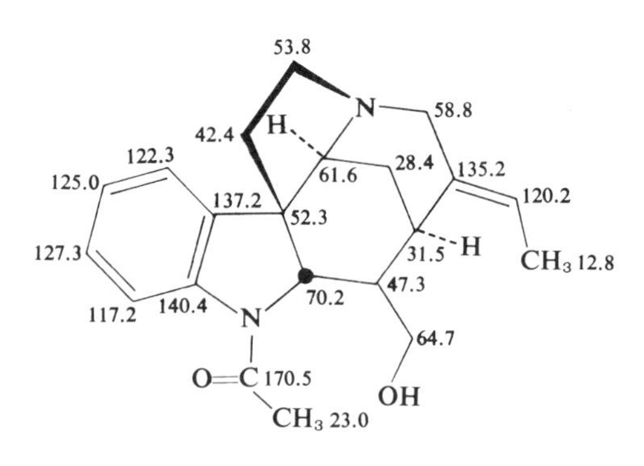

Ref. 132

680 10-Methoxy-*O*-demethylsilanine

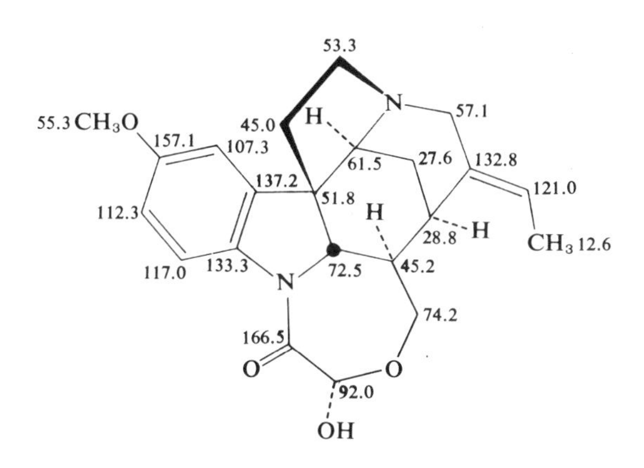

Ref. 132

681 Strychnine *N*-oxide

(DMSO-d_6)

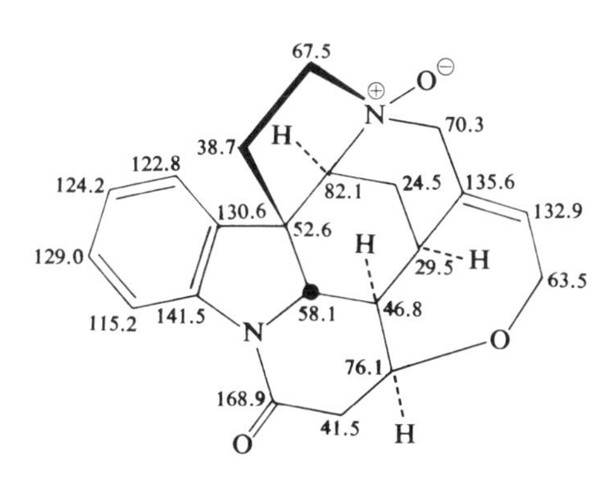

Ref. 132

682 Toxiferine-I

(D_2O–MeOH, 1:5)

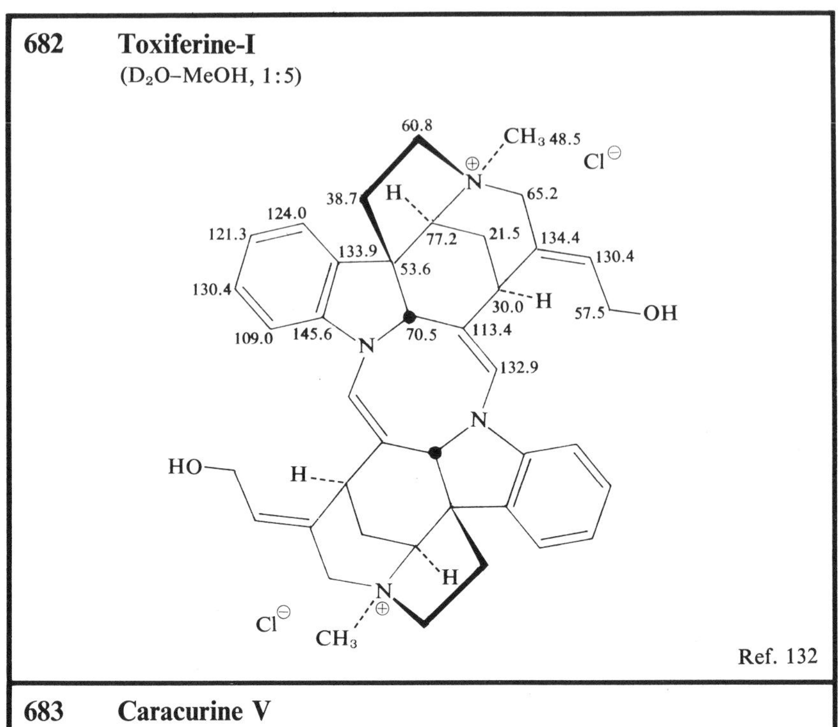

Ref. 132

683 Caracurine V

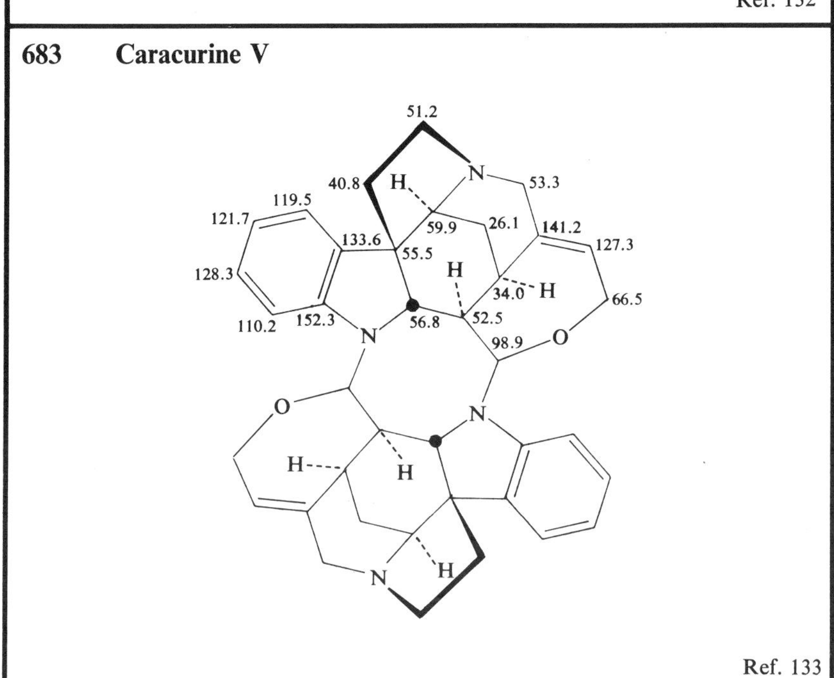

Ref. 133

684 Caracurine V di-*N*-oxide

($CDCl_3$–CD_3OD, 7:3)

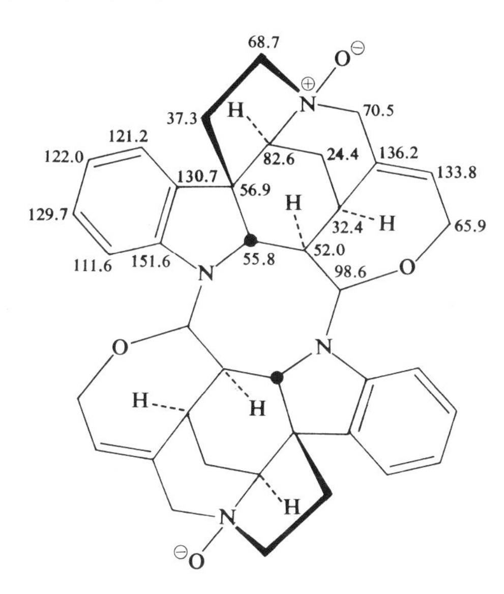

Ref. 133

685 Ajacusine

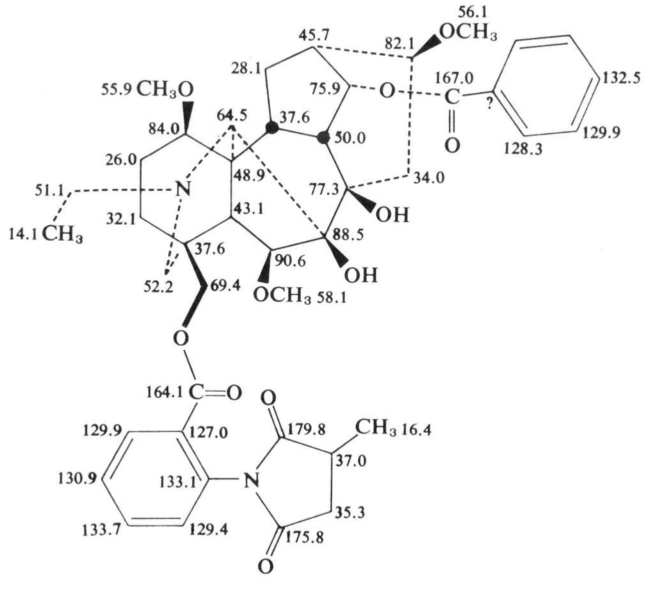

Ref. 134

686 Ajadine

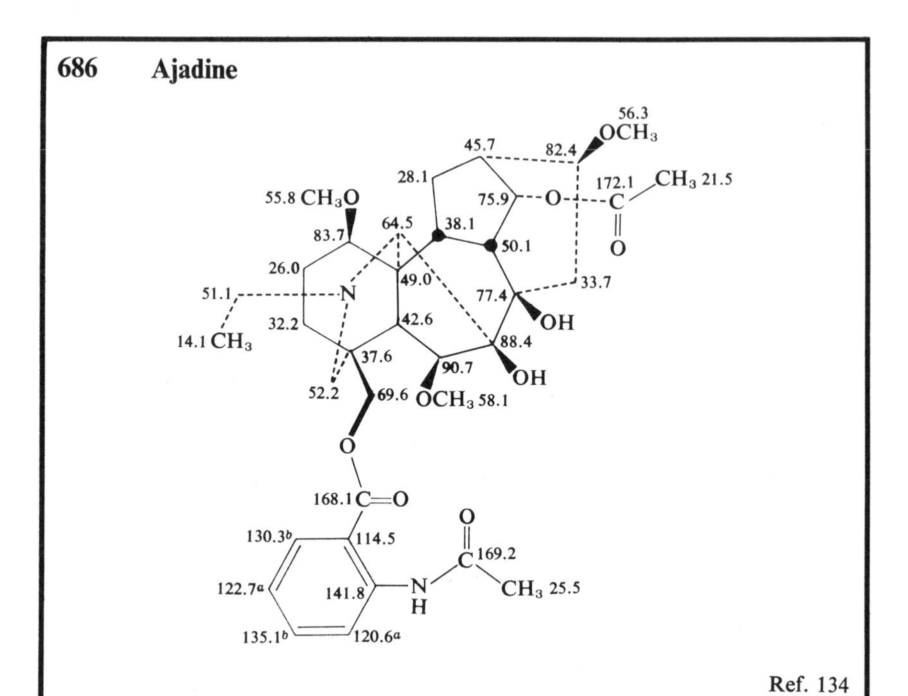

Ref. 134

687 Frangulanine

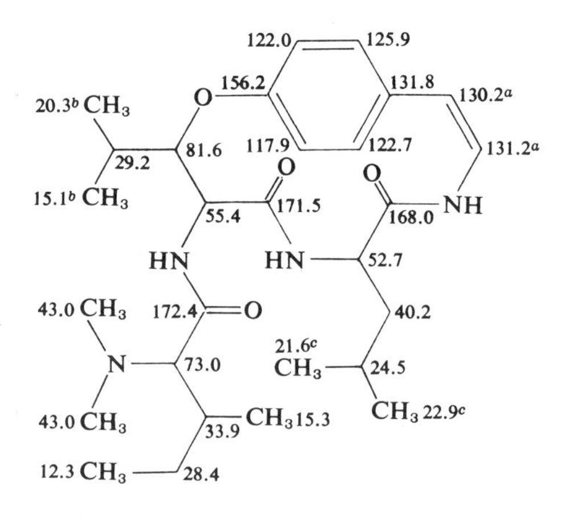

Ref. 135

REFERENCES AND NOTES

1. For additional data on the cmr spectra of pyrroles, see R. J. Cushley, R. J. Sykes, C.-K. Shaw, and H. H. Wasserman, *Can. J. Chem.* **53**, 148 (1975).
2. R. J. Abraham, R. D. Lapper, K. M. Smith, and J. F. Unsworth, *J. Chem. Soc. Perkin II*, 1004 (1974).
3. E. Lippmaa, M. Mägi, S. S. Novikov, L. I. Khmelnitski, and A. S. Prihodko, *Org. Magn. Reson.* **4**, 153 (1972).
4. D. G. Hawthorne, S. R. Johns, and R. I. Willing, *Aust. J. Chem.* **29**, 315 (1976).
5. W. R. Begg, J. A. Elix, and A. J. Jones, *Tetrahedron Lett.*, 1047 (1978).
6. H. Casal, J. Altamirano, and P. Moyna, *Gazz. Chim. Ital.* **107**, 361 (1977).
7. For additional data on the cmr spectra of pyridines, see: (a) G. Miyajima, Y. Sasaki, and M. Suzuki, *Chem. Pharm. Bull. Tokyo* **20**, 429 (1972); (b) H. L. Retcofsky and R. A. Friedel, *J. Phys. Chem.* **71**, 3592 (1967); (c) H. L. Retcofsky and R. A. Friedel, *J. Phys. Chem.* **72**, 290 (1968); (d) H. L. Retcofsky and R. A. Friedel, *J. Phys. Chem.* **72**, 2619 (1968); (e) H. Iwamura, M. Iwamura, M. Imanari, and M. Takeuchi, *Bull. Chem. Soc. Japan* **46**, 3486 (1973); and (f) R. D. Lapper, H. H. Mantsch, and I. C. P. Smith, *Can. J. Chem.* **53**, 2406 (1975).
8. F. A. L. Anet and I. Yavari, *J. Org. Chem.* **41**, 3589 (1976).
9. L. Kozerski, *Org. Magn. Reson.* **9**, 395 (1977).
10. R. J. Cushley, D. Naugler, and C. Ortiz, *Can. J. Chem.* **53**, 3419 (1975).
11. Y. Takeuchi and N. Dennis, *Org. Magn. Reson.* **8**, 21 (1976).
12. A. T. Balaban and V. Wray, *Org. Magn. Reson.* **9**, 16 (1977).
13. U. Vögeli and W. von Philipsborn, *Org. Magn. Reson.* **5**, 551 (1973).
14. V. Galasso, *Mol. Phys.* **26**, 81 (1973).
15. R. P. Thummel and D. K. Kohli, *J. Org. Chem.* **42**, 2742 (1977).
16. F. Coletta, A. Gambaro, and G. Rigatti, *Spectrosc. Lett.* **9**, 469 (1976).
17. W. M. Litchman, A. E. Zune, and U. Hollstein, *J. Magn. Reson.* **17**, 241 (1975).
18. E. Leete, *Bioorg. Chem.* **6**, 273 (1977).
19. H.-J. Sattler and W. Schunack, *Arch. Pharmaz.* **309**, 222 (1976).
20. For additional data on the cmr spectra of piperidines, see: (a) D. Wendisch, H. Feltkamp, and U. Scheidegger, *Org. Magn. Reson.* **5**, 129 (1973); (b) A. J. Jones, C. P. Beeman, A. F. Casy, and K. M. J. McErlane, *Can. J. Chem.* **51**, 1790 (1973); (c) P. Geneste, J. M. Kamenka, and C. Brevard, *Org. Magn. Reson.* **10**, 31 (1977); and (d) E. Eliel and D. Kandasamy, *Tetrahedron Lett.*, 3765 (1976).
21. E. Wenkert, J. S. Bindra, C.-J. Chang, D. W. Cochran, and F. M. Schell, *Acc. Chem. Res.* **7**, 46 (1974).
22. G. Ellis and R. G. Jones, *J. Chem. Soc. Perkin II*, 437 (1972).
23. H. Booth and D. V. Griffiths, *J. Chem. Soc. Perkin II*, 842 (1973).
24. E. Wenkert, D. W. Cochran, E. W. Hagaman, F. M. Schell, N. Neuss, A. S. Katner, P. Potier, C. Kan, M. Plat, M. Koch, H. Mehri, J. Poisson, N. Kunesch, and Y. Rolland, *J. Am. Chem. Soc.* **95**, 4990 (1973).

25. N. J. Bach, H. E. Boaz, E. C. Kornfeld, C.-J. Chang, H. G. Floss, E. W. Hagaman, and E. Wenkert, *J. Org. Chem.* **39**, 1272 (1974).
26. F. Bohlmann and R. Zeisberg, *Chem. Ber.* **108**, 1043 (1975).
27. E. Wenkert, E. W. Hagaman, B. Lal, G. E. Gutowski, A. S. Katner, J. C. Miller, and N. Neuss, *Helv. Chim. Acta* **58**, 1560 (1975).
28. A. J. Jones, A. F. Casy, and K. M. J. McErlane, *Can. J. Chem.* **51**, 1782 (1973).
29. E. Wenkert, B. Chauncy, K. G. Dave, A. R. Jeffcoat, F. M. Schell, and H. P. Schenk, *J. Am. Chem. Soc.* **95**, 8427 (1973).
30. P. R. Srinivasan and R. L. Lichter, *Org. Magn. Reson.* **8**, 198 (1976).
31. A. Rabaron, M. Koch, M. Plat, J. Peyroux, E. Wenkert, and D. W. Cochran, *J. Am. Chem. Soc.* **93**, 6270 (1971).
32. E. Wenkert, D. W. Cochran, E. W. Hagaman, R. B. Lewis, and F. M. Schell, *J. Am. Chem. Soc.* **93**, 6271 (1971).
33. G. Van Binst and D. Tourwé, *Org. Magn. Reson.* **4**, 625 (1972).
34. E. Wenkert, D. W. Cochran, H. E. Gottlieb, E. W. Hagaman, R. B. Filho, F. J. de A. Matos, and M. I. L. M. Madruga, *Helv. Chim. Acta* **59**, 2437 (1976).
35. E. Leete, *Phytochemistry* **16**, 1705 (1977).
36. A. J. Jones and M. M. A. Hassan, *J. Org. Chem.* **37**, 2332 (1972).
37. Y. Takeuchi, *J. Chem. Soc. Perkin II*, 1927 (1974).
38. M. G. Ahmed and P. W. Hickmott, *J. Chem. Soc. Perkin II*, 838 (1977).
39. D. Tourwé, G. Van Binst, S. A. De Graaf, and U. K. Pandit, *Org. Magn. Reson.* **7**, 433 (1975).
40. R. J. Molyneux and R. Y. Wong, *Tetrahedron* **33**, 1931 (1977).
41. V. I. Stenberg, N. K. Narain, and S. P. Singh, *J. Heterocycl. Chem.* **14**, 225 (1977).
42. P. Hanisch, A. J. Jones, A. F. Casey, and J. E. Coates, *J. Chem. Soc. Perkin II*, 1202 (1977).
43. For additional data on the cmr spectra of benzoquinolizidines and benzoquinolines, see G. Van Binst, G. Laus, and D. Tourwé, *Org. Magn. Reson.* **10**, 10 (1977); and G. Van Binst, D. Tourwé, and E. De Cock, *Org. Magn. Reson.* **8**, 618 (1976).
44. M. Sugiura and Y. Sasaki, *Chem. Pharm. Bull. Tokyo* **24**, 2988 (1976).
45. Y. Arata, T. Aoki, M. Hanaoka, and M. Kamei, *Chem. Pharm. Bull. Tokyo* **23**, 333 (1975).
46. R. T. LaLonde, T. N. Donvito, and A. I.-M. Tsai, *Can. J. Chem.* **53**, 174 (1975).
47. Y. Itatani, S. Yasuda, M. Hanaoka, and Y. Arata, *Chem. Pharm. Bull. Tokyo* **24**, 2521 (1976).
48. T. N. Nakashima, P. P. Singer, L. M. Browne, and W. A. Ayer, *Can. J. Chem.* **53**, 1936 (1975).
49. For additional data on the cmr spectra of quinolines, see P. A. Claret and A. G. Osborne, *Spectrosc. Lett.* **8**, 385 (1975).
50. S. R. Johns and R. I. Willing, *Aust. J. Chem.* **29**, 1617 (1976).
51. P. A. Claret and A. G. Osborne, *Org. Magn. Reson.* **8**, 147 (1976).
52. L. Ernst, *Org. Magn. Reson.* **8**, 161 (1976).
53. P. A. Claret and A. G. Osborne, *Org. Magn. Reson.* **9**, 167 (1976).
54. P. A. Claret and A. G. Osborne, *Org. Magn. Reson.* **10**, 35 (1977).
55. A. M. Nadzan, and K. L. Rinehart, Jr., *J. Am. Chem. Soc.* **99**, 4647 (1977).
56. E. Wenkert, C.-J. Chang, A. O. Clouse, and D. W. Cochran, *Chem. Commun.* 961 (1970).
57. E. L. Eliel and F. W. Vierhapper, *J. Org. Chem.* **41**, 199 (1976).
58. For additional data on the cmr spectra of benzoquinolines see D. Tourwé, L. Vandersteen, and G. Van Binst, *Bull. Soc. Chim. Belg.* **86**, 603 (1977).

59. D. M. Hindenlang and M. Shamma, unpublished results.
60. D. W. Hughes, H. L. Holland, and D. B. MacLean, *Can. J. Chem.* **54**, 2252 (1976).
61. E. Wenkert, B. L. Buckwalter, I. R. Burfitt, M. J. Gašić, H. E. Gottlieb, E. W. Hagaman, F. M. Schell, and P. M. Wovkulovich, in *Topics in C-13 NMR Spectroscopy, Vol. 2*, G. C. Levy, ed., Wiley-Interscience, New York (1976), pp. 105–110.
62. J. B. Bremner and Le van Thuc, *Aust. J. Chem.* (1978), in press.
63. G. S. Ricca and C. Casagrande, *Org. Magn. Reson.* **9**, 8 (1977).
64. L. M. Jackman, J. L. Moniot, J. Trewella, M. Shamma, R. L. Stephens, E. Wenkert, M. Lebœuf, and A. Cavé, in press. See also M. Shamma, in *Specialist Periodical Reports, The Alkaloids, Vol. 7*, M. F. Grundon, Senior Reporter, The Chemical Society, London (1977), p. 163; and M. Shamma and J. L. Moniot, *Isoquinoline Alkaloids Research 1972–1977*, Plenum Press, New York (1978), p. 153.
65. S. Kano, Y. Takahagi, E. Komiyama, T. Yokomatsu, and S. Shibuya, *Heterocycles* **4**, 1013 (1976).
66. C. Tani, N. Nagakura, S. Hattori, and N. Masaki, *Chem. Lett.*, 1081 (1975).
67. N. Takao, K. Iwasa, M. Kamigauchi, and M. Sugiura, *Chem. Pharm. Bull. Tokyo* **25**, 1426 (1977).
68. R. H. F. Manske, R. Rodrigo, H. L. Holland, D. W. Hughes, D. B. MacLean, and J. K. Saunders, *Can. J. Chem.* **56**, 383 (1978).
68a. T. Kametani, K. Fukumoto, M. Ihara, A. Ujiie, and H. Koizumi, *J. Org. Chem.* **40**, 3280 (1975).
69. T. T. Nakashima and G. E. Maciel, *Org. Magn. Reson.* **5**, 9 (1973).
70. D. W. Hughes, B. C. Nalliah, H. L. Holland, and D. B. MacLean, *Can. J. Chem.* **55**, 3304 (1977).
71. C. Tani and K. Tagahara, *J. Pharm. Soc. Japan* **97**, 93 (1977).
72. A. Buzas, R. Cavier, F. Cossais, J.-P. Finet, J.-P. Jacquet, G. Lavielle, and N. Platzer, *Helv. Chim. Acta* **60**, 2122 (1977).
73. A. M. Ismailov, M. K. Yosynov, and X. A. Aslanov, *Khim. Prir. Soedin.* **13**, 422 (1977).
74. Y. Terui, K. Tori, S. Maeda, and Y. K. Sawa, *Tetrahedron Lett.*, 2853 (1975).
75. F. I. Carroll, C. G. Moreland, G. A. Brine, and J. A. Kepler, *J. Org. Chem.* **41**, 996 (1976).
76. S. P. Singh, S. S. Parmar, V. I. Stenberg, and S. A. Farnum, *Spectrosc. Lett.* **10**, 1001 (1977); as modified by C. D. Hufford, private communication.
77. W. O. Crain Jr., W. C. Wildman, and J. D. Roberts, *J. Am. Chem. Soc*, **93**, 990 (1971).
78. L. Zetta, G. Gatti, and G. Fuganti, *Tetrahedron Lett.*, 4447 (1971).
79. E. Rosenberg, K. L. Williamson, and J. D. Roberts, *Org. Magn. Reson.* **8**, 117 (1976).
80. R. G. Parker and J. D. Roberts, *J. Org. Chem.* **35**, 996 (1970).
81. V. Galasso, G. Pellizer, and G. C. Pappalardo, *Org. Magn. Reson.* **9**, 401 (1977).
82. V. Galasso, G. Pellizer, H. Le Bail, and G. C. Pappalardo, *Org. Magn. Reson.* **8**, 457 (1976).
83. T. Nozoye, T. Nakai, and A. Kubo, *Chem. Pharm. Bull. Tokyo* **25**, 196 (1977).
84. H. Fritz and T. Winkler, *Helv. Chim. Acta* **59**, 903 (1976).
85. P. A. Crooks, B. Robinson, and O. Meth-Cohn, *Phytochem.* **15**, 1092 (1976). Alternative assignments for physostigmine are given by V. I. Stenberg, N. K. Narain, S. P. Singh, R. H. Obenauf, and M. J. Albright, *J. Heterocycl. Chem.* **14**, 407 (1977).
86. C. Poupat, A. Ahond, and T. Sévenet, *Phytochem.* **15**, 2019 (1976).
87. G. W. Gribble, R. B. Nelson, G. C. Levy, and G. L. Nelson, *Chem. Commun.*, 703 (1972).

88. E. Wenkert, C.-J. Chang, H. P. S. Chawla, D. W. Cochran, E. W. Hagaman, J. C. King, and K. Orito, *J. Am. Chem. Soc.* **98**, 3645 (1976). For the revision of the stereochemistry of alloyohimbine, see L. Töke, K. Honty, L. Szabó, B. Blaskó, and C. Szántay, *J. Org. Chem.* **38**, 2496 (1973).
89. R. H. Levin J.-Y. Lallemand, and J. D. Roberts, *J. Org. Chem.* **38**, 1983 (1973).
90. M. Damak, A. Ahond, P. Potier, and M.-M. Janot, *Tetrahedron Lett.*, 4731 (1976).
91. E. Bombardelli, A. Bonati, B. Gabetta, E. Martinelli, G. Mustich, and B. Danieli, *Phytochem.* **15**, 2021 (1976).
92. E. Wenkert and H. E. Gottlieb, *Heterocycles* **7**, 753 (1977).
93. P. Rasoanaivo and G. Lukacs, *J. Org. Chem.* **41**, 376 (1976).
94. A. Ahond, M.-M. Janot, N. Langlois, G. Lukacs, P. Potier, P. Rasoanaivo, M. Sangaré, N. Neuss, M. Plat, J. Le Men, E. W. Hagaman, and E. Wenkert, *J. Am. Chem. Soc.* **96**, 633 (1974).
95. J. Bruneton, A. Cavé, E. W. Hagaman, N. Kunesch, and E. Wenkert, *Tetrahedron Lett.*, 3567 (1976).
96. J. Le Men, M. J. Hoizey, G. Lukacs, L. Le Men-Olivier, and J. Lévy, *Tetrahedron Lett.*, 3119 (1974).
97. A. Cavé, J. Bruneton, A. Ahond, A.-M. Bui, H.-P. Husson, C. Kan, G. Lukacs, and P. Potier, *Tetrahedron Lett.*, 5081 (1973).
98. N. Neuss, H. E. Boaz, J. L. Occolowitz, E. Wenkert, F. M. Schell, P. Potier, C. Kan, M. M. Plat, and M. Plat, *Helv. Chim. Acta* **56**, 2660 (1973).
99. E. Bombardelli, A. Bonati, B. Gabetta, E. M. Martinelli, G. Mustich, and B. Danieli, *Tetrahedron* **30**, 4141 (1974).
100. A. Ahond, A.-M. Bui, P. Potier, E. W. Hagaman, and E. Wenkert, *J. Org. Chem.* **41**, 1878 (1976).
101. B. C. Das, J. P. Cosson, G. Lukacs, and P. Potier, *Tetrahedron Lett.*, 4299 (1974).
102. Z. Votický, E. Grossman, J. Tomko, G. Massiot, A. Ahond, and P. Potier, *Tetrahedron Lett.*, 3923 (1974).
103. R. Verpoorte, P. J. Hylands, and N. G. Bisset, *Org. Magn. Reson.* **9**, 567 (1977); and J. Leung and A. J. Jones, *Org. Magn. Reson.* **9**, 333 (1977).
104. E. Wenkert, C.-J. Chang, D. W. Cochran, and R. Pelliciari, *Experientia* **28**, 377 (1972).
105. G. Tóth, K. Horváth-Dóra, O. Clauder, and H. Duddeck, *Justus Liebigs Ann. Chem.*, 529 (1977).
106. L. Zetta and G. Gatti, *Org. Magn. Reson.* **9**, 218 (1977).
107. L. Zetta and G. Gatti, *Tetrahedron* **31**, 1403 (1975).
108. C. G. Moreland, A. Philip, and F. I. Carroll, *J. Org. Chem.* **39**, 2413 (1974).
109. F. I. Carroll, D. Smith, M. E. Wall, and C. G. Moreland, *J. Med. Chem.* **17**, 985 (1974).
110. F. I. Carroll, A. Philip, and M. C. Coleman, *Tetrahedron Lett.*, 1757 (1976).
111. M. Daudon, M. H. Mehri, M. M. Plat, E. W. Hagaman, F. M. Schell, and E. Wenkert, *J. Org. Chem.* **40**, 2838 (1975).
112. M. Daudon, M. H. Mehri, M. M. Plat, E. W. Hagaman, and E. Wenkert, *J. Org. Chem.* **41**, 3275 (1976).
113. For additional data on the cmr spectra of dimeric indole alkaloids and also some monomeric indoles, see: (a) M. Damak, C. Poupat, and A. Ahond, *Tetrahedron Lett.*, 3531 (1976); (b) B. C. Das, J.-P. Cosson, and G. Lukacs, *J. Org. Chem.* **42**, 2785 (1977); (c) E. Bombardelli, A. Bonati, B. Danieli, B. Gabetta, E. M. Martinelli, and G. Mustich, *Experientia* **31**, 139 (1975); (d) E. Bombardelli, A. Bonati, B. Gabetta, E. M. Martinelli, G. Mustich, and B. Danieli, *J. Chem. Soc. Perkin I*, 1432 (1976);

(e) D. E. Dorman and J. W. Paschal, *Org. Magn. Reson.* **8**, 413 (1976); and (f) E. Wenkert, E. W. Hagaman, B. Lal, G. E. Gutowski, A. S. Katner, J. C. Miller, and N. Neuss, *Helv. Chim. Acta* **58**, 1560 (1975).

114. M. C. Koch, M. M. Plat, N. Préaux, H. E. Gottlieb, E. W. Hagaman, F. M. Schell, and E. Wenkert, *J. Org. Chem.* **40**, 2836 (1975).
115. Y. Rolland, N. Kunesch, J. Poisson, E. W. Hagaman, F. M. Schell, and E. Wenkert, *J. Org. Chem.* **41**, 3270 (1976).
116. S. Yamamura, H. Irikawa, Y. Okumura, and Y. Hirata, *Bull. Chem. Soc. Japan* **48**, 2120 (1975).
117. For additional data on the cmr spectra of diterpenoid alkaloids see A. J. Jones and M. H. Benn, *Can. J. Chem.* **51**, 486 (1973).
118. S. W. Pelletier, Z. Djarmati, and N. V. Mody, *Tetrahedron Lett.*, 1749 (1976).
119. S. W. Pelletier, N. V. Mody, A. J. Jones, and M. H. Benn, *Tetrahedron Lett.*, 3025 (1976).
120. S. W. Pelletier and Z. Djarmati, *J. Am. Chem. Soc.* **98**, 2626 (1976).
121. S. W. Pelletier, N. V. Mody, R. S. Sawhney, and J. Bhattacharyya, *Heterocycles* **7**, 327 (1977).
122. M. Sangaré, F. Khuong-Huu, D. Herlem, A. Milliet, B. Septe, G. Berenger, and G. Lukacs, *Tetrahedron Lett.*, 1791 (1975).
123. R. Radeglia, G. Adam, and H. Ripperger, *Tetrahedron Lett.*, 903 (1977).
124. R. J. Weston, H. E. Gottlieb, E. W. Hagaman, and E. Wenkert, *Aust. J. Chem.* **30**, 917 (1977).
125. L. H. Zalkow, L. Gelbaum, and A. Keinan, *Phytochem.* **17**, 172 (1978).
126. F. Cavagna and H. Pietsch, *Org. Magn. Reson.* **11**, 204 (1978).
127. C.-K. Wat, A. G. McInnes, D. G. Smith, J. L. C. Wright, and L. C. Vining, *Can. J. Chem.* **55**, 4090 (1977).
128. J. E. Arrowsmith, M. J. Cook, and D. J. Hardstone, *Org. Magn. Reson.* **11**, 160 (1978).
129. E. Graf and W. Lude, *Arch. Pharm.* (*Weinheim*) **311**, 139 (1978).
130. S. P. Singh, S. S. Parmar, V. I. Stenberg, and S. A. Farnum, *Spectrosc. Lett.* **11**, 59 (1978).
131. S. J. Gould and C. C. Chang, *J. Am. Chem. Soc.* **100**, 1624 (1978).
132. E. Wenkert, H. T. A. Cheung, H. E. Gottlieb, M. C. Koch, A. Rabaron, and M. M. Plat, *J. Org. Chem.* **43**, 1099 (1978).
133. R. Verpoorte and A. B. Svendsen, *J. Pharm. Sci.* **67**, 171 (1978).
134. S. W. Pelletier and R. S. Sawhney, *Heterocycles* **9**, 463 (1978).
135. E. Haslinger, *Tetrahedron* **34**, 685 (1978).
136. D. Tourwé and G. Van Binst, *Heterocycles* **9**, 507 (1978).
137. R. M. Acheson and G. Procter, *J. Chem. Soc. Perkin I*, 1924 (1977).
138. J. K. Baker and R. F. Borne, *J. Heterocycl. Chem.* **15**, 165 (1978).
139. M. Sugiura, N. Takao, K. Iwasa, and Y. Sasaki, *Chem. Pharm. Bull.* **26**, 1168 (1978).
140. C. Olieman, L. Maat, and H. C. Beyerman, *Recl. Trav. Chim. Pays-Bas* **97**, 31 (1978).

INDEX

The numbers next to the compound names are the spectral assignment diagram numbers, not page numbers.